OUR UNDISCOVERED UNIVERSE

Introducing Null Physics

The Science of Uniform and Unconditional Reality

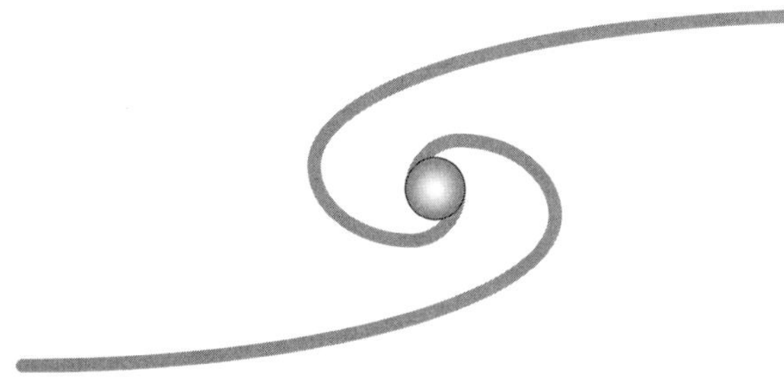

Terence Witt

ARIDIAN PUBLISHING CORPORATION
MELBOURNE

To Ginny, the brightest star in my universe

Our Undiscovered Universe, Introducing Null Physics

Book design by Terence Witt. All graphs and calculations used in this volume were produced on numerous laptop computers (2.2-3.4 GHZ) using software written by the author. The galactic image on the front cover is courtesy of NASA/Hubble Heritage Project. The image of the Milky Way's galactic core region on the rear cover is courtesy of NASA/CXC/MIT/F.K. Baganoff et al. This book is typeset in *perpetua*, designed by Eric Gill and made available to the public by Monotype in 1932.

Library of Congress Control Number: 2006911267

ISBN-10: 0-9785931-3-8
ISBN-13: 978-0-9785931-3-1

PRINTED IN THE UNITED STATES OF AMERICA

FIRST EDITION

Contents

Preface

In questions of science the authority of a thousand is not worth the humble reasoning of a single individual.

Galileo Galilei, 1564-1642

The breadth of scientific concepts has expanded in fits and starts throughout the brief history of civilization, each time leaving our perception of the universe progressively more immense and refined. Along the way our tools have kept pace with our gestating worldview. In less than four hundred years telescopes evolved from Galileo's first effort, barely as powerful as binoculars, to the massive Hubble and Keck. Yet after all the progress - after Socrates, Aristotle, and Lucretius paved the way for Newton, Planck, and Einstein; after tens of thousands of scientists across the world have probed deeply into nature - no one has a clue why the universe is here, how it really works, or what it is ultimately composed of. Not the slightest clue. Cosmologists promote the Big Bang without the most meager understanding of what could have caused it or came before it. Particle physicists labor at their accelerators year after year, decade after decade, never finding the truths behind matter's subatomic facade. Why should protons be composed of quarks? Why are protons, antiprotons, positrons, and electrons the only stable elementary particles? No one knows, and no one has even come close to answering questions like these. We measure the universe; we do not understand it. At the current state of scientific development there are far more questions than answers, and they aren't even the right questions.

The investigation of the universe in its entirety has always been hampered by a desperate lack of certainty. Indeed, cosmology's inherent and greatest challenge is the distance and history it encompasses. Universal processes move so slowly in relation to a human lifespan they might as well be static. Our modern instruments give us exquisite snapshots of an ancient cosmic cinema written across the heavens, but they do not foster comprehension, and scientists are left to wander through forests of data without a map, capturing new measurements while never fully explaining or understanding the previous ones.

A breakthrough theory doesn't need more data, nor does it need more detailed data. We are drowning in data. What is missing is a comprehensive explanation of measurements made over a century ago. There lies the glaring incompleteness. We have counted so many trees; it is time to take a hard look at the forest and keep looking at it until it starts to make sense. It is simply not possible to understand anything of importance about the universe if we fail to understand its most fundamental mystery: ***Why does it exist***? This is the question most cosmologists are convinced is beyond their reach now and in the foreseeable future. Many would claim it unknowable. "*Why does the universe exist?*" casts a shadow of doubt across all of the physical sciences. *It is the gaping void in science itself.*

The enigma of our existence has tormented thinkers for thousands of years. It causes philosophers to doubt the power of reason; it causes scientists to doubt the rationality of nature. It convinces theorists the universe has an impenetrable barrier to knowledge and understanding. But the time has finally come to move to a new level of understanding. *Null Physics* tells us why the universe exists in no uncertain terms. The reason for existence is the most important scientific question there is, and I'm convinced the theory presented in this volume answers it rationally, completely, and unequivocally. A wealth of compelling evidence gives me reason to believe it is the first cohesive explanation of the underlying basis of the physical world.

The problem is how to present my findings, and here the challenge is twofold. First, the concepts are scientific, yet entirely new to science. This is not catastrophic; science is often the testing ground for novel ideas. The second and more difficult issue is that my cosmology is diametrically opposed to the Big Bang. It is not that the Big Bang is eloquent or insightful; it falls short on both counts. The problem is no scientific consensus has been refuted in nearly a century. One of the *most recent* broadly held scientific concepts that was eventually recognized as fictitious was that the Milky Way constituted the sum total of the celestial sphere (~1915). Science has, in its own estimation at least, been unremittingly correct since then. Wrangling about details continues unabated, but not about the big picture or the process used to investigate it.

Solving the existential impasse is one thing; convincing scientists a major paradigm is wholly off the mark is quite another. The theory of relativity won acceptance not only as a result of its sweeping eloquence but also because it was a refinement of earlier work. Newton was not *wrong*; his theory is an approximation of a weak gravitational field. In stark contrast, the Big Bang theory is a gross misinterpretation of the nature of the universe and is utterly, irreconcilably false. Yet it correlates reasonably well with observations and is based on one of the most successful theories of all time, General Relativity.

Although a number of books take the Big Bang to task both observationally and theoretically, what they tend to overlook is that for the past sixty years *it has been the best, albeit incomplete, explanation of the large-scale universe.* The Big Bang is one of the few cosmologies to even address the origin of the universe. It tells us it came from a primordial cataclysm, *it just doesn't tell us why.* Yet if someone asks a cosmologist where the universe came from, the immediate answer is "an inconceivably energetic expansion about 13.7 billion years ago." Herein lies the primary utility of the Big Bang; it provides a short answer to the question humankind has asked since before the Stone Age. The answer is irredeemably wrong, but is at least mercifully brief and easy to relate to. People are born; kittens are born; even stars are born. The notion the universe was born and is very old certainly seems reasonable.

While a small number of scientists are convinced there was no Big Bang, no one has a reasonable "short answer" to replace it. Saying the universe has existed forever isn't the short answer unless it is also possible to say why this is the case. Ironically, the reason why the Big Bang is so popular is the same reason why it is wholly erroneous. There is no partial credit for solving the existential dilemma because there is no partial answer. Either you know why the universe exists or you do not. A cover story about what happens the first moments after it magically arrives avoids the basic question, and in the process avoids a basic truth that has remained undiscovered until now.

A few scientists have pursued alternatives to the Big Bang at great risk to their professional careers. Their theories tend to be thoughtful but ultimately inconclusive. None are capable of providing the entire solution to the existential puzzle. Nor are they consistent with many of the observed properties of the heavens. They try to explain selected aspects of the universe, but not where it came from. This is the defining riddle. Alternative cosmologies invariably reference observations contrary to the Big Bang's premise, yet fail to provide better explanations themselves.

Null Physics is, for the first time in the history of science, both a complete answer to the riddle of our existence and a quantitative theory of universal properties. It does not divert the question of our existence to the murky past of some nascent version of the universe. It does not depend on unseen mechanisms or unknowable precursors. This book answers the question of why the universe exists fully and finally, using nothing except logic and reasoned extensions of known physics. It leads to testable conclusions and surprising results, precisely the way real science is supposed to work.

This investigation is not and cannot be limited to cosmology. The universe is a single, vast machine. Its substructure defines its mass density as well as its elementary particles. In order to properly reflect this level of unification, our theoretical development will follow the path of discovery set before it by principles defining the existence of reality, moving smoothly from the utter desolation of empty space down to the cores of protons, then returning to the stars to show us the structure and function of galaxies. All of the steps along this progression are as inevitable as existence itself, and are subjected to rigorous empirical validation.

The reader has no doubt heard much of this before. Alternative cosmologies invariably make bold claims about revealing the secrets of the universe and are quick to attack the Big Bang's soft philosophical underbelly. Yet not one of these perspectives provides even remotely plausible explanations of the observations leading to the Big Bang in the first place. They decry the current paradigm while offering nothing of substance to replace it. *Null Physics'* standards are much higher, and the intergalactic redshift, Cosmic Microwave Background (CMB) radiation, and other relevant measurements receive the detailed attention they merit. My ultimate purpose is to supplant the Big Bang with a more comprehensive, *quantitative* perspective of reality. This is a natural consequence of the evolution of scientific thought and can only be achieved through a systematic approach. The Big Bang is over sixty years old; it was the first widely accepted scientific cosmology. The time has come to recognize its conceptual incompleteness and move on, not cling to it wishing it were true.

As of this writing, cosmologists are firmly convinced the latest high-resolution CMB measurements all but "prove" that the Big Bang is the correct model of the universe's birth, and that it occurred 13.7 billion years ago (±1%). But this is only true if we begin by assuming in the first place that (a) the universe *had* an origin and (b) intergalactic redshift represents universal expansion, neither of which is supported by *known physical laws*. The only reason the Big Bang fits the Wilkinson Microwave Anisotropy Probe (WMAP) data so well now is because it has been adjusted to this data since the WMAP project began, and the only reason the CMB correlates to the universe's large-scale material distribution is because this material is immersed in the CMB, and has been, for quite some time. In the final analysis, the Big Bang concept invokes phantasmagorical entities such as dark energy yet reveals nothing of any real substance about the universe's nature. The Big Bang's relentless departure from known physics is unwarranted as well as intellectually unsatisfying.

Physics is currently in a sad, sad state. Theorists have, upon their global failure to provide coherent solutions to nature's toughest questions, chosen to relinquish much of their responsibility. They have reached the internally-agreed-upon consensus that their job is not to *explain* reality, their one and only job is to *describe* it, using models that need not even reference real, physical entities. Yet at the same time, many physicists go to great lengths to *explain* why they needn't *explain* certain things. Further, in what might be best referred to as physics chauvinism, interested bystanders are told that the only reason they ask questions such as "Why does the universe exist?" is because they lack the formal training and expertise of scientists. The bystander may also learn, if they request further clarification, that a number of questions about nature, especially those driven by common sense, are meaningless and without merit. Such hubris! Other fields of science, from molecular biology to geology, still seek a deeper understanding of the phenomena they study. Physics, from the realm of subatomic particles to galactic superclusters, is the only branch of science that claims an unwarranted exclusion from the pursuit of understanding. Where else does not knowing the answer make a question irrelevant?

Suppose a primitive native, with no prior contact with modern civilization, found a digital watch on a jungle trail. Being the shaman of his village, he studies this object and soon recognizes patterns in the symbols it displays. Eventually, he develops a model of the precession of these symbols, and wows his tribesman by predicting the appearance and moment of arrival of the next cipher. Yet he has no idea what the watch is or why it was laying on the trail in the first place. These are insignificant details, he tells his ignorant compatriots, because he knows what the next symbol is going to look like and approximately when it will appear, and this remarkable foreknowledge transcends all other considerations. As physics creeps into the twenty-first century, its methodology bears an uncanny resemblance to the approach used by our friend with the digital watch. Scarier still, many physicists would not see this as a problem. We can do better, far better.

Anyone who has ever wondered why the universe exists and how it functions on the largest scale; anyone with a critical mind who wants to know the truth and is weary and dissatisfied with the prevailing quantum and cosmological dogma; is absolutely going to love this book. Welcome to the new world ... the world where reality is *real*, not the winner of a science fiction contest.

ACKNOWLEDGEMENTS

I would first like to thank my wife, Ginny, for her unwavering support. She shared the joy of my discoveries and retained her optimism no matter how many times I struggled well into the wee hours, beaten bloody by Mother Nature. And thanks to my daughter Pharaba and so many good friends who helped me put the final draft together and suffered through tortuous readings of my initial efforts. Finally, I owe a special debt of gratitude to Will Hurley and Roy Burton for their editorial insight and suggestions. Any errors or omissions in this text are entirely my own.

Organization and Structure

A streamlined method of reading *Null Physics* has been included in order to provide a quick overview and foster curiosity. The last section of every chapter is entitled *General Conclusions*. It is a short list of pivotal results that can be read out of context, making it possible to skip from chapter to chapter and scan the entire book in about ten minutes. At face value many of the conclusions will probably seem surprising or perhaps even nonsensical, but anyone who has studied the universe should begin to see a hint of the truth and be encouraged to travel some of the roads leading to these intriguing discoveries.

Also note that the four parts of this volume can be reviewed separately, depending on the reader's specific interests.

- Part I: To find out why the universe exists, read the first chapter. For a number of unprecedented discoveries about the nature of space and time, read the rest.

- Part II: To explore the substructure of energy, read this part. It begins with an innovative nonstatistical resolution of wave-particle duality and ends with the connection between the four-dimensional size of a photon and the four-dimensional size of the universe.

- Part III: To examine the ultimate composition of elementary particles and their relationship within atomic nuclei, read this part. It begins with the derivation of the core size of protons and electrons and ends with the state and distribution of matter in the interior of black holes.

- Part IV: To learn how the large-scale universe operates and where galaxies and black holes fit into the cosmic engine, read this part.

Parts II through IV begin with short lists of prerequisite concepts from earlier parts. The lists are brief as all parts are designed to be as self-contained as possible.

Although every attempt has been made to make *Null Physics* amenable to the widest possible audience, several sections are abstract, others require first-year calculus to follow the derivations, and in general no section is particularly easy. Please don't let the algebra obscure the governing concepts. Mathematical expressions are included in this work for only two reasons:

- To summarize the text.
- To perform supporting calculations for various derivations.

Logic is the tool for deriving the reality of nature; algebra is what is used to demonstrate that it is the correct reality. Most calculations use a number of intermediate steps in order to maximize clarity, and the more rigorous mathematical work has been included in a wealth of appendices. Also, a detailed glossary can be found near the end of this volume. It contains definitions of a number of standard scientific terms as well as many new terms specific to *Null Physics*.

STRUCTURE

The body of *Null Physics* is composed of an interrelated network of four different formalisms used to mark our path as well as provide an organized development. Each is symbolized and described as follows:

- **§ Definition**

  Used for clarification purposes only. It formalizes but does not constitute a progression of theoretical development. Symbolically, a connecting element.

- **Φ Axiom**

  A major tenant acknowledged as a self-evident truth. *Null Physics* has only a single axiom. An axiom allows entry into the contextual circle of existence from which all other findings are derived, deduced, induced, or inferred. Symbolically, a closed, complete premise.

- **Ψ Theorem**

  A premise derived from axioms, other theorems, and/or independent reasoning. It represents a preponderance of evidence and certainty. Symbolically, a foundation for future growth.

- **Ω Hypothesis**

  Similar to a theorem, but more speculative. It is derived from theorems and axioms but is not sufficiently certain to form the basis of subsequent theorems or hypotheses. A hypothesis is a work in progress that will hopefully attain further substantiation as the investigation proceeds. Symbolically, a partially open issue.

All formalisms are separately numbered in each chapter. The third theorem in Chapter 4, for instance, is (Ψ4.3).

PART I: FOUNDATION
OUT OF THE DARKNESS

The substructure and nature of unbounded nothingness

- *Competing cosmological theories*
- *Derivation of the universe from nothingness*
- *Null Axiom*
- *Geometric basis of the quantization of energy and matter*
- *Relationship between space and time*
- *Uniqueness of three-dimensional space*
- *Cosmic ultrastasis*

Nothing can be created from nothing.

Titus Lucretius, 99-55 BC

Figure (I) Gravitational lensing in the Hubble Deep Field
(Courtesy NASA/Hubble Heritage Project)

1. SOMETHING FROM NOTHING

1.1 COSMOLOGY PARADE

The Big Bang is by no means the only hypothesis offered to account for the universe's existence. There are a number of other explanations, referred to by mainstream theorists as *alternative cosmologies*, that have aspirations of competing with the current paradigm. Let's begin our investigation by comparing and contrasting all of the most prominent theories available prior to the introduction of *Null Physics*. They are (in descending order of popularity):

- Big Bang Cosmology (**quantitative**)
- Quasi Steady State Cosmology (**quantitative**)
- Plasma Cosmology (qualitative)
- Machian Cosmology (qualitative)

Cosmologies are identified as either quantitative or qualitative theories, based on their approach. Quantitative theories are the hallmark of science; here calculations can be performed based on formal premises and compared to experimental results. Qualitative theories focus on the philosophical incompleteness and observational failures of quantitative theories, and generally have a foundation too poorly defined to allow for a definitive assessment of their own premises. Speculation is easy; there are countless possibilities for every reality. The hard part is matching an idea to nature.

Let's review all of these theories from the standpoint of a critical outsider with no vested interest other than curiosity about the true nature of reality. In the comparison to follow, responses of each theory to key cosmological observations are included, as well as the contemporary physical principles they violate (something common in theories requiring a primordial universe where the rules are different from our current universe).

One of the most prevalent limitations of modern cosmology is the uncertainty of its data. The Hubble constant is thought to contain as much as 50% error, but the actual error range is unknown.⟨1.2⟩ Other universal parameters, like average matter density, are barely known to the nearest order of magnitude. The only cosmological measurement with any reasonable degree of accuracy is the intensity of the Cosmic Microwave Background (CMB) radiation field. This being the prevailing environment, our comparison among different cosmologies will focus on *observationally known* aspects of the universe - phenomena whose existence is beyond a reasonable amount of doubt. Dark energy, relic neutrinos, and other hypothetical entities will not be addressed.

As it turns out, the indisputable facts of the large-scale universe are few:

CMB: There is a diffuse background of microwave radiation permeating the known limits of space. Its spectrum is that of a perfect 2.724 °K blackbody with slight intensity fluctuations (~18 μK) distributed randomly across the entire sky. This field also has a dipole anisotropy (opposing blueshift/redshift) corresponding to Earth's motion through it at 370 km/s:⟨36⟩

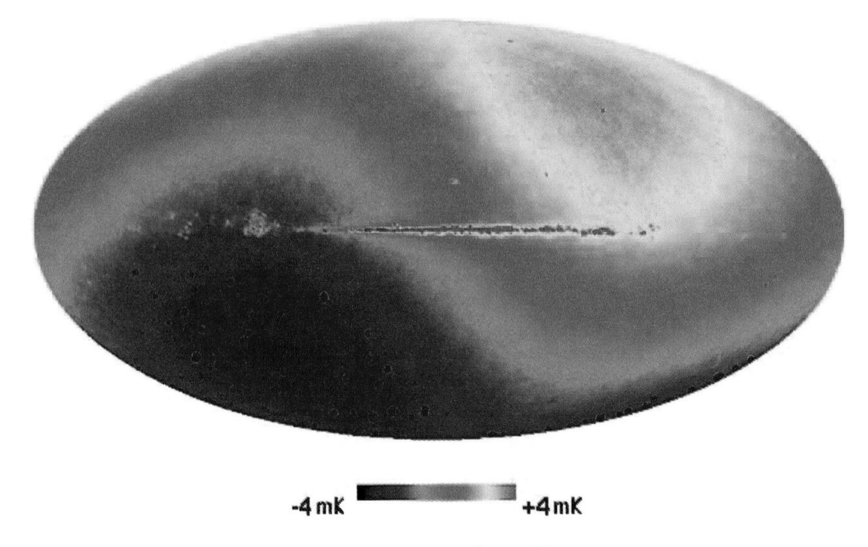

Figure (1.1) WMAP with Earth's motion
(Courtesy NASA/WMAP Science Team)

The horizontal red streak in Figure (1.1) is the microwave emission along the plane of our own galaxy, the Milky Way.

Elements: The universe's material is composed of a wide variety of elements, 92 of which occur naturally on Earth. Hydrogen makes up the majority of the *luminous* universe, ~70% by mass, helium is the second most abundant, at ~25%, and all of the other elements comprise the remaining ~5%.⟨1.1⟩

Redshift: The light from distant objects is redshifted in (roughly) direct proportion to the distance of the object. The Hubble constant defines the ratio of redshift to distance.⟨7.1⟩

Structure: Most of the universe's material is concentrated in galaxies of various types, ~30% of which are spiral like our own. Ellipticals and lenticulars are the most common galactic morphologies, and tend to congregate into clusters and superclusters, which in turn form the walls of a foam-like structure of interconnected cells stretching throughout the observable universe. These cells are about 100 to 300 Mly (million light years) in diameter, as shown below:

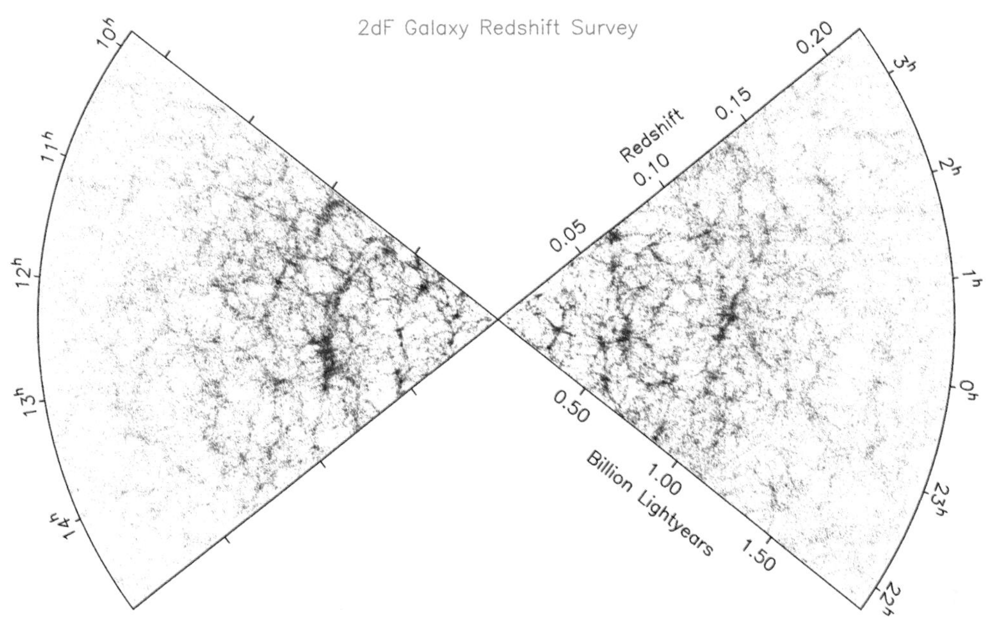

Figure (1.2) Large-scale structure of the universe - the ~80,000 galaxies of the 2dF redshift survey (Courtesy 2dF GRS Team)

The interiors of these intergalactic cells consist of empty intergalactic space sprinkled with a few isolated galaxies.⟨8.1⟩

Olbers: Named after the famous astronomer Henrick Olbers, Olbers' paradox is a problem of astrophysical thermodynamics. The luminous energy given off by stars heats space. Any cosmology supporting an infinitely old universe must deal with the unchecked accumulation of this energy.⟨6.6⟩

How do our cosmologies account for these basic facts?

BIG BANG COSMOLOGY

Reference: *Principles of Physical Cosmology*, P.J. Peebles.

Theorist(s): George Gamov, Ralph Alpher, Robert Herman, 1946.

Premise: The universe is a solution to the field equations of General Relativity. It emerged as a rapidly expanding mote of pure energy/space about 13.7 billion years ago. Its frenetic initial growth is known as the inflationary period. This was followed by a slowly accelerating expansion that resulted in a steady decrease in energy density. The sequence of events associated with the universe's adiabatic expansion is (1) separation of the four universal forces from a single unifying force; (2) formation of elementary particles from pure energy; (3) formation of selected light isotopes, such as helium; and (4) formation of initial stars/proto-galaxies, galaxies, and large structures. The continued and accelerating expansion of the universe is powered by a cosmic abundance of dark energy.

Violation(s): Conservation of energy and gravitation (antigravitational dark energy).

<u>CMB</u>: Microwaves are the relic radiation generated from the formation of helium during the first few moments of the universe's birth. They have a thermal distribution because when the universe was young, matter was so dense that photons were in thermal equilibrium with it. Universal expansion has occurred in just such a way as to preserve this thermal spectrum even though the wavelengths of its component photons have expanded by a factor of over a thousand.

<u>Elements</u>: Selected light elements arose immediately after the initial formation of matter. All other elements are produced in stellar interiors as a normal consequence of fusion, by supernovae, or through the collision of cosmic rays with other elements.

<u>Redshift</u>: Energy loss in light from distant celestial objects is caused by the recession velocities of universal expansion. This expansion is uniform, so an observer located in any galaxy would see the same redshift effect in all directions.

<u>Structure</u>: Quantum fluctuations in the primordial energy density.

<u>Olbers</u>: The paradox is avoided by the expansion of the universe into progressively larger volume, reducing the energy density of the light released by stars.

Comments: This is a complicated theory. The reason gravity is listed as a violation is because the antigravity required for universal expansion has never been experimentally verified. Energy conservation is listed as a violation because no known interactions create or destroy energy. Cosmologists claim the Big Bang doesn't violate conservation; they maintain that the universe's total energy is offset by its own negative gravitational potential. This is fantasy. The creation of even the smallest amount of energy, along with a comparable amount of negative gravitational potential, *has never been observed*. The other way to resolve the conservation problem is for the Big Bang to originate from a prior universe, but this only leads to additional theoretical fabrication. And the issues keep coming. The origin of the elements is a problem, as it is difficult to envision why only a few specific light elements were created in the beginning. Current stellar models require heavy elements to initiate fusion, so one wonders how the first stars were able to ignite in their absence. Also, since there are billions of galaxies containing billions of stars, all burning hydrogen into helium, the ratio of primordial helium to recently minted helium is yet another complication and its value is arbitrary. Finally, there are no cases in our contemporary universe where a redshifted thermal spectrum remains thermal. A detailed explanation is given for how the CMB remains perfectly thermal *after a redshift of over a thousand*, but this just adds to the artificiality and complexity of this model. The Big Bang is an ad hoc, retrospective theory. In the final analysis it doesn't even address causation, so it doesn't explain why the universe exists. Everything else is incidental.

QUASI STEADY STATE COSMOLOGY

Reference: *A Different Approach to Cosmology*, Fred Hoyle, Geoffrey Burbidge, Jayant Narlikar.

Theorist(s): Hermann Bondi, Thomas Gold, Fred Holye, 1948.

Premise: The universe is a solution to the field equations of General Relativity. It has cyclical periods of expansion and is of indeterminate age and size. Matter is continuously created as either a by-product of, or driving force behind, universal expansion. New matter accumulates into stars and galaxies, expanding outward under the influence of antigravitational potential.

Violation(s): Conservation of energy and gravitation (antigravitational component).

CMB: Cosmic microwaves are an equilibrium concentration of the luminous energy given off by fusion. They are thermalized into the CMB spectrum by their interaction with the tiny iron whiskers that supernovae explosions release throughout space.

Elements: All elements are formed in stellar interiors or through the collision of cosmic rays with other elements.

Redshift: Energy loss in light from distant celestial objects is caused by the recession velocities of universal expansion. This expansion is uniform, so an observer located in any galaxy would see the same redshift effect in all directions.

Structure: Built slowly over hundreds of billions of years (or longer) by the interaction of matter in deep space.

Olbers: Expansion of the universe into larger volume maintains universal temperature.

Comments: This theory is in some ways identical to the Big Bang and in other ways entirely divergent. It relies on a hypothetical antigravitational force for universal expansion, but also needs an additional unknown mechanism for the continuous creation of matter. Observationally, the known fraction of universal helium corresponds fairly well to the CMB's energy density. In other words, if hydrogen were continuously created and burned, the energy density it generated would be close to the CMB's current value. The iron whisker hypothesis for thermalizing the CMB is suspect but more scientific than the Big Bang's explanation because it is at least testable in a contemporary universe. The continuous creation of matter is difficult to support either theoretically or empirically. Like the Big Bang, this theory provides no explanation for why the universe exists.

PLASMA COSMOLOGY

Reference: *The Big Bang Never Happened*, Eric Lerner.

Theorist(s): Hannes Alfvén (*Worlds-Antiworlds*, 1966), Oskar Klein, 1961.

Premise: The universe is governed by electromagnetism, not gravitation. It is infinitely large and expanding. Its expansion is caused by a universal interaction between matter and antimatter. Electric currents define the universe's large-scale structure because plasma interactions cause the formation of stars, galaxies, and superclusters. The internal motion of galaxies is more in keeping with electromagnetic force than gravitation. The universe could not possibly be 13.7 billion years old since many massive celestial structures such as the Great Wall required hundreds of billions of years to form.

Violation(s): None.

<u>CMB</u>: This is the ambient radiation given off by the cosmic power grid. The solar wind of the universe's stars creates huge electrical currents in deep space. These currents, in turn, form galaxies and emit and absorb microwave radiation. This radiation is thermalized by scattering with microfilaments (plasma artifacts) or iron whiskers in deep space.

<u>Elements</u>: Not specifically referenced.

<u>Redshift</u>: Energy loss in light from distant celestial objects is caused by the recession velocities incurred during a cyclical and violent interaction between matter and antimatter.

<u>Structure</u>: Created by condensation and distribution of intergalactic plasma filaments.

<u>Olbers</u>: Not specifically referenced.

Comments: This theory seems more a counterpoint to specific aspects of the Big Bang than a cosmology of its own merit. The CMB, which can be measured more accurately than any other cosmic parameter, is only given incidental consideration. One is given to wonder how "microfilaments" of plasma can exist in deep space when there is essentially no material there. Plasma requires the motion of charged particles along magnetic and electric fields. Certainly such fields exist in deep space but charged particles are few and far in between. No calculations are presented that relate universal currents with the power needed to control the "nongravitational" motion of galaxies. As was the case with the other cosmologies, no reason is forthcoming about why the universe exists.

MACHIAN COSMOLOGY

Reference: *Seeing Red: Redshifts, Cosmology and Academic Science*, Halton Arp.

Theorist(s): Halton Arp, 1980.

Premise: The universe is arbitrarily large and not expanding. Matter is continuously created in areas of high density, such as the cores of stars and galaxies, and its elementary character changes over time. New particles emerge with a mass of zero and acquire their mass through a light-speed interaction with pre-existing matter (based on Ernst Mach's concept of inertial mass). At a certain age, matter becomes unstable and disintegrates.

Violation(s): Conservation of energy.

<u>CMB</u>: Microwaves are an equilibrium concentration of the light energy given off by fusion. They are thermalized by some operator (perhaps iron whiskers) in extragalactic space.

<u>Elements</u>: Not specifically referenced.

<u>Redshift</u>: Energy loss in light from distant celestial objects is caused by the relationship between the age and mass of elementary particles. The farther the object, the younger the object was when it emitted the light we observe. Since the mass of particles has increased over time, objects farther away emit spectra of lower energy, appearing uniformly redshifted.

<u>Structure</u>: New matter, continuously created, acts like a superfluid with eddies of structure.

<u>Olbers</u>: Not specifically referenced.

Comments: This theory is certainly original. To say the mass of particles spontaneously increases over time, thereby changing the spectra of elements, is a novel approach for explaining intergalactic redshift. The only law it really violates is energy conservation, but the mechanics of having protons and electrons proportionately gain mass makes this theory in some ways more complex than the Big Bang. It eliminates universal expansion because redshift is directly proportional to the transit time of celestial light, yet doesn't reference

Olbers' paradox, and any steady state theory ought to. There is little to no formal development of equations for any of the processes cited in the theory, and the CMB is mentioned in passing as if it were an unimportant phenomenon. Quasars are treated as local objects ejected from galactic cores. The entire theory is qualitative and for the most part untestable. It fails to explain why the universe exists in the first place.

In summary:

Pre-Null Cosmologies						
Theory	**Violation(s)**	**CMB**	**Elements**	**Redshift**	**Structure**	**Olbers**
Big Bang	Conservation, Gravity	Relic Light	Bang/Stars	Recession	Quantum Fluctuation	Expansion
Quasi Steady State	Conservation, Gravity	Iron Whiskers	Stars	Recession	Cosmic Cycles	Expansion
Plasma	None	Plasma Glow	---	Recession	Plasma Filaments	---
Machian	Conservation	Iron Whiskers	---	Aging	Superfluidity	---

Table (1.1) Cosmological alternatives prior to the introduction of *Null Physics*

Not a single model provides a meaningful reason for the existence of the universe or its more interesting properties, and most require the violation of physical laws. The overriding mystery that a complete cosmology must confront is *why the universe exists*. None of the most popular cosmologies even address this enigma, let alone resolve it. Nor do any of these theories provide reasonable explanations for the intergalactic redshift and CMB. For the redshift, either space itself is expanding (into what?) or matter changes mass spontaneously as it ages. For the CMB, deep space microwaves are either (a) in equilibrium with matter (what matter?) or (b) have somehow retained an ideal thermal spectrum after a uniform redshift expanded them to over a thousand times their original wavelengths. The latter is tantamount to a radio signal reflecting around the Earth's ionosphere a hundred times with zero signal degradation.

The bottom line is no currently available cosmology addresses the riddle of our existence or provides cogent explanations of simple, irrefutable cosmic properties. Moreover, by violating physical laws, they fail to even achieve the status of *science*. Science is not a case of trying to explain the universe of today by abandoning its governing laws. Science is to derive the universe of today, along with its laws, and demonstrate the relationship between the two. *The reason the universe has laws is not a matter of convenience; it has these laws because they are a necessary requirement for its existence.*

Alternative cosmologies attack the Big Bang because, among other things, the random distribution of material throughout space shows no evidence of a uniform expansion. This is particularly evident in the 2dF galactic redshift survey that was shown earlier in Figure (1.2). But empirical discrepancies are not where the Big Bang truly fails. It fails because it *does not satisfactorily reconcile the existence of the universe with the conservation of energy*. The negative gravitational potential of the universe's matter does not even come close to balancing its total energy content. But even if it did, the spontaneous emergence of energy has never been observed, so any theory that relies on such an emergence *violates energy conservation, regardless of arguments to the contrary*. This creates a glaring inconsistency, an indelible flaw invalidating all subsequent analysis. The laws the universe must obey, its core essence, apply at each moment in time throughout all of space because they are the *rules of existence*. These laws are as immutable as the universe itself. They are the rules it has to use. When the Big Bang reinterprets energy conservation in order to make our existence possible it becomes self-contradictory. The universe is not free to choose the laws it obeys and the laws it ignores. Neither is a scientific cosmology.

1.2 PRIME QUESTIONS

As a new century of scientific development begins, the most important questions about the nature of existence remain unsolved:

- Why does the universe exist?
- Where did it come from?
- What is it made of at the smallest level?

Physics has countless other mysteries, such as why unit charge has the value it does, or why electrons are so much lighter than protons, but these are mere details compared to questions about *existence*. Existence questions form the undercarriage of reality. They are the questions children ask and never get good answers to. They are the questions that started with cavemen and endured past Einstein. They are the foundation for all questions about physical nature, and will be called *Prime Questions* because they cannot be divided into smaller, easier problems.

All of the cosmologies presented in the last section have a common thread. They do not tell us why the universe exists. The reigning paradigm, the Big Bang, does not even address the existence questions. It focuses on a detailed description of an emergent universe, not the cause of its origin. The calculation of primordial helium formation is considered science; what happens before is considered philosophy. This kind of approach creates more questions than answers.

If the universe began in an expansion 13.7 billion years ago, then:

- What caused it?
- What existed before the beginning?
- Where did all of the universe's material come from?
- Why did the event happen when it did; does time predate the Big Bang?

Indeed, even if the Big Bang were adjusted to have a perfect correlation with the age and distribution of matter in space and all other known observations, it still wouldn't tell us *why* it happened in the first place. Redshift is measured for distant galaxies and interpreted to mean they are all moving away from us. The theory attributes this to a cosmic expansion beginning billions of years ago, but the pivotal question is the *cause* of the event. The Big Bang is not an *explanation* of anything; it is a vacuous recitation that lacks even the hint of a coherent justification or philosophy. By its very nature it is incomplete, and worse than that, it lacks the capacity for completeness.

Suppose a prototype race car engine explodes during a test run. Pieces of metal litter the smoldering remains of the engine block and surrounding floor much as galaxies are thought to be scattered across space after the Big Bang. Are the design engineers going to measure the location of all the pieces, carefully plot the results, and calculate their distance/mass curve? Of course not. The single most pertinent aspect of the event is its *cause*. The results are incidental; what everyone wants to know is why the engine flew apart. The Big Bang obscures the original question of existence with a drama of cosmic transmutation. *Why does the universe exist?* That is the question - a question people have been asking since the dawn of human intelligence. We have stories, not explanations. We have data, not information. The search for the truth begins now.

*** Preview ***

The existence of the universe is a paradox to us because we do not understand its basic nature, not because its basic nature cannot be understood.

1.3 UNIVERSE PROBLEM

The paradoxical nature of our existence is embodied by a set of Prime Questions:

- **Why does the universe exist?**
- **Where did it come from?**
- **What is it made of at the smallest level?**

These questions constitute the *universe problem*. There is a universe out there with billions of stars in each of untold billions of galaxies, and no one has a clue as to where it came from or why it exists. It is like someone having to explain to an income tax auditor why they suddenly have two hundred trillion dollars in their bank account. A few extra dollars might be easy to track down, but a "discrepancy" as large as the universe has got to have an interesting explanation. It does.

The solution to the existential paradox is the universe, which clearly exists. To resolve the paradox is not to find the universe; it's right in front of us. The trick is to know enough about it so the fact of its existence changes from a paradox to a foregone conclusion. The process is purely conceptual and no additional measurements are necessary. The universe existed long before anyone understood why and will continue to exist long after the reason it does is generally accepted. The transition is ours to make.

THE UNIVERSE'S SIZE

As introduced by our accounting analogy, one of many things that is disconcerting about the universe problem is the *size* of the problem. Visualizing unending reaches of star-filled space is difficult; understanding an origin to it all is even more so. Yet it is here now, regardless of how it got here. The question is whether or not the size of the universe compounds the problem of *why* it exists. This would certainly seem to be the case, but let's take a closer look.

The spontaneous generation of matter has been proposed in contemporary physics in terms of *virtual particles*. These phantasms are thought to exist near charged particles where they suddenly emerge as matter-antimatter pairs, take place in interactions, and then dissolve back to empty space. If enough energy is transferred into a reaction, virtual particles become real particles in the form of equal amounts of matter and antimatter. Energy conservation mediates the transition from virtual to permanent. Virtual particles are in keeping with the Copenhagenist view that reality itself has a quantum statistical nature. This idea, originating in the early 1900's, is usually called by the more contemporary name of *quantum reality*. It holds that within certain guidelines, (the Uncertainty Principle) the very existence of matter is a function of probability. If this were true (and it is not, as demonstrated in Part II), space would never be empty because it would be a seething foam of virtual energy constantly fading in and out of existence - a perpetual state of micro creation/dissolution.

While many physicists are comfortable with particles created from nothing over very brief intervals, pulling a *permanent* universe from the void is a clear violation of energy conservation. This is the *size of the universe* problem in modern physics, implicit in the Copenhagenist idea of scale versus reality. Its premise is that things like rocks and chairs are *real*, but the elementary particles of which they are composed *are not*. This is why individual particles can momentarily wink in and out of existence while macroscopic objects have a more substantive, stable presence. If there were some way to generate a universe on the scale of virtual particles, a conversion from the unreal (subatomic particles) to the real (people, planets, stars) might be possible.

Enter the Big Bang. Here the universe begins life as a submicroscopic particle of unimaginable density: a *universe particle*. This reduces the universe's size to a hypothetical (and by definition unobservable) realm even smaller than virtual particles. In so doing it connects it to the mysterious province of quantum reality where existence is nothing more than a statistical wave function. By its own admission, however, quantum physics has no underlying basis, so the virtual particle concept is a dead end, and the Big Bang focuses on the universe's birth as a problem of energy distribution, not energy conservation. But the size problem remains, unscathed by the paucity of the origin rationale. Either the universe came from *somewhere*, or it never came from *anywhere*. A sobering accounting problem is present in either case.

As initially suspected, the universe's size seems to make the problem of its origin quite a bit more difficult.

*** Preview ***

The universe's size actually makes the explanation for its existence easier, not more difficult. Like everything else, there is a reason the universe is so vast, and as it turns out the only way it can exist is to be as large as it is.

UNIVERSAL CONSERVATION

The conservation of energy is the cornerstone of modern physics, yet is blatantly violated by a universal origin from nothing - *ex nihilo*. There are four ways to address this problem:

- Vacuum fluctuations. Many cosmologists believe that the universe's origin did not violate energy conservation because gravitational energy is intrinsically negative and all of the universe's matter/light energy is balanced by its total negative gravitational energy. This is an interesting idea, but *there is absolutely no evidence to support it*. If it were possible to produce matter from empty space and account for the difference with negative gravitational energy, why has this never been observed? Why is it not happening now? What limits the process? Why can't matter vanish and take a comparable gravitational potential with it? And there is a more basic inconsistency to consider. The Big Bang allegedly produces space through expansion. How can there be vacuum fluctuations without a vacuum?

- Not in effect. In this scenario, energy conservation came into existence *after* the universe emerged. While it is true that conservation is meaningless in the absence of something to conserve, it is just as true that the two are inseparable.

- Approximation. Perhaps energy conservation is only an approximation of the true situation, and the total energy in any interaction varies by a tiny, currently undetectable amount. But even if this were the case, it has no bearing on the salient issue. Energy conservation has been documented with sufficient precision to preclude the spontaneous emergence of even the smallest traces of matter, and the material scattered across deep space constitutes far more than a trace amount.

- Eternal energy, periodic universal renewal. Here the energy of which the universe is composed is eternal, and the Big Bang is just the latest cosmic renovation. This satisfies energy conservation but fails to explain why the universe exists *instead of nothingness*. Philosophers attack this issue with the disingenuous counter argument that "somethingness" could well be the universe's *natural* state, and we have no way of knowing otherwise. The last time this line of reasoning was used to mask a general ignorance of a phenomenon was when Archimedes explained gravitational attraction as an object's innate tendency to return to its *natural* place. This was over 2200 years ago! We ought to be able to do better by now.

Speculation about exceptions to fundamental physical laws is entertaining, but *the universe exists now, there is a reason why this is the case, and no violation of energy conservation has ever been observed.* Although physicists have only documented a small part of the universe, they have measured it carefully. Electromagnetic spectra from distant regions of space indicate our corner of reality is much the same as the rest of the universe. The signatures of elements are the same in ancient light, billions of years old, as they are in the light from our sun, only minutes old. This means the laws of physics *have not changed in billions of years*. If this isn't a good enough track record for their stability, it is difficult to envision a better one. The machinery of existence everywhere, in the form of particles, photons, and interactions, depends on energy conservation. It's what keeps the universe's stars burning.

Origin ideas attempt to explain the universe of today (the conserved universe; the isotropic universe) in terms of various incarnations of the universe of yesterday. Whenever these narrations fail to explain a characteristic of the current universe, changes are made to the primordial universe, which by its very nature is inaccessible. The true solution to the universe problem is to *explain* the universe of today constrained by the rules of today.

*** Preview ***

At no moment in the universe's history has energy conservation been violated in any way - not a billion years ago, not a trillion years ago, never. It is the quantification of existence. Like immense size, energy conservation is a vital universal characteristic.

TIME BEFORE TIME

The idea that the universe came into being in the distant past is as perplexing as how it emerged from nothingness in the first place. If it began at some particular moment, what transpired prior to this? If no events preceded the Big Bang, how could time be reckoned at all? This is the *time before time* problem. If time predates the universe, energy emerged at some time t_0 in what will be called the *post-time model*:

SPACE-TIME → t_0 → SPACE-TIME-ENERGY

This doesn't really make the time before time problem any easier. When was primordial space-time born in this model? Does it even *need* an origin? Either primordial space-time has been around forever or it had an origin itself. If the latter is true, space-time begins at t_0 and energy follows at t_1, yielding the *two-stage model*:

NOTHING → t_0 → SPACE-TIME → t_1 → SPACE-TIME-ENERGY

As noted earlier, universal origin theories often take the case of $t_0 = t_1$, consolidating all of the time-related dilemmas into a single transition, the *single-stage model*:

NOTHING → t_0 → SPACE-TIME-ENERGY

The origin of time remains a quandary in all of these models because the problem of *first origin* remains. Either time is infinite and has been around forever or it had a beginning moment. The cases are mutually exclusive and equally incomprehensible. If time had no origin why should the rest of the universe? If time had an origin what were its initial conditions?

CHANGE OF STATE

As is evident from the models in the last section, the origin of space-time from nothingness and the origin of energy from space-time are two forms of the same origin problem. Time may or may not be present, but any origin is a *change of state*. This can occur in one of two ways: an instantaneous change with no initial conditions or a gradual change brought about by a certain set of initial conditions.

Gradual transitions infer some sort of preliminary change in the beginning state to establish initial conditions for the ending state. Perhaps space-time underwent a gradual warping and finally snapped to yield a tremendous amount of energy. But consider the instant when the *first change occurred* in the post-time model. In the beginning there is pure space-time. It remains pristinely uniform until it experiences an infinitesimal fluctuation. This might be only a precursor, but it is still a transition of state and can therefore be defined as energy's *origin*. The same is true of the origin of space-time from nothing. In this way, the origins of energy from space-time and space-time from nothing are both *instantaneous. They are the moment of the very first infinitely small change of state*. This is their intrinsic temporal boundary. It is the instant where the before and the after are different from each other by an infinitely small amount. *The beginning of any transition is by definition instantaneous.*

Herein lies the vexing problem with time before time. An instantaneous origin from nothing is not viable because there can be no *cause* for the event. An origin with no initial conditions has no relationship between the before and after. It is a discontinuity between different states. If, for instance, space-time changes to space-time-energy, the change must occur instantaneously by definition, and this means there can be no continuous connection between the two. No *cause* can exist in space-time for space-time-energy because, prior to the first infinitely small transition, space-time at any given moment is identically equal to any other moment, stretching back into infinite history.

First origins are inherently unviable and any cosmic origin is by its nature a first origin with an instantaneous beginning. Origins like these can have no causation because the state preceding them is perfectly homogeneous. They are the cosmic embodiment of the chicken and the egg. Chickens laying eggs and eggs hatching into chickens provide a consistent system because no discontinuous transition is required. Chickens cause eggs and eggs cause chickens continuously over time. Regardless of how finely time is sliced, change persists throughout. The Big Bang is postulated as a *first origin*, so it is the egg that never came from a chicken. Either it had no cause or is not a first origin, only one of infinitely many universal expansions and contractions. This is the motivation behind the quasi steady state cosmology described earlier. It avoids the paradox of a first origin but has its own issues. If the universe had no beginning why is it here now instead of just nothingness? There are trillions of stars out there and no one knows why. Even though the *time before time* problem is removed with an infinite history, the problem of the *source of the universe* remains. Until this is resolved, there's no way to know if the universe even had an origin.

*** Preview ***

There is a causal explanation for the universe and our investigation is slowly marching toward it. What this also means, by the above reasoning, is the universe had no origin and has existed forever. As it turns out, the knowledge of why untold billions of stars and galaxies exist makes the universe's eternal nature a foregone conclusion. This, like its size and conservation, is mandated by its fundamental essence.

SOURCE OF THE UNIVERSE

Clearly, a universal origin has some serious scheduling problems, but it provides hints for where to look next. Investigating *when* energy came from space or *when* space came from nothing leads to the more essential problem of *how*. If the universe had an origin how did it *come from nothing?* If the universe had no origin why is it here now *instead of nothing?* The time before time depends on the mechanism driving its emergence. This means the universe's source is a more profound cosmological problem.

One and only one question must be answered to understand why the universe exists. It will be called the *Prime Paradox*:

PRIME PARADOX

HOW IS EXISTENCE DERIVED FROM NONEXISTENCE?

where *existence* is a formal term for the entirety of the universe and *nonexistence* is the state of its absence, nothingness. The terms *existence*, *universe*, and *something* will be used interchangeably to refer to reality, while *nonexistence*, *nothing*, and *nothingness* are synonymous for the lack thereof.

Compared to the Prime Paradox, the universe's origin and size are mere details. All of the Prime Questions revolve around this solitary enigma - it is the unfathomable riddle implicit in each of them. The relationship between something and nothing is so pervasive that it surfaces in literature as often as it does in science. Lucretius and Einstein contemplated it, and so did Shakespeare:

Nothing will come of nothing.

(King Lear, Act I, Scene I) Ironically Shakespeare came as close to the truth as Einstein did, perhaps closer. But no one has been able or willing to take the next logical step.

The Prime Paradox touches every aspect of the Prime Questions and the universe problem they represent. It is the unified mystery of *why the universe exists*. But what a question! *How is existence derived from nonexistence?* No wonder everyone went off to look for black holes and quarks. Solving this paradox is the ultimate parlor trick. While no one has proven such a derivation is impossible, no one has ever come close to doing it, either. Not to worry though, *if there was no solution, the universe wouldn't exist and we wouldn't be here asking these questions*.

1.4 EXISTENCE, DEFINED

We have to know where the universe came from before we can possibly begin to understand it, and the answer lies in the Prime Paradox and nowhere else. The universe's conservation and size are important clues about its underlying nature, but they only serve to deepen the mystery of its existence. However, universal properties, regardless of how enigmatic they might be, are not paradoxical in and of themselves. The real reason the Prime Paradox is a paradox is because *something and nothing are perceived to be radically and irreconcilably different from each other*. Matter and energy are significantly different from empty space; space is significantly different from a dimensionless void. In either case an insoluble disparity exists. Consider for a moment:

TEST PARADOX 1: How is something derived from something?

Trivial. Any chemical reaction converts one thing into another. Or:

TEST PARADOX 2: How is nothing derived from nothing?

Also trivial. *Nothing from nothing leaves nothing.* There is no contention in either of these examples as there is no enormous difference to overcome. This is not the case in the Prime

Paradox. A void is empty of space, energy, time, and all else, while the universe contains massive amounts of solid material. *Real* material. To get one from the other is like turning a vacuum into steel. The mindbender isn't the process of one thing changing into another; it is the *perceived magnitude of the required transition*. Getting smoke from air seems reasonable since they are similar forms of the same thing. Pulling steel from a vacuum is far more impressive.

Since the root of the problem is the scope of the cosmic transmutation, let's consider it as a set of two smaller steps:

<GENERATING A UNIVERSE>

NOTHING ⇒ SPACE-TIME

SPACE-TIME ⇒ SPACE-TIME-ENERGY

If this looks familiar it should; it is a reiteration of one of the models used in the time before time problem. The context now is the basic *difference* between the states in each transition, not the time they occurred. At which step in the above process does "something" come from "nothing"? Is space-time a "something" like energy? Or is empty space-time just a form of nothing and energy is the only thing of consequence? It is not possible to characterize the difference between something and nothing without cohesive definitions of both. Let's see what we can do to identify the bedrock beneath reality.

QUANTITY

The Prime Paradox is a closed system. The step from nonexistence to existence can be understood only by knowing what each of these states ultimately represents. Can existence be characterized without knowing how it is related to nonexistence? Absolutely. Description and understanding are two entirely different things.

Existence is often defined as our ability to measure, where reality is strictly limited to the known empirical realm. If it can be observed it exists; if it cannot it does not. This is a far too limited perspective. The existence of a thing does not depend on our ability to measure it. Just because our telescopes can't detect planets orbiting stars in other galaxies doesn't mean they don't exist. The aliens looking back at us probably can't see us, either. Protons are too small to see with our microscopes, but they exist nonetheless. There are limits on the macro and micro scale for any instrument. Our ability or inability to measure a thing is of little consequence to the universe, and certainly not a requirement for existence. To exist is to have an attribute that (if sufficiently large) *could be measured.*

The binding energy of deuterium has been measured as 2.224589 ± 0.000002 MEV (million electron volts).⟨18.4⟩ In reality, it might be 2.224589625284... MEV. Does this mean the additional 0.000000625284... MEV does not exist? No. This fraction exists in the same way as 2.224589 MEV; it simply falls below our detection threshold. The existence of a thing is related to the quantity measured, not the size of said quantity.

Physics' International System of units (SI) is a specific reflection of our working relationship with existence. Its base units can express quantities of anything in nature, and are listed below:

SI Units	
Measurement	**Unit**
Length	meter
Mass	kilogram
Time	second
Current	ampere
Temperature	Kelvin
Numerical count	mole
Light intensity	candela

Table (1.2) International system of units (SI)

There are a number of other measurement standards, but the SI is the most widely used. All of the units in Table (1.2) have one thing in common - they are all expressions of *quantity*, the primary feature of existence. The definition of existence used in *Null Physics* follows from this:

§ DEFINITION 1.1 - EXIST

TO HAVE QUANTITY

§ DEFINITION 1.2 - EXISTENCE, [∃]

THE SET OF ALL QUANTIFIABLE THINGS

Quantity is in turn defined as a value on a continuous scale, something expressible as a real number. Definitions of nothingness are the converse of quantity:

§ DEFINITION 1.3 - NONEXIST

TO HAVE NO QUANTITY

§ DEFINITION 1.4 - NONEXISTENCE, ℵ

THE SET OF ALL THINGS WITH NO QUANTITY

If these definitions seem simplistic it is because the deepest level of reality is simple; *less is more*. If a thing has a quantity of any type it exists. Space exists in terms of volume, time in terms of seconds, and energy in terms of joules. Some values are a ratio of quantities, such as velocity. It exists by virtue of the existence of time and distance. The universe consists of things with quantity and the relationships between those things.

One of the far-reaching implications of our definition of existence is that the difference between nonexistence and existence rests on only a single characteristic. There is no other distinction as to whether something exists or not. Space exists, time exists, matter exists, and photons exist because they all have quantity.

SPACE - AN ETHEREAL YET INDESTRUCTIBLE FABRIC

Our definition of existence diverges from the beaten path by maintaining the physical existence of space. This is not widely accepted even though General Relativity requires spatial curvature. In modern physics, space provides a framework for reality; it is not a constituent of reality itself (at least on par with energy). The exclusion of space from the set of *something* comes about because it is not possible to directly interact with it. Space is the most inert medium imaginable. Herein lies more of the confusion that pervades contemporary physics. Space is not real, yet is curved around massive objects. Matter is solid, yet is ~100% empty space. Einstein even described matter as *curved empty space*. If space is not a real, physical thing, what is it? Space is just as real as energy because *volume* is a quantity of space in precisely the same way that *joule* is a quantity of energy. Energy density is energy/volume. If space did not exist, how is it possible to compare it so directly to something that does?

Space is a thing just as energy is a thing. They differ from each other, just as waves on a pond differ from the water they move through, but they have comparable existence, and therefore comparable *realness*. Interaction and existence are two entirely different considerations. The two most powerful lasers in the world will pass through each other with no detectable interaction. This doesn't mean the light in the beams is an illusion; it merely demonstrates the weakness of photon-photon interaction. It is too small to measure. Each of the laser beams exist because they contain energy with quantity. Their realness could be easily demonstrated by tossing a wad of paper into their path and watching it burst into flame. Whether one laser interacts with the other is of no consequence. Existence is quantifiable, not circumstantial. Since some forms of energy do not noticeably interact with each other, the lack of interaction between energy and space is certainly irrelevant to its existence.

Perhaps the best way to incorporate space into the set of existence is to look closer at the meaning of *interaction*. The dynamic interaction between forms of matter or energy is measured as an exchange of energy or momentum. This type of interaction is not possible between energy and empty space. Space is space and energy is energy. But this does not mean space and matter do not interact *geometrically*. The presence of space allows for the distribution of energy. Without this interaction, reality would be a single point of infinite energy density.

Space is not a material thing, but is still a real, physical thing. Although this might currently be an unpopular interpretation, it is not at odds with relativity. Einstein's theory maintains that space has no effect on the physics of matter and is therefore impossible to isolate. It is not a statement of spatial nonexistence. While this might seem a minor semantic, it is a crucial distinction in the grand geometric scheme of things. However, this is not to say that no distinction will be made between the existence of space and that of matter. The terms *material existence* and *material universe* will be used to refer to the sum total of the universe's matter and antimatter, which is a subset of all of existence.

Reality is synonymous with existence, so it follows that the universe is composed exclusively of real things. Electrons are as real as chairs and photons as real as planets. The existence of space, energy, or anything else is not dependent in any way on human observation, interaction, or interpretation. In short, the sum total of quantity *is* reality. A number of things may have no measurable interaction with each other, but if they have quantity of any form they are *real*. Quantity and reality are inseparable as well as indistinguishable, and no unified description of our universe can possibly succeed without a full recognition of this fact.

REALNESS

The emergence of existence from nonexistence is incomprehensible due to the staggering difference between the two states, yet energy and space share a similar level of *realness*. This begs the question of just how real anything is. Getting steel from empty space is an impressive trick, but:

How hard is steel and how thin is space?

The struggle to this point will not end with the trite claim that getting something from nothing is easy because nothing really exists. While this is sometimes postulated in philosophies that deny reality itself (relativism), the question of how real is real or how solid is solid must still be addressed in any theory about reality. Since our existence is perceived to be a paradox, its inherent nature has clearly been misinterpreted. Any theory attempting to answer the Prime Paradox must redefine the worldview in order for our existence to be a logical consequence of reality. The trick is not to do this by denying reality; the trick is to do this by embracing it. Unfortunately, the task of figuring out just what is meant by *reality* remains.

The universe provides a few clues about its true essence:

1. Matter is for the most part empty space. The search into the innermost regions of elementary particles yields no fixed boundaries, only fields and interactions.

2. Matter and energy have wave-particle features consistent with their distributed nature.

3. Matter and energy are two expressions of the same substance - energy. In *Null Physics* the energy equivalent of a particle's mass (mc^2) will be called its *rest energy*.

4. Gravitational fields are best described as curved regions of space-time. Einstein called matter and energy empty spatial curvatures.

5. The only thing the *solidity* of matter can be compared to is other matter. This means that solidity is entirely self-referencing.

Suppose, for the sake of argument, energy and matter were tiny curvatures of space-time and the greater the energy density, the greater the curvature. This would be consistent with all of the above properties. Energy as curvature makes space-time the tapestry upon which the universe is written. So, if steel is the result of bending space into another dimension:

How hard is steel and how thin is space?

Perhaps the derivation of a universe is not so impossible after all. First determine how to get space from nothingness. This might be doable since it is just an empty vacuum anyway. Next bend it here and there to yield energy and matter.

* Final Preview *

The universe contains only two things: space-time and curvatures of space-time. This is why energy is as real as space; they are both composed of the same thing: space. Energy's geometric structure is somewhat more complicated than the curvature described in Einstein's General Theory of Relativity, but it is still nothing more than curved space.

1.5 EVERYTHING FROM NOTHING

We are now in a position to solve the Prime Paradox. Existence is derived from nothingness as follows:

1. Assume nonexistence is the prevailing state of reality. Nothing exists: no stars, no planets, no space, and no time. It is an utter void.

2. Existence is a state *different* from nonexistence. Generally it is defined as the converse of nonexistence, but for our purposes here it will suffice to say it is *different*.

3. If nonexistence is the complete state, the only state different from it is a subset, or incomplete nonexistence. Nothingness is by definition composed of nothing and incomplete nonexistence is a subset of this composition.

4. There are only two existential states, existence and nonexistence. Similarly, nothingness is either complete or incomplete.

5. It follows that existence is the state of *incomplete nonexistence*, as it is the only thing different from total nonexistence.

The thought process arrives at the same conclusion if the starting point is existence. Consider a universe filled with air, water, and light. Now turn off the lights. Then drain out the water. Then let the air escape. Each step is a step closer to nothingness. The completeness of nothingness can be judged by the degree to which things are lacking. The lack of air, water, and light is a more complete void than just the lack of water or air.

In general, the lack of part of existence is an incomplete form of the lack of all of existence, which is defined as the state of nonexistence.

Existence is by definition incomplete nonexistence.

Let's look at this argument graphically.

Suppose existence consists of the union of three sets:

Figure (1.3) Existence as three regions

This can be symbolized as:

$$\exists \equiv \exists_1 \cup \exists_2 \cup \exists_3 \tag{1.1}$$

Nonexistence is the converse of this:

$$\overline{\exists} \equiv \overline{\exists_1 \cup \exists_2 \cup \exists_3} \equiv \overline{\exists}_1 \cap \overline{\exists}_2 \cap \overline{\exists}_3 \tag{1.2}$$

The nonexistence of this universe is therefore the simultaneous nonexistence of each of its elements.

But the *nonexistence* of any given element is the *existence* of the rest:

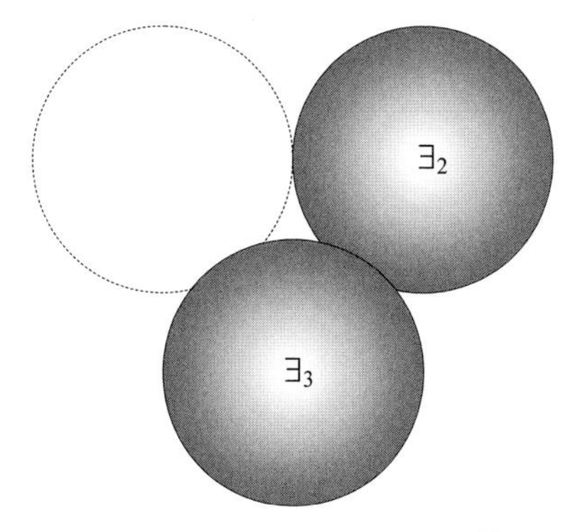

Figure (1.4) Nonexistence of $\exists_1$

Or symbolically:

$$\overline{\exists}_1 \equiv \exists_2 \cup \exists_3 \tag{1.3}$$

This demonstrates a valid yet odd relationship between the nonexistence of a given object and the existence of others. Existence is a case of *incomplete* nonexistence; it is innately contextual. The nonexistence of one thing is directly related, or dependent, on the existence of others. But this isn't the relationship needed to solve the Prime Paradox. To get something from nothing is to demonstrate the relationship between the existence and nonexistence *of the same object*. Equation (1.3) only shows us existence is by definition incomplete, or *partial* nonexistence.

PARTIAL NONEXISTENCE, THE LIMITING CASE

Suppose the universe's material existence (matter and antimatter) is an infinite union of the form:

$$\exists \equiv \bigcup_{i=1}^{\infty} \exists_i \tag{1.4}$$

where $\exists_i$ are the rest energies of its elementary particles.

The nonexistence of this universe is expressed as:

$$\overline{\exists} \equiv \bigcap_{i=1}^{\infty} \overline{\exists}_i \tag{1.5}$$

and the nonexistence of the rest energy of any particle j is:

$$\overline{\exists}_j \equiv \bigcup_{\substack{i=1 \\ i \neq j}}^{\infty} \exists_i \tag{1.6}$$

This means, strangely enough, that the existence of the sum total of all of the universe's material is equivalent to the nonexistence of its smallest conceivable element. If, for example, $\exists_j$ represents a single particle, then Equation (1.6) indicates that the physical nonexistence of a single particle is the same as the existence of *all of the matter and antimatter of the universe minus said particle*.

Getting a particle from nothing is the difference between its existence and nonexistence. In order to calculate this difference, the magnitude of material existence will be defined as:

$$|\exists| = \sum_{i=1}^{\infty} |\exists_i| \tag{1.7}$$

where individual elements are taken to be nonoverlapping in keeping with energy conservation. The magnitude of the Earth's existence, for example, is the sum of the energy of all of its constituent particles.

Applying this to the nonexistence of an individual particle, Equation (1.6), yields:

$$\left|\overline{\exists}_j\right| = \sum_{\substack{i=1 \\ i \neq j}}^{\infty} \left|\exists_i\right| = \sum_{i=1}^{\infty} \left|\exists_i\right| - \left|\exists_j\right| \tag{1.8}$$

where the universe's matter (energy) is immense in comparison to a single particle. So while the existence of something can be quantified in terms of an amount of energy, the magnitude of its nonexistence is instead a contextual relationship scaled by the size of the universe in which the object exists. There are two terms in Equation (1.8), a summation and a single particle. The summation is the contextual (external) component of the particle's nonexistence. The particle is the intrinsic (internal) component. Changing the universe's size has no effect on the size of the particle's intrinsic component. Note that this is true whether its universal context is finite or infinite.

So how much energy does it take to create an electron out of nothing? It is the difference between the *existence and nonexistence of an electron*. Using Equation (1.8), treating the electron as the jth element of existence, we have:

$$\left|\exists_{e^-}\right| - \left|\overline{\exists}_{e^-}\right| = \left|\exists_{e^-}\right| - \left(\sum_{i=1}^{\infty} \left|\exists_i\right| - \left|\exists_{e^-}\right|\right) = 2\left|\exists_{e^-}\right| - \sum_{i=1}^{\infty} \left|\exists_i\right| \tag{1.9}$$

The difference between the existence and nonexistence of an electron is equal to *twice an electron's energy minus the energy of the entire universe* - an interesting result.

The amount of energy needed to create an electron should be comparable to its rest energy, not the energy of the universe. In fact, we know it to be precisely twice its rest energy since electrons are always created with their antimatter twin, the positron.

Thus:

$$\left|\exists_{e^-}\right| - \left|\overline{\exists}_{e^-}\right| = 2\left|\exists_{e^-}\right| - \sum_{i=1}^{\infty} \left|\exists_i\right| = 2\left|\exists_{e^-}\right| \tag{1.10}$$

There is only one case where the above is true:

$$\sum_{i=1}^{\infty} |\exists_i| = |\exists| = 0 = \aleph \qquad (1.11)$$

The sum of all material existence is zero (nonexistence). If the universe can be derived from nothing it must be a form of nothing, and so its sum total is nothing. What Equation (1.11) indicates, which seems like a bold-faced contradiction, is that the amount of energy necessary to generate a universe from scratch is zero. How can energy be required to build electrons while building an entire universe is free? *Because the nature of existence is contextual.* Equation (1.11) is a requirement for the sum of all existence, not the sum of a few particles. It is no coincidence that matter cannot be created without an equal creation of antimatter. The two must balance on a universal scale in order to express the true nature of existence as a form of nothing. Zero energy is required for the existence of the universe *because nothingness is all there is to work with in the first place*. This is true *regardless of the universe's size.*

Please note that this does not indicate antimatter has negative energy in terms of conservation. Antimatter is the mirror image of matter, its negative existence. An electron is space with negative curvature; a positron has the opposite curvature. Matter and antimatter sum to zero in the context of the entire universe, but energy is always conserved within it.

The true relationship between the existential states will be called the *Null Axiom*:

Φ AXIOM 1.1 - NULL AXIOM

EXISTENCE SUMS TO NONEXISTENCE

This is the solution to the Prime Paradox, and is precisely the opposite of how the reality question has always been approached. Theorists throughout history have been so enthralled with existence that nonexistence languished in near obscurity. All the while, the relationship between the existential states remained a paradox because the physical implication of *zero* was never recognized. Existence is by definition an incomplete state of nonexistence and must therefore sum to nonexistence. The unequivocal evidence of this has been taunting us for centuries.

THE REAL ZERO

A number of questions need to be answered before the Null Axiom's physical reality becomes fully evident. The first is how, precisely, can existence be composed of nothing?

The solution to this question begins with the composition of nothingness. Nothingness can have only one type of component, itself:

NOTHINGNESS ≡ ... ∪ NOTHINGNESS ∪ NOTHINGNESS ∪ ...

Or symbolically:

$$\varnothing \equiv \cdots \cup \varnothing \cup \varnothing \cup \varnothing \cup \varnothing \cup \varnothing \cup \varnothing \cup \cdots \quad (1.12)$$

where ∅ is the empty set.

A profound relationship has just emerged. On the left side of the equation a null set prevails and on the other side a relationship between null sets exists. Where before there was nonexistence there is now a relationship between nonexistent sets. *This relationship is what existence is.* The only component nonexistence can have is nonexistence, as it is a total void. *Yet the fact that it can be a component of itself is sufficient to provide a relationship, and this relationship is existence in its purest form.*

Mathematicians come tantalizing close to this concept when they construct number systems directly from combinations of empty sets.⟨13.1⟩ The difference between these efforts and the Null Axiom is the realization that at the most austere level (even though mathematicians might be the first to disagree), *mathematics is more physical reality than human invention.* Math is, as is everything else, derived from the universe. There is no other source, because the universe provides the entirety of the input that human consciousness uses to create math in the first place. As such, empty sets have a *physical* analog, and so does their relationship to each other. Although it is certainly possible to build number systems using empty sets, the far more sweeping perspective is that *they are also the bedrock of reality's construction.*

Existence sums to nonexistence so *nonexistence is composed of existence.* As the union of empty sets, *nonexistence is composed of itself.* The only way nonexistence can be composed of both states is if existence is also composed of nonexistence:

Ψ THEOREM 1.1 - COMPOSITION OF EXISTENCE {Φ1.1}

EXISTENCE IS COMPOSED OF NONEXISTENCE

Everything is constructed of the same thing: nothing. There is no logical alternative. In the preceding sections it became clear that the universe's source was nothingness: the lack of the universe. Whether this happens with or without an origin is immaterial. In the final analysis, the universe is derived from nothingness as there is no other conceivable basis, so the only material it could possibly be built of is nothing. The Null Axiom is a statement of the limitations for a source of existence. Existence is a form of nothing because nothing is literally the only available ingredient. The Null Axiom solves the problem of the source of existence as well as the problem of its composition, and it does so with absolute inevitability. The universe is not a paradox; it is an inescapable reality because *nothingness must be a component of itself.*

The Null Axiom solves the time before time problem as well. The universe had no origin from nothingness; it is just another expression of it. *Parts do not come from a whole; they are its necessary constituents.* An origin of existence from nonexistence would mean existence *was not composed of nothingness.* This is not possible as no other source is available.

The two sides of Equation (1.12):

$$\varnothing \equiv \cdots \cup \varnothing \cup \varnothing \cup \varnothing \cup \varnothing \cup \varnothing \cup \varnothing \cup \cdots$$

are equal as well as indistinguishable. The only way to have an origin is if they were not, because an origin represents a difference between one state and another. The equality specifies that there is no difference and this eliminates the possibility of an origin.

A hypothetical transition of the form:

$$\varnothing ? \rightarrow \cdots \cup \varnothing \cup \varnothing \cup \varnothing \cup \varnothing \cup \varnothing \cup \varnothing \cup \cdots \tag{1.13}$$

would suggest that the original global empty set no longer exists. This is not valid because the union of all empty sets is the immutable relationship between them. Their union still has a value identically equal to the original nonexistent set. It is present whether the expansion of the individual empty sets is shown or not, so there is no transition and therefore no origin. The overall union exists by default and the subsets within it exist because they are a logical and unavoidable consequence. *Any empty set is composed of the union of an infinite number of empty sets.*

We think the universe came from somewhere because we cannot see its existential sum. From our finite perspective, only a small part of nonexistence is visible. The relationship between existence and nonexistence is one of perspective and composition, not process. There never was a universal origin. It is simply not possible, because there is no difference between nothing and a universe whose sum is nothing. *The Big Bang is quite literally the twenty-first century's version of the flat earth concept. It couldn't be further from the truth, yet nearly everyone believes it.*

FROM SETS TO CONTINUUM

The expressions used to support the Null Axiom thus far have been sets and subsets. While these are valid, they fail to fully express the nature of existence as a set of quantifiable objects. Quantity is a continuum that is more consistent with the mathematical operations of addition and subtraction than intersection and union. Space is continuous, energy distributions are continuous, and the key aspect of reality reflected in the real number system is its continuous, dimensionally structured distribution.

The universe can therefore be accurately described as a zero-sum equation of the form:

Ψ THEOREM 1.2 - ZERO EQUATION {Φ1.1}

$$0 = \ldots + 0 + e^{+} + 0 + p^{-} + 0 + \gamma^{(+-)} + 0 + 0 + e^{-} + p^{+} + 0 + \gamma^{(+-)} + \ldots$$

On the right are an infinite number of positive and negative quantities dispersed throughout a similarly unbounded collection of zero quantities. On the left is zero. The right represents the infinite distribution of matter, antimatter, and photons scattered across space. The left represents the state of nonexistence itself. They are inseparable and always have been. As indicated by their superscripts, photons are neutral by virtue of the combination of their internal positive and negative fields, not by their lack of polarity.

There is literally no limit to the amount of nothing within nothing. *The universe, as the physical manifestation of the zero equation, must be infinite to express this fact*. The universe is not expanding for the same reason the state of nonexistence is not expanding; it cannot. It is already infinitely large. Any increase in size would mean the preexisting universe was smaller and therefore an incomplete (invalid) expression of nonexistence. Nothingness defines the universe's full extent, and it is immeasurably vast:

Ψ THEOREM 1.3 - EXTENT OF EXISTENCE {Φ1.1}

EXISTENCE IS INFINITE AND ETERNAL

There is no limit to space or energy or time. The universe has existed forever and will continue to exist for eternity. The billions of galaxies visible to our most powerful telescopes don't even scratch the surface of nature's immensity. Pick a direction, any direction. A spaceship leaving in that direction could travel forever, continually finding new galaxies and uncharted space *regardless of how far it went*. A boundary on any quantifiable aspect of the entire universe is a boundary on the state of nonexistence. Nonexistence is not a trillion light years in diameter. It does not have a mass of a trillion quadrillion suns. Quantities are finite; they characterize the state of existence, not nonexistence.

Nonexistence is the province of unboundedness. This is precisely the opposite of how its magnitude has always been portrayed in mathematics: *zero*. From our local perspective, a statement such as:

A photon has no mass

certainly seems unambiguous. It does not imply infinitely large mass; if anything it suggests *infinitely small mass*. Similarly, if a scientist said an electron has *no* volume, this means it has a volume infinitely smaller than an atom. The apparently equivalent relationship between no quantity and infinitely small quantity is the way zero has always been perceived. Mathematicians may disagree about the difference between a null quantity and an infinitely small quantity, but this difference has never been recognized as having physical significance or being particularly important. Nothing could be further from the truth. This difference is everything:

Ψ THEOREM 1.4 - CONTEXTUAL EXISTENCE {Φ1.1}

THE DIFFERENCE BETWEEN INFINITE SMALLNESS AND ZERO'S EXTERNAL PERSPECTIVE IS INFINITE LARGENESS

Recall that the nonexistence of an electron is equivalent to the existence of all of the universe's energy less a single electron. Now extend this down to the nonexistence of an infinitely small portion of a single electron's rest energy. The difference between no energy and infinitely small energy is whether or not this tiny bit of electron exists, and the smaller it is, the closer this difference comes to the entirety of the universe's energy.

There are two perspectives to nonexistence, internal and external, and they are completely isolated from each other. The internal perspective is only available to the finite beings within nonexistence that measure their surroundings on scales comparable to themselves. Reality appears immense to them, and they in turn consider themselves gigantic in comparison to infinitely small regions. The external perspective is not available through observation, but it governs the universal properties these beings witness, such as the curious symmetry between matter and antimatter.

1.6 NULL EMPIRICISM

Since no Big Bang occurred and the universe is not expanding, the redshift of distant objects and the CMB that permeates space require alternate explanations. Both will be explored in detail in Part IV, resulting in some surprising discoveries. For now our focus remains on the universe's basic properties.

The relentless truth of the Null Axiom is everywhere:

1. Matter and antimatter are always created in equal, yet opposite amounts, whose electrical sum is **zero**.

2. Positive and negative electric fields sum to a neutral universe with **zero** net electrical charge.

3. Energy is conserved in all interactions; the magnitude of the universe's energy has **zero** change.

4. Space is a collection of points, little bits of nothingness itself, embodiments of the geometric **zero**.

5. Charge must be conserved in particle interactions; the sum of charge differences is **zero**.

6. Momentum is conserved, so the universe's net momentum remains constant at **zero**.

All around us are clues to the true nature of existence and many have been there for hundreds of years - clues telling us what the sum of everything really is. As scale reaches cosmic proportions, every charge sums to zero and every vector quantity sums to zero. *The universe is infinite and eternal because nonexistence contains no boundaries to limit it in space or time. Its material sums to neutrality because as a component of zero no other state is possible.*

Reality certainly seems to have substance. The idea that it is composed entirely of pieces of geometric nothingness might be unsettling. Yet there is no other possible conclusion. The only material available to build the universe from is nothingness, so what else could it be composed of? The universe's components are real to each other; but there is no absolute realness; it all vanishes in the cosmic summation. Everything is composed of the same form of nothingness, so reality itself is uniform and unconditional:

Ψ THEOREM 1.5 - UNIFORM AND UNCONDITIONAL REALITY {Φ1.1}

ALL OF THE UNIVERSE'S COMPONENTS HAVE THE SAME UNIFORM AND UNCONDITIONAL LEVEL OF REALNESS BECAUSE THEY ARE ALL COMPOSED OF NOTHINGNESS

Reality is a complete system. Excluding the realness of any of its constituents prevents a full comprehension of the whole. No parts are optional, nor are some less real than others. The human predisposition of establishing hierarchies of realness is far and away the greatest obstacle between contemporary science and a deep understanding of the natural world.

The universe is the *most complex, perfect equation possible*. It is simple; it adds to zero; but as we will discover in Part III, there is more information in a single electron than the sum total of electronic storage on Earth. All the things of our local environment - chairs, windows, desks, trees, mountains - are intricate distributions of curved space. So are the tens of billions of galaxies, each with tens of billions of stars, that our telescopes are currently able to detect. The sum total of our observable universe is the tiniest bit of the four-dimensional equation we call existence. It is all *just an equation*. All of the particles, all of our instruments, the planets and the stars, are part of an equation so indescribably precise and beautiful it adds up to zero down to the smallest fraction of a single electron's rest energy. There are no rounding errors in this equation; there is no fuzziness or uncertainty. It is perfection personified.

Universe as equation isn't an unprecedented concept. The idea of a cosmic wave function has surfaced in a number of theories, where every interaction in the universe is determined by a single all-encompassing probability distribution. It even evaluates to the same value as the zero equation of (Ψ1.2):

$$\psi_U(x, y, z, t) = 0$$

The difference between this and our premise is that unlike this wave function, the zero equation expresses the physical implications of zero and its invariant connection to universal composition. This relationship is not and cannot be statistical.

As mentioned earlier, some cosmologists have postulated that the Big Bang doesn't violate the conservation of energy because the total energy of the universe is in fact zero, where negative gravitational energy cancels the energy of matter and light throughout space. Consider for a moment how tantalizingly close this speculation is to the Null Axiom. One of its premises is wrong:

- zeroing energy *within* the universe would violate conservation, so gravitational energy cannot offset matter and light energy,

and another is missing:

- the realization that nothingness as a state of zero energy *is identical to a universe whose energy sums to zero*. If the universal sum before and after the "origin" is zero energy, where is the transition called the origin? Without an origin, positive and negative energy become intrinsic components of nothing … the Null Axiom.

So close and yet so far.

ANTIMATTER'S SYMMETRY

One of the theoretical issues the Big Bang has unsuccessfully tried to overcome is the universe's apparent predisposition toward matter. The pure energy of its alleged emergence ought to have condensed into equal parts matter and antimatter, which in turn would have reacted, producing light and a small residue composed of equal parts matter and antimatter. These opposites would continue to obliterate each other, eventually leaving a universe filled with nothing but light. In order to resolve this serious conundrum, scientists have painstakingly searched for any form of asymmetry between matter and its inverse.

In 1964, Val Fitch and James Cronin of Princeton discovered a difference of about 0.2% in the decay modes of positive and negative kaons. A similar finding was reported for B mesons in 2001. As it turns out, the observed asymmetries are far too small to support matter's perceived universal dominance. More to the point, however, they *categorically do not represent a violation of the flawless symmetry between matter and antimatter.* Unstable elementary particles decay after brief average lifespans; they are particularly brief in the case of the B meson. These events occur because of internal change in the particle. Like the universe, elementary decay is causal. It is a product of initial conditions and interactions. One kaon decays into an electron and photon, another into a muon and photon. This is not a case of chance; it's a case of circumstance.

Unstable particles interact with their environment during their brief lifespan, and the environment in our laboratories is composed *entirely of matter.* The only way matter-antimatter asymmetry could be truly isolated is to study particles and antiparticles with detectors composed of matter and antimatter. The extraordinary volatility of antimatter makes this impractical. Matter and its inverse are utterly symmetric in every sense of the word; any asymmetries we detect in local experiments are the product of an environment disproportionately composed of matter.

CONSERVATION

A number of universal properties start to make sense in light of the Null Axiom. One of these is *conservation*. Nonexistence is conserved because by definition it cannot be removed. Existence, as a partial form of this state, must be conserved as well:

Ψ THEOREM 1.6 - CONSERVATION OF EXISTENCE {Φ1.1}
FUNDAMENTAL QUANTITIES OF EXISTENCE ARE CONSERVED

The universe must exist as a consequence of nonexistence, and conservation is the physical realization of the relationship between the existential states. It connects the immutability of unbounded nothingness to the finite amounts of energy stored in matter and light.

Bits of existence fluctuate in many ways, but the nature of their underlying quantity never varies. To know what the universe is really made of is to look at its conserved quantities. Everything else is cosmetic. Energy is conserved because (as demonstrated later) it is vital to the universe's dimensional configuration. Space is conserved because the universe's volume cannot change and each and every cubic meter within it must be preserved as parts of the greater invariant whole. Indeed, the only reason gravitational fields can distort space is because its volume is a conserved quantity. *If spatial volume could vary when it was distorted by matter, curvature would not exist and neither would gravity.*

Three different types of quantities of existence can be identified based on the nature of their conservation:

- Quantities conserved by magnitude: *energy and space.* The total quantity of energy present in any interaction is conserved. Similarly, although space can be displaced by gravitational fields, its total volume remains constant.

- Quantities conserved by summation: *charge and momentum.* The *net* momentum and charge within any isolated group of particles are both conserved. This form of conservation preserves the universe's sum, but does not address its substance.

- Variant quantities: *pressure, force, etc.* These are either cosmetic characteristics of energy forms or the relationship between more substantial quantities. They exist but are not directly linked to reality's composition.

1.7 INFINITE OVERVIEW

The Null Axiom tells us why the universe is present (it has always coexisted with nothingness). It tells us what its net charge is (neutral). It tells us the universal ratio between matter and antimatter (unity). It tells us how big the universe is (infinite). All of the universe's truly mysterious features are bound to nonexistence in one way or another. The paradox of getting something from nothing, the riddle of time before time, the enigmatic nature of infinity - all of these are found in the same relationship that our love of substance and solidity has obscured since the dawn of rational thought. Let's take a final look at the Prime Questions, and use the Null Axiom to fully answer each in turn.

1. Why does the universe exist?

If the universe didn't exist, nonexistence would be the prevailing state. *Yet the universe is still its internal structure.* Parts coexist with their whole. It is not possible to have nonexistence without the universe as they are two forms of the same thing. The universe exists because there is no other alternative.

2. Where did it [the universe] come from?

It didn't come from anywhere; it has always existed. Nothingness does not transform itself into reality - the relationship between the two is fixed. There is no existential transmutation of any kind in nature. Existence is partial nonexistence. Parts do not come from the whole, they are its substructure. This is why there can never be a violation of the conservation of energy or the conservation of space. And what holds true for energy and space holds true for the existential soup of which they are composed.

3. What is it [the universe] made of at the smallest level?

Nothing. What else could it be made of? The universe is a distribution of nothing. It is unthinkably huge; it is inconceivably perfect, yet it is also ultimately no more real than an equation. To the happy inhabitants of Earth, the universe could not be more substantial. But the only comparisons we can draw are made within the universe and they are of necessity self-referencing. How small is a point, how large is space? Matter is built of spatial curvature, space is built of geometric points, and points are nothingness incarnate. This means matter is composed of *points*. The infinite universe full of matter and energy of every conceivable form is a scintillating array of geometric points. When the dots are connected and the universe is viewed from the outside, the summation vanishes into the solitary state of nothingness.

Every aspect of the universe as a whole:

- Apparently boundless space;
- Overall electric neutrality;
- Conservation of energy, momentum, and charge;

leads inexorably to nothingness. The zero equation is the reason the universe *is the way it is*, the reason why the universe *must be the way it is*, and the reason why it *is*. The universe is *permanent*. It is not some cosmic toy doomed to run down and stop in a few billion years.

Nor is it the result of a random explosion in the distant past. It can't go away for there is quite literally no place for it to go. We can worry about Earth's destiny, but the universe it floats through is here to stay.

So take a moment and look at this page as a collection of tiny spatial curvatures called particles, assembled into atoms, and bound by weaker forces into molecules. The molecules are connected by even weaker electrostatic forces. The book is held down by the curvature of gravity. The sobering truth is the book, the Earth, and the universe and all of its contents are the solution to a single equation - nothing more, nothing less. There is no unreality; there is only comprehension. We are the lucky ones. After three billion years of wrenching evolutionary order out of chaos we are finally smart enough to know the truth. The universe, along with the struggling, occasionally sentient beings within it, exists *because there is no other possible alternative*.

NONEXISTENCE

The term *nonexistence* was originally defined as the set of all things with no quantity. While this remains true, the Null Axiom provides a more accurate formalization of this state:

Ψ THEOREM 1.7 - THE STATE OF NONEXISTENCE {Φ1.1}

NONEXISTENCE IS THE TOTALITY AND OVERVIEW OF EXISTENCE: $\aleph$

Nonexistence is symbolized as $\aleph$ per its original definition and will be referred to equivalently as *the overview*, *the void*, *nothingness,* or *totality*.

Although little is known about the nature of nonexistence, it is primarily characterized by the properties it lacks. Nonexistence is unbounded because it has no boundary. This gives it the simplest form of symmetry:

Ψ THEOREM 1.8 - SYMMETRY OF NONEXISTENCE {Ψ1.3, Ψ1.7}

NOTHINGNESS IS INHERENTLY SYMMETRIC

Nonexistence is symmetric just as zero is symmetric. Asymmetry is a net quantifiable attribute; the void has no such distinction.

1.8 REDEFINING EXISTENTIAL

The study of existence is the relationship between the universe and the lack thereof. It has nothing to do with inner subjective experience, though unfortunately this is how the term *existential* has historically been defined. The discovery of the Null Axiom calls for a redefinition of this term. Any form of the word *existence* used in *Null Physics*, such as *existential*, will refer specifically to fundamental physical characteristics.

§ DEFINITION 1.5 - EXISTENTIAL
AN ENTITY'S MOST BASIC PHYSICAL QUALITIES

Although the term *existential* will be used throughout our investigation, the term *existentialism* is not referenced because the study of the existential aspects of space, energy, particles, and photons is more properly called *Null Physics*. The designation *metaphysics* is customarily used to describe similar lines of inquiry, but this term is a misnomer based on the flawed idea that standard scientific and mathematical approaches do not work on reality's substrate. As will be demonstrated in the chapters to follow, the rationality of classical physics and math extends, uninterrupted, from the farthest reaches of space all the way down to the very core of existence.

1.9 GENERAL CONCLUSIONS

The guiding premise of *Null Physics* is the Null Axiom. It is an entirely new scientific paradigm - a concept that redefines our relationship to the physical world:

- The relationship and difference between something and nothing have been misinterpreted in the mathematical and physical sciences since their inception. Infinitely small quantity and zero quantity are vastly different things.

- The universe did not come from the void; it is a distributed form of nothing that sums to nothing and is composed of nothing.

- All of the universe's global qualities; its unboundedness, neutrality, perpetuity, and ultimately its very existence only make sense by viewing it as the inevitable and omnipresent substructure of nothingness.

- The universe is a perfect zero-sum equation.

2. FINITE HYPERSPACE

The previous chapter provided the underpinning for a universe composed of nothing. The next step is to build it from nothing, one level at a time.

2.1 SPACE FROM NOTHING

Reality begins with space, and space, like all of the other components of the universe, is a distribution of nothing. Consider the series:

$$0 = (\ldots + 0 + 0 + 0 + \ldots) \tag{2.1}$$

Nothing plus nothing ad infinitum yields nothing. This is one-dimensional space. The zeros in the series represent geometric points. Points are not abstractions; they are the physical incarnation of nonexistence. A point has no mass, charge, or any other property, except for its position relative to other points. It is the ideal minimalist representation of the contextual nature of existence.

The derivation of space is not a process whereby space emerges from nothingness; space is by Equation (2.1) an intrinsic component. On the left side of the equation there is nothing, and in the series there is a distribution of bits of nothing. There is no difference between these two forms. *Existence is the relationship of nonexistent objects.*

Now suppose each element of one-dimensional space is expanded into a series of its own, to:

$$\begin{array}{c} + \\ 0 \\ + \\ 0 = \ldots + 0 + 0 + 0 + 0 + \ldots \\ + \\ 0 \\ + \end{array} \tag{2.2}$$

or in a more compact form:

$$0 = (\ldots + 0 + 0 + 0 + \ldots)^2 \tag{2.3}$$

Equation (2.3) is, given the appropriate evaluation of its summed zero elements, an accurate representation of *two-dimensional space*. Repeated expansion results in spaces of higher dimensions, such as three-dimensional space:

$$0 = (\ldots + 0 + 0 + 0 + \ldots)^3 \tag{2.4}$$

All of these summations have no limit to the number of terms within them, thereby reflecting the boundlessness of space itself. *There is no limit to the amount of nothing within nothing.*

An expansion of zero is not infinite as a matter of chance; it must be infinite to accurately represent the state of nonexistence. The sum of a finite number of zeros doesn't represent nothingness:

$$0 \neq (0 + 0 + 0 + 0 + 0 + 0 + 0) \tag{2.5}$$

because nothingness contains no information to differentiate between internal elements and boundary elements (0_b):

$$0 \neq (0_b + 0 + 0 + 0 + 0 + 0 + 0_b) \tag{2.6}$$

Nor does it specify a certain number of terms. An accurate portrayal of zero's internal composition must contain an unlimited number of terms in order to express its *completeness*.

It is true that:

$$(0 + 0 + 0) = (0 + 0 + 0) \tag{2.7}$$

because both sides of the equation have the same finite number of terms. But this is only a small part of zero. The infinite expansion of Equation (2.1):

$$0 = (\ldots + 0 + 0 + 0 + \ldots)$$

expresses both the extent and composition of zero. The left side (nothingness) contains no information about the number of terms on the right and the only way to express this correctly is with an infinite series. *Points are cheap, space is big.* There is a marked distinction between the mathematical use of zero and the physical reality of zero. Treating zero as nothing more than an accounting necessity obscures its connection to the Null Axiom. *At the foundation of reality the simplest things are the most important things.*

2.2 SPATIAL STRUCTURE

Scientists have been trying to make the universe finite for as long as they have gazed into the starry sky. As comforting and human as finiteness may be, it simply isn't the nature of reality. Space and the material distributed therein are infinite in all directions. Lucretius knew this over two thousand years ago, he just didn't know why.

Suppose for a moment the universe was of a certain size, perhaps fifty billion light years in diameter. If this were true, what exists outside of it, a hundred billion light years away?

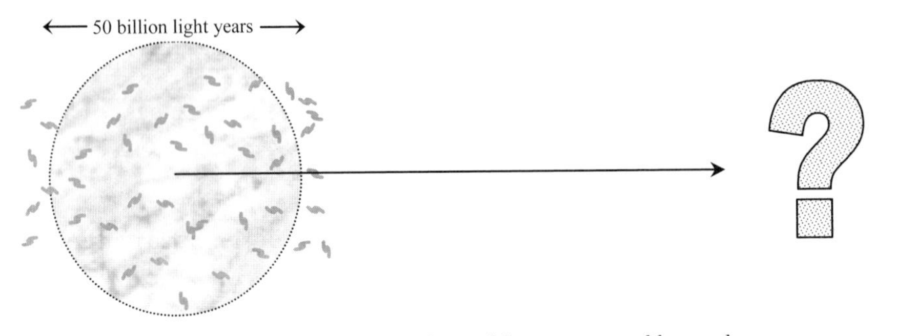

Figure (2.1) Observable universe and beyond

Similarly, if space-time were curved into a finite four-dimensional sphere, are other spheres floating near it, a few billion light years away through four-dimensional space?

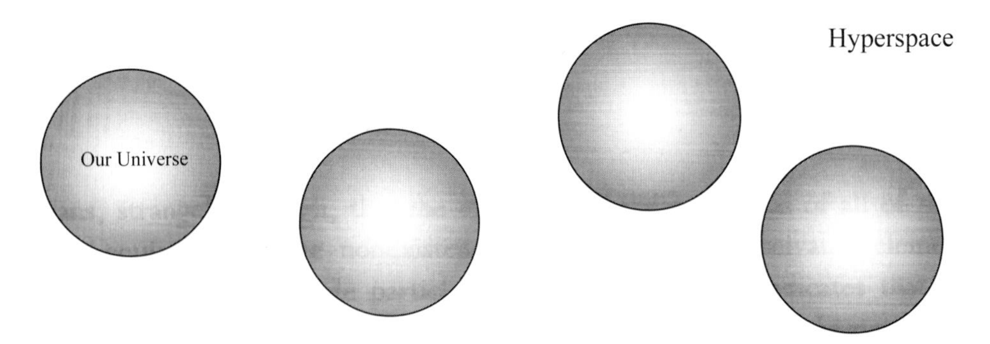

Figure (2.2) Rendering of multiple universes in a 4-D continuum

If so, is four-dimensional space infinite?

A finite universe is simply problematic. How can a seamless continuum be bounded? This is the same intractable issue as a first origin. Boundaries are inconsistent with the universe's basic nature because there is simply no limit to the amount of nothing within nothing. Distance is linear and metric. It allows extrapolation beyond or around any boundary possible to envision. This is one of the many difficulties the Big Bang approach never satisfactorily answers. Its inflation concept has space-time rushing outward in a rapid initial expansion, but expanding into what, exactly? How can space expand into a region already present by definition?

Unbounded space is usually referred to by mathematicians in terms of *closure*. A coordinate system is closed under vector addition, for instance, when a vector of any length can be added to another vector and the result still lies within the coordinate system. This is true for the space of our universe. If an object anywhere in the universe moves any finite distance in any direction, it is still in the universe. The reason for this arrives directly from nonexistence. There can be no boundaries in space because a boundary is a quantity, quantity is existence, and existence is *incomplete* nonexistence. Boundaries are local distances such as fifty billion light years. Space is forever. In *Null Physics* the lack of a boundary to space will be referred to as *unboundedness* or *completeness*, not closure. Infinite space is closed in a way far more profound than its size.

UNBOUNDEDNESS

The first thing any investigation of space requires is a consistent mathematical framework for handling infinities. Since the universe is the only infinite reference we have, its extent will be used to define infinity.

The universe's diameter is the invariant width of nonexistence. It constitutes *a fixed, exact level of linear largeness, and is therefore the absolute metric of unboundedness.*

Ψ THEOREM 2.1 - INFINITY, ∞ {Ψ1.3}

INFINITY IS THE UNIVERSE'S INVARIANT DIAMETER

The universe is the source of all data and all knowledge. The only meaningful generalizations within this context are not human creations; they are direct reflections of reality's underlying structure. The most accurate definition of infinity is therefore a property of the universe, not an assertion of our mathematics.

The relationship between infinite and finite numbers is one of scale, not substance. The universe's diameter is composed of an infinite number of meters. If a single meter were removed from space, the remainder would be *incomplete*. Each and every meter of the universe's diameter is required to fully define the linear breadth of nothingness. In general, infinite numbers are *composed* of finite ones just as the universe's diameter is composed of individual meters.

This property will be referred to as *unbounded scalability*.

Ψ THEOREM 2.2 - UNBOUNDED SCALABILITY {Ψ2.1}

INFINITE QUANTITIES CAN VARY BY FINITE OR INFINITE AMOUNTS AND HAVE FINITE OR INFINITE RATIOS

Finite numbers do not vanish when they are combined with infinity as contemporary math would have us believe. Although finites and infinites cannot be directly compared, their magnitude is preserved under addition and subtraction:

$$\infty + 1 > \infty \tag{2.8}$$

as well as multiplication and division:

$$\infty(\infty + 1) = \infty^2 + \infty \tag{2.9}$$

Theorem (Ψ2.2) also supports infinite numbers scaled by finites:

$$\infty > \frac{\infty}{2} \tag{2.10}$$

Even so, the only case where two values can be directly compared is when they have a comparable level of unboundedness:

$$\frac{\infty}{\left(\frac{\infty}{2}\right)} = 2 \tag{2.11}$$

Any mathematical operations done on finite numbers can also be done on infinite numbers with the caveat that the results will often be infinite and irreducible. Modern math tends to lump infinities of all sizes into a quagmire of countable and uncountable sets. Nature is far more eloquent, as the correct definition of infinity should reflect.

ACROSS THE EDGELESS BOUNDARY

One of the conceptual problems with the Scalability Theorem (Ψ2.2) is that it makes it seem like it would be possible to go to the edge of the universe and take a step beyond it, as expressed by Equation (2.8):

$$\infty + 1 > \infty$$

This is not how the relationship between finites and infinites works. The sum of an infinite and finite value cannot be *evaluated* because infinite numbers don't have specific boundaries. Although ($\infty + 1$) is in fact one unit larger than ∞, the two terms in the expression ($\infty + 1$) have two different levels of scale and no conversion or merger is possible. However, this boundary problem can be resolved by converting the finite term to an infinite product:

$$\infty + 1 = \infty\left(1 + \frac{1}{\infty}\right) > \infty \qquad (2.12)$$

Adding a finite to the universe's size crosses no boundary; it merely increases the size of its individual linear components by an infinitely small amount. The purpose of the Scalability Theorem is not to provide a means of comparing finites directly to infinites. It is presented to establish the physical reality of infinites composed of finites and *to allow calculations using various levels of unboundedness*.

MAXIMAL EXISTENCE

The universe's size can be calculated directly from our interpretation of infinity. Its three spatial dimensions are equivalent because a difference between them denotes asymmetry and nothingness is by definition symmetric (Ψ1.8). *Shapelessness is the most eloquent form of symmetry*. Thus the universe's diameter is the same *in any of its three dimensions*. Since its diameter is by definition ∞, its volume is given by:

$$V_U = \infty^3 \qquad (2.13)$$

The universe's volume is significantly larger than its diameter *and its diameter is infinite*. Reality's unbounded dimensional nature demands the existence of *distinct levels of infinite largeness*. The universe's size is the largest possible *quantity of existence*.

The universe has the same volume today as it did a trillion years ago, and it will have the same volume a trillion years from now. Its volume is fixed throughout time.

Ψ THEOREM 2.3 - VOLUME OF THE UNIVERSE {Ψ2.2}

THE UNIVERSE'S VOLUME IS THE LARGEST QUANTITY OF EXISTENCE;
IT IS INVARIANT AND IS EQUAL TO ∞^3

This is the endpoint - the maximum amount of nothingness in the universe. While its full extent is not possible to imagine, the relationship between the diameter and volume of the universe is similar to the relationship between the diameter and volume of a baseball. The primary difference is one of scale.

Contemporary cosmology often refers to space-time's geometry and cites three possible shapes for it:

- Flat (rectilinear or homoloidal).
- Open (saddle shaped).
- Closed (spherical).

The size of the first two is thought to be infinite whereas the closed version typically has a finite radius but might float within an infinite four-dimensional framework. All of these geometries treat time as a dimension having an extent similar to space, albeit existing along a fourth dimension. The only model approximating the truth is the flat one, as it is a case of no shape (symmetry) and this resembles nonexistence more than the others.

An object can be a sphere or a cube when a boundary surface defines its topology. The totality of the universe is cubic from the standpoint of symmetry in three dimensions but is not a cube as such because it has no boundary. A similar situation exists for geometric points. Is a point a cube or a sphere? Just as the universe lacks a boundary in the realm of infinite largeness, a point is formless as a consequence of infinite smallness.

Since space vanishes to zero in the overview one might wonder if it is even meaningful to talk about its infinite volume. It is. The contextual nature of existence means its largest physical form is infinite space less any arbitrarily small volume ($\infty^3 - \delta$). Since the missing element can exist anywhere within infinite space, the maximal existence every spatial context shares is ∞^3, even though it is only visible within the universe. For the purposes of our calculations, the liberty will be taken of being able to see both sides of the existential coin simultaneously. Whenever references are made to unbounded characteristics or quantities of the universe, it assumes a region arbitrarily close to the size of the entire universe. The universe less an electron or photon still, for all intents and purposes, has a volume of ∞^3. This is the contextually asymptotic size of reality when measured in relation to any finite volume.

2.3 EXISTENTIAL SINGULARITY

Geometric points occupy one end of the cosmic size scale and infinite volume marks the other. How are the two related? Although the three-dimensional series of Equation (2.4):

$$0 = (\ldots + 0 + 0 + 0 + \ldots)^3$$

is a good approximation for space, it is only preliminary because it doesn't show how an infinite number of points *constitutes* distance, or how distance forms the substructure of volume. The universe's volume is ∞^3. What is not as clear is how many points exist in a unit cube. How many infinite summations exist between a geometric point and the entirety of the firmament? Where does summation stop and the structure of distance begin? To fully understand space is to find the relationship among:

- Geometric points.
- Finiteness.
- The volume of the universe.

TWO PATHS TO INFINITY

The curious thing about Equation (2.1):

$$0 = (\ldots + 0 + 0 + 0 + \ldots)$$

is the sum is precisely equal to any *single* element of the series. Subtract one point from each side:

$$0 - 0 = (\ldots + 0 + 0 + 0 + \ldots) - 0 \qquad (2.14)$$

Behold. The difference between a single point and all of space is the same as the difference *between a single point and itself*. The only possible conclusion is:

In the limiting case, the large is the same as the small.

This is the arithmetical consequence of contextual existence (Ψ1.4). From our finite human vantage there are only two directions away from finiteness, infinite largeness and infinite smallness. *But both paths lead to the same state, nothingness, because they both lead away from existence and there is only one other existential alternative.*

At the level of totality two existential states are no longer available; there is only nonexistence. It is an existential singularity. It is where infinite smallness and infinite largeness merge into the same state. *Existence is the abstraction produced from an incomplete perspective; it is the partial formulation of reality bounded by the same vacant truth on both ends of the size scale.*

Reality is a self-referencing proposition. A rock is hard, cotton is soft, but there is no absolute solidity beyond this. Since the infinitely large is too large to observe, the infinitely small is our only perspective of the totality of existence. Yet its apparently small size makes us fail to realize that it is just as inaccessible as infinite largeness. Common sense tells us our surroundings are vastly larger than geometric points and provides the false impression that points are tiny and insignificant. Nothing could be further from the truth. There is no ceiling to the universe, but just as importantly, there is no floor. Our finite existence lies suspended above the submicroscopic world of the infinitely small and below the immensity of the infinitely large. Why is this? What keeps our reality from shrinking into a single point? What keeps finiteness hanging between infinite smallness and largeness? What defines its location on the cosmic size scale? *Existential singularity.* Finiteness is a certain distance removed from the scale of the singularity of nonexistence. It is perched between infinite largeness and smallness, roughly the same distance from each *because they are the same thing.*

Astrophysicists and particle physicists often work together to try to find a theoretical connection between the macro and micro universe, but the two are more inseparable than they would dare imagine. It is no wonder why zero and infinity cause many of the same problems in algebra; they are two different viewpoints of the same thing - the existential certainty of nonexistence.

Ψ THEOREM 2.4 - EXISTENTIAL SINGULARITY {Ψ1.4}

INFINITE LARGENESS AND INFINITE SMALLNESS ARE INVERTED VIEWS OF THE SINGULARITY OF NONEXISTENCE

From an observational standpoint, it seems abundantly clear that infinitely small length is smaller than finite length, which is in turn smaller than infinitely large length. But this is an incomplete perspective. *Infinite length, large or small, cannot be directly observed so a definitive comparison is simply not possible.* In the self-referencing framework of space, there is a clear order to the size of things, but this is only because finite size is the only size available for us to measure, mired as we are at the midpoint of existential scale.

The cosmic size scale is so apparent to a pure empiricist that there is no reason for further inspection:

$$\infty \text{ small} << \text{finite} << \infty \text{ large} \tag{2.15}$$

But it takes more than empiricism to account for the *basis* of the empirical. The premise of existential singularity is the universe has two appearances, the *single* infinite sum of its infinitely many parts and an infinitely small *single* point. If an absolute comparison were possible, their equality would be obvious. From within a finite vantage, their equivalence is so much less than obvious that it has evaded discovery since the time we did our math on cave walls. The sum of reality is either too large or too small to see, but the truth is the universe has no ceiling and it has no floor, and there is only one place to go from finiteness just as there is only one place to go from existence.

This is not to say there is no order to size in the finite realm. Atoms are certainly smaller than planets. Existential singularity applies to the relationship between existence and nonexistence, not the relationship between various finite sizes. Atoms are smaller than planets, and planets are smaller than galaxies, but when the scale moves from planets to atoms or planets to galaxies it is one step closer to nothingness. There is no difference between a single point and all of space, less one point. Either case is infinitely close to a complete state of nothingness and is approached by any departure from finiteness.

Blaise Pascal came close to the truth 330 years ago (Pensées, Sec II, para 72):

> *For in fact what is man in nature?*
> *A Nothing in comparison with the Infinite,*
> *an All in comparison with the nothing,*
> *a mean between nothing and everything.*

The only thing he didn't understand is the All and the Nothing are synonymous perspectives of the same thing. Nothingness is the only form that completeness can take.

COMPOSITION AND CONTINUITY

According to existential singularity, individual points are equivalent to the sum total of existence. This implies that every point of space has a universe inside of it composed of points, which in turn contain entire realities. This is true with one clarification. The only states of reality are existence (finite) and nonexistence (infinite), so the universe existing within each point of space is the *same* universe, not some other level of reality. Points of space are related because they all belong to the same universe and thereby represent *its* totality. This is their common connection.

Yet how can the multitude be the same as the one? *It is possible because the superimposition of any number of points at the same position is indistinguishable from a single point.* All a point has is its position relative to other points. It has no intrinsic properties. This means any given position in space essentially contains an immeasurable number of points. Points have an intrinsically unbounded duplicity.

From within the universe, the composition of nothingness as points is self-evident because only a portion of the summation is present. From the outside it vanishes.

Ψ THEOREM 2.5 - COMPOSITIONAL SINGULARITY {Ψ2.4}
SPACE IS COMPOSED OF POINTS BECAUSE A POINT IS COMPOSED OF POINTS

Existence is the relationship between nonexistent objects. At the perspective of totality all of the separate nonexistent objects coalesce into a *single* nonexistent object. Totality is not a process or merger event. Nonexistence is by definition composed of itself regardless of the form of its distribution. *Nonexistence is a singularity composed of itself such that the number of points at a given position in space is the same as the number of positions in space itself.*

Is our galaxy an atom in the next level of scale? Do entire galaxies exist within every elementary particle? No and no. The true situation is far more profound. *Our entire universe exists within each geometric point within it.* All of the points of space scattered across untold trillions of light years are copies of the same singularity - *the same cosmic point.* From within our universe legions of copies of nothingness expand to form the celestial realm; in infinite largeness and smallness they collapse back to a self-contained nothingness. This is the physical reality of Equation (2.1):

$$0 = (\ldots + 0 + 0 + 0 + \ldots)$$

There are an unbounded number of points on the left side of this expression as well; it is just not obvious because they all have the same location. This only becomes apparent in the internal perspective.

Composition and continuity are closely related properties. Continuity is the statement of homogenous and complete composition. This is also derived in terms of a point.

Ψ THEOREM 2.6 - CONTINUOUS SINGULARITY {Ψ2.4}
SPACE IS CONTINUOUS BECAUSE A POINT IS CONTINUOUS

Spatial continuity is an interesting proposition. One might think no gaps larger than a single point could exist because any such gap constitutes distance, distance is composed of points, and any distance within space belongs to the same universe. After all, a position cannot exist within space without also being related to space, because *position is the operative relationship*. While this is certainly true for empty space, the situation is not quite so straightforward in the curved space of matter. Space is composed of points at a density sufficient to support infinite smallness. But when it is distorted by matter the metric and reality of distance is distorted along with it. It is therefore possible to have voids in space as long as *they are not internal to our universe*. Space is a closed system just like existence; any discontinuities within it are by default *extraspatial*. As it turns out, this is not an esoteric consideration. As will be demonstrated in Part III, matter *could not exist* without its associated spatial discontinuity, and elementary particles actually provide a glimpse of the universe's four-dimensional *exterior*.

2.4 DIMENSIONALITY

If the composition and continuity of space can be traced to properties of a geometric point, what about its dimensionality? *Geometric points represent infinitely small size, not the complete absence of size.* Space is infinitely large in all three of its dimensions regardless of how large each extent might be. Similarly, infinite smallness is infinitely small in three dimensions regardless of how small this might be. Space was originally given by Equation (2.4):

$$0 = (\ldots + 0 + 0 + 0 + \ldots)^3$$

where the product of three infinite linear summations describes its three-dimensional nature. While this expression is true, it is more a statement of the metric of the reference frame as a whole than the configuration of the space within it. Taking the entire infinite series to a power limits the three-dimensional perspective of space's fabric to the level of the universe's diameter. Space is a distribution of points but also of volume, and volume exists on scales far below totality. This can be expressed by expanding Equation (2.4) to:

$$0 = (\ldots + 0^3 + 0^3 + 0^3 + \ldots) \tag{2.16}$$

Equation (2.16) is a more accurate rendition of spatial substructure because it shows *dimension across scale*. Since this scale also includes totality, dimensionality is coincident with both composition and continuity. *Space is composed of and continuous in multi-dimensional units.* Not only is distance continuous, volume is as well.

Ψ THEOREM 2.7 - SPATIAL SUBSTRUCTURE {Ψ2.4}

DIMENSION IS CONCURRENT WITH COMPOSITION AND CONTINUITY; POINTS OF SPACE HAVE THE SAME SPATIAL DIMENSIONS AS SPACE AND ARE CONTINUOUS IN LENGTH, AREA, AND VOLUME

Although at present the meaning of dimension isn't entirely clear, the connection between properties of the totality of existence and nothingness is indisputable. Since:

- Space is composed of points.
- The totality of space is a point.

It follows that the dimensionality of the universe and the points of which it is composed are largely indistinguishable. All a point has is its relationship to other points, and this relationship is by necessity dimensional.

Dimension is a reflection of the way nothingness is composed of nothingness. Distances can be infinitely long yet divisible into infinitely small lengths. *The balance between these two infinities results in finite dimension just as the balance between infinite space and energy produces finite energy density.* Earlier it was noted that the number of points coexisting at a single position in space is the same as the number of positions in space. Dimension is the purest expression of this parity.

2.5 CLOSURE

Several spatial properties have been introduced and related to nonexistence, all dealing with the unboundedness of points and the totality of the universe. Yet space's principal characteristic, *quantity*, remains mysterious. Quantity cleanly separates what exists from what does not. How is a unit cube related to a point? How is it related to infinity? To understand space is to connect these three levels of size. Let's begin with finiteness.

Finiteness comes from the balance between infinite largeness and smallness. It is the realm equidistant from both. Infinitely large and infinitely small are inverted viewpoints of the same totality, and *finiteness is the magnitude of size lying precisely halfway between them.* Infinite smallness is to finiteness as finiteness is to infinite largeness. This relationship is expressed as a product of the two ends of infinity and will be called *closure.*

Ψ THEOREM 2.8 - CLOSURE {Ψ2.4}

THE PRODUCT OF INFINITE LARGENESS AND INFINITE SMALLNESS IS UNITY

Closure places existence squarely between the infinite limits of scale. The relationship between finiteness and infinite smallness/largeness is *simultaneous* and is therefore the *product* of the two limits. Since the universe's size is invariant, its inverse of infinite smallness is also invariant. Their product is unity, which is also invariant. *Finiteness is located exactly midway along the scale of size and unity is the standard by which it is measured.*

Theorem (Ψ2.8) is consistent with the treatment of infinity and zero in the Riemann sphere:⟨12.1⟩

$$\infty \cdot 0 = 1 \tag{2.17}$$

But our premise is far more comprehensive - *all of the terms in Equation (2.17) are physical realities and their relationship holds the key to the structure of the universe.* Riemann geometry was developed to generalize complex number space, but by demonstrating a direct relationship between zero and infinity it makes a sweeping statement about the quantifiable structure of nonexistence. However, as it turns out, ($0 = 1/\infty$) is not the relationship between zero and infinity in our universe; *it is their relationship in one-dimensional space.*

Think of closure in terms of finite distances. Suppose largeness is defined as a million meters and smallness as a micron. Their product is a square meter:

$$K = (1{,}000{,}000)(0.000001) = 1^2 \tag{2.18}$$

These size limits are expressed as distances, so their product K is an area that represents the size of the entire system. Smallness and largeness are related through unity but still exist simultaneously, which in this case requires two dimensions to describe. The square root of K is the distance hovering midway between largeness and smallness in this model: 1 meter. It is a million times larger than a micron and a million times smaller than a million meters. Regardless of how large infinite largeness becomes, infinite smallness can be defined as comparatively small, so their product remains unity. The universe is closed because largeness and smallness extend to infinity with equivalent magnitudes in a self-referencing reality.

2.6 CLOSURE CONSTANT

Infinite smallness and largeness have an inverted relationship; the question now is the form these limits take in a three-dimensional universe. Infinite largeness has already been unambiguously identified as space's volume, ∞^3. The equivalence between quantity and existence leaves no reasonable alternative. Reality's other limit, infinite smallness, is the *inverse* of ∞^3.

From (Ψ2.8), the universe's closure relationship can be written:

$$1^{M} = \left(\frac{1^{M}}{\infty^{3}}\right)\left(\infty^{3}\right) \tag{2.19}$$

where:

Infinite largeness = Size of space = ∞^3

and:

Infinite smallness = Size of a point = $(1^M/\infty^3)$

The value 1^M will be referred to as the *universal closure constant*. It encapsulates the full breadth of reality and therefore represents *totality's boundary condition*. Space and infinite smallness are components of this boundary. How many dimensions does it have? Let's look at the general relationship between boundaries and dimension.

SURFACE BOUNDARIES

In general, any (N+1)-dimensional space can have an N-dimensional surface subject to the following related criteria:

Ψ THEOREM 2.9 - CLOSURE BOUNDARY {Ψ2.4, Ψ2.8}

(A) ANY (N+1)-DIMENSIONAL REGION CAN BE BOUNDED BY AN INFINITELY THIN N-DIMENSIONAL SURFACE

(B) THE MAXIMUM DIMENSION AN INFINITELY THIN N-DIMENSIONAL SURFACE CAN BOUND IS (N+1)

A circle is bounded by an infinitely thin line; a sphere is bounded by an infinitely thin area. Similarly, a line is too dimensionally small to form a spherical boundary. In reference to our universe, no boundaries of any kind exist along its three spatial dimensions. *This means space is not a bounded interior region; it is a bounding surface.*

Look again at Equation (2.19):

$$1^{M} = \left(\frac{1^{M}}{\infty^{3}}\right)\left(\infty^{3}\right)$$

Totality is the simultaneous product of infinite smallness and infinite largeness, exhibiting their *combined dimensional content*. Infinite smallness lies *external* to the dimensions of infinite largeness. It is the only way the two can coexist as equivalent paths to nonexistence. What this means is that *the space of our universe is the boundary surface of its own totality*:

Ψ THEOREM 2.10 - TOTALITY BOUNDARY {Ψ2.9}

UNBOUNDED SPACE IS THE BOUNDARY SURFACE OF ITS OWN TOTALITY

This in turn means, in accordance with (Ψ2.9B), that the maximum dimension that can be bound by three-dimensional space is the fourth. It follows that the universal closure constant is 1^4 (M=4).

Although lower-dimensional analogs fail to convey the full reality of space and closure, consider the relationship between the surface and interior of a sphere. Here the interior and exterior share the same dimensions, but most significantly the bounding surface is not in itself bounded. A sphere's surface has no beginning or end, yet serves to delineate a precise limit to the interior region it encapsulates. In much the same way, the space of our universe is a membrane in a higher-dimensional framework. Its infinitely small extent constitutes the thickness of this membrane along its *interior* dimension. Although we naturally think of a bounded region's interior as more significant and expansive than its surface, *the universe's interior region is infinitely small as an inescapable consequence of closure*. The full extent of reality consists of an infinitely large exterior (infinite largeness) that bounds an infinitely small interior (infinite smallness).

Closure is the level of existence where the large becomes indistinguishable from the small, both merging into the singularity of totality. The totality boundary is the geometric interpretation of this. Imagine an enormous sphere in space, growing progressively more immense. As long as its volume remains finite, it is a three-dimensional object with a two-dimensional surface. When its volume reaches infinity, its three-dimensional interior *is now the surface by which it is bound*. The sphere becomes inverted just as infinite smallness and largeness are inverted.

The singular nature of interior dimensionality will be called the *closure limit*:

Ψ THEOREM 2.11 - CLOSURE LIMIT {Ψ2.10}

THE INFINITELY SMALL INTERIOR EXTENT OF ANY CLOSED, UNBOUNDED N-DIMENSIONAL SPACE IS LIMITED TO A SINGLE DIMENSION

Universal closure has the following parameters:

Infinite largeness:	three-dimensional reality.
Infinite smallness:	diameter of a point.
Finiteness:	four-dimensional unity.

These are related by:

$$1^4 = \left(\frac{1^4}{\infty^3}\right)\left(\infty^3\right) \quad (2.20)$$

This sustains an inverse relationship between infinite largeness and infinite smallness while simultaneously providing geometric closure. It will be referred to as *dimensional closure*.

2.7 DIMENSIONAL CLOSURE

A more general approach can be used to arrive at the same conclusion as the previous section, as follows.

Infinite largeness takes three different forms in a three-dimensional universe - one for each dimension. The universe's length, area, and volume are all infinitely large. The largest example of infinite largeness is the volume of the universe because it is the unbounded quantity having the maximum number of dimensions. It encompasses all of space and everything in it.

Infinite smallness also occurs in length ($1^4/\infty^3$), area ($1^4/\infty^2$), and volume ($1^4/\infty$). For any bounded totality, the smallest form of infinite smallness is infinitely small length because it has the maximal power of infinity in the denominator as well as the minimum number of dimensions. The only way to encapsulate the universe with closure is to span the full range of existence, so it follows that *dimensional closure is the relationship between the largest of the infinitely large and the smallest of the infinitely small.*

Dimensional closure spans the full range of cosmic scale:

Ψ THEOREM 2.12 - DIMENSIONAL CLOSURE {Ψ2.4, Ψ2.11}

THE SPACE OF THE UNIVERSE IS CLOSED BY $1^4 = (1^4/\infty^3)(\infty^3)$

This is the relationship between the infinite volume of space and the smallest length in the universe. The smallest length, in turn, is the ultimate divisibility of linear distance - its *linear continuity*. Closure takes us all the way from the universe's volume down to the minimum distance between two points of space:

$$\ldots 0 + 0 + 0 < 0 > 0 + 0 + 0 \ldots$$

Existence is the relationship between nonexistent objects, and the distance between two points is as small as this relationship can get. Dimensional closure defines the hierarchy between individual points, finiteness, and the volume of the universe, and it governs our relationship to zero:

Ψ THEOREM 2.13 - ZERO (INTERNAL PERSPECTIVE) {Ψ2.12}

THE DIAMETER OF A GEOMETRIC POINT IN THREE-DIMENSIONAL SPACE IS ZERO; ITS SIZE IS GOVERNED BY THE VOLUME OF SPACE: $0 = (1^4/\infty^3)$

The area and volume of a point in three-dimensional space are given by the square and cube of zero, $(0^2 = 1^8/\infty^6)$ and $(0^3 = 1^{12}/\infty^9)$, respectively.

Just as there is a limit to space's dimensionality, there is a corresponding limit to its divisibility and they are directly and inversely related to each other. Continuity is the inverted perspective of dimension just as infinite smallness is the inverted perspective of infinite largeness. Both are a consequence of the distribution of nothing in nothing. While there is no limit to the total amount of nothing within nothing, there are limits to its density within itself. Nothingness provides infinite largeness and infinite point density and *the ratio between the two creates finite dimensionality*.

Spatial continuity is often described as:

A space is continuous if there is always a point between any two other points.

In the physical space of our universe, if two points are less than or equal to $(1^4/\infty^3)$ apart, there is no other point between them. Space's continuity can be defined either in terms of the diameter of an individual point or the distance between two adjoining points:

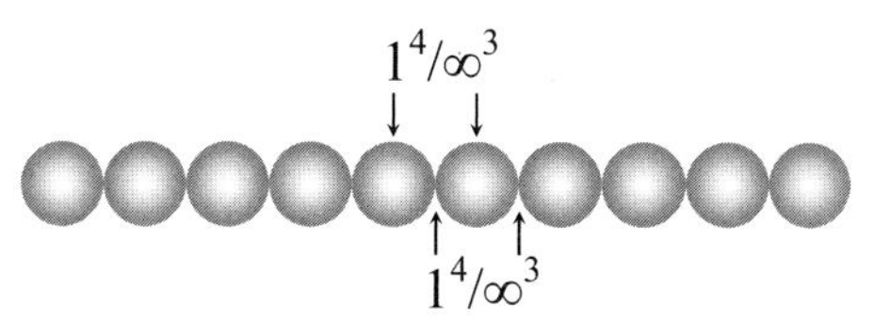

Figure (2.3) Linear continuity as point diameter or distance between adjoining points

This leads directly to *point density*:

Ψ THEOREM 2.14 - SPATIAL CONTINUITY {Ψ2.13}

THREE-DIMENSIONAL SPACE HAS A LINEAR DENSITY OF ∞^3 POINTS PER ABSOLUTE FINITE DISTANCE

Absolute finite distance is derived from the physical interpretation of the universal closure constant, as revealed in the next section.

2.8 UNIT HYPERVOLUME

The universe's volume has the same total size as ∞^3 finite unit cubes (1^3). Its linear density is ∞^3 points per finite distance. This means our entire reality could be stacked at its linear point density inside a four-dimensional cube of size 1^4, the quadric constant of its dimensional closure. *The universe has a constant, finite, four-dimensional size.* This is what the universal closure constant 1^4 in Equation (2.20) represents:

$$1^4 = \left(\frac{1^4}{\infty^3}\right)\left(\infty^3\right)$$

Our infinite three-dimensional universe fits inside a finite hypercube. Moreover, this is the *only* finite representation of its size. 1^4 is the gold standard of finiteness - a sharp line drawn in the otherwise seamless continuum of infinity. *Space's featureless, unending expanse has a global four-dimensional boundary condition.*

Think of the relationship between three-dimensional space and its finite four-dimensional size in terms of a two-dimensional analog. Suppose a plane were sectioned into ∞^2 finite unit areas (1^2), which were then stacked into a cube as shown below:

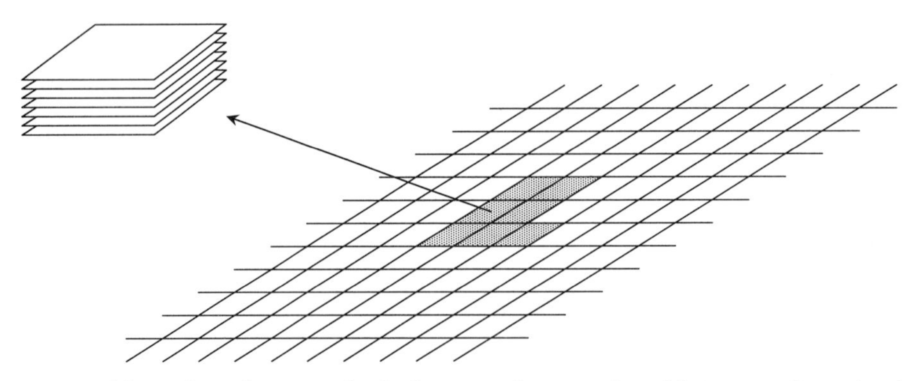

Figure (2.4) Building a three-dimensional cube from an infinite number of finite, two-dimensional areas

This cube's *height* has one of three possible values:

- Infinitely small.
- Infinitely large.
- Finite.

The first is too small, since each of the ∞^2 segments used to build this cube also have infinitely small height. The second is too large, as a plane is smaller than a region with infinite volume. The last choice is the only viable option. *A finite cube is the three-dimensional size of a plane, just as a finite hypercube is the four-dimensional size of space.*

Dimensional closure demonstrates that *relative to totality, space is infinitely thin along the fourth dimension*. Space is constructed of points with three spatial dimensions, as noted earlier, but they also contain a single infinitely small *extraspatial* metric. Modern string theories propose spatial dimensions too small to measure; as many as eleven in some versions. It is interesting that such a concept should arise as physicists grapple with the nature of matter. In reality space only contains a single additional dimension beyond its self-evident extent. Four-dimensional space will be referred to equivalently as *hyperspace* or *hypervolume*.

The universe's finite hypervolume defines an absolute size scale - *it is why atoms behave differently than galaxies.* On the macroscopic scale matter and energy appear smooth and continuous, but on the atomic and nuclear scale their discrete structure emerges. The operator creating granularity in energy relative to scale is the same operator defining our absolute location between infinite smallness and largeness - the four-dimensional volume of the universe. *The inevitable conclusion is the finite four-dimensional size of the universe is the boundary condition for energy quantization.* It was introduced as the closure constant, 1^4, but will henceforth be referred to as *unit hypervolume*.

Ψ THEOREM 2.15 - UNIT HYPERVOLUME, $\Diamond_4$ {Ψ2.12}

THE UNIVERSE'S FOUR-DIMENSIONAL VOLUME, UNIT HYPERVOLUME, IS FINITE

The universe has a fixed, albeit infinite, three-dimensional volume and therefore a fixed, finite, four-dimensional volume. This in turn defines a universal standard unit of distance that will be called the *absolute meter*, m_a.

Ψ THEOREM 2.16 - ABSOLUTE METER, m_a {Ψ2.15}

THE FOURTH ROOT OF THE UNIVERSE'S FINITE FOUR-DIMENSIONAL VOLUME IS THE ABSOLUTE UNIT OF FINITE LENGTH - THE ABSOLUTE METER

Symbolically:

$$1\ \mathrm{m}_a = \sqrt[4]{\Diamond_4} \tag{2.21}$$

This is the universe's absolute unit of distance. It exists precisely halfway between infinite largeness and infinite smallness. In accordance with (Ψ2.14), space has a linear density of ∞^3 points per absolute meter. Note that unit hypervolume is symbolized as $\Diamond_4$ instead of 1^4 because *it is only a power of unity when expressed in units of absolute meters.* Its value in standard physical units (SI) will be derived shortly.

QUANTIZATION BOUNDARY

Take the infinite, three-dimensional space of the universe, divide it into innumerable finite cubes, and lay these onto each other, infinitely close together, along the fourth dimension. The resulting finite hypercube is the one and only boundary condition that can be derived *directly from the seamless unbounded vastness of empty space.* Yet only one is necessary. This single value defines finiteness and at the same time defines the precise relationship between energy, packaged and discrete, and space, continuous and unbounded. The universe's finite hypervolume governs the quantization of energy; it is the only boundary condition infinite space has to offer. *Energy is constrained within the limitations of space and is ultimately packaged by it as well.* Unit hypervolume is how it is possible to have a *universal* constant for microscopic processes like the creation of particles and emission of photons. It has the same value anywhere in the universe, yet can also have tiny linear extents.

Since unit hypervolume is the only true cosmic standard, it ought to betray its presence in the universal constants used in physics, and it does. Table (2.1) lists key physical constants, most of which are considered universal:[5.2]

Universal Physical Constants				
	Name	**Value**	**Units**	**Δ(ppm)**
q	Elementary Charge	$1.602176462(10)^{-19}$	coulomb	0.039
c	Speed of Light	299,792,458	meter/second	Exact
h	Planck's Constant	$6.62606876(10)^{-34}$	joule-second	0.078
k	Boltzmann Constant	$1.380650324(10)^{-23}$	joule/Kelvin	1.7
G	Gravitational Constant	$6.673(10)^{-11}$	$\mathrm{meter}^3/(\mathrm{kg\text{-}second}^2)$	1500
ρ_U	Universal Energy Density[3.1]	$9(10)^{-12}$	$\mathrm{joule/meter}^3$	10^7

Table (2.1) Key physical constants

None of the units of these constants are obviously four-dimensional, but on closer inspection one certainly stands out from the rest: *Planck's constant*. It can easily be converted into a four-dimensional quantity using only the values in this table. Its units are energy-time. Universal energy density has units of energy/distance^3 and the speed of light has units of distance/time. Combining these expressions with Planck's constant yields a four-dimensional constant. As it turns out, this constant *is* the physical manifestation of unit hypervolume, and is directly proportional to it:

$$\lozenge_4 = \Lambda_\gamma \left(\frac{hc}{\rho_U} \right) = \Lambda_\gamma \left(\frac{E_\gamma \lambda}{\rho_U} \right) \tag{2.22}$$

where Λ_γ is the unitless constant of proportionality defined by the way the energy of individual photons is geometrically distributed into space. It will be called the *photon hyperscaling factor*.

Note that the hypervolume of Equation (2.22) has units of *quadric* distance (distance^4). *Planck's constant is the boundary condition for the quantization of energy*, just as would be expected of the hypervolume of the universe.

ENERGY AS VOLUME

Planck's constant is four-dimensional *because energy is three-dimensional*. This requires no great leap of intuition. Energy is spread throughout space in the form of an endless and variable density distribution. On one end of this scale is deep space with its virtually nonexistent energy density. On the other are atomic nuclei, with an incredible energy concentration dozens of orders of magnitude more intense. Yet throughout the entire scale, average energy density remains *finite*, and the only way the ratio of energy to space can be finite is if both share a similar dimensionality. This will become self-evident as our analysis proceeds.

The absolute meter is the fourth root of unit hypervolume. Substituting Equation (2.22) into Equation (2.21):

$$1\,\mathrm{m_a} = \sqrt[4]{\lozenge_4} = \sqrt[4]{\Lambda_\gamma} \left(\sqrt[4]{\frac{hc}{\rho_U}} \right) \tag{2.23}$$

Most geometric factors, such as π, tend to be much closer to unity than physical constants. If Λ_γ has a magnitude comparable to π or $1/\pi$, for instance, its fourth root will be close to unity:

$$\sqrt[4]{\Lambda_\gamma} \approx 1 \quad (2.24)$$

Applying this approximation to Equation (2.23) produces an absolute meter of ~0.4 mm using values from Table (2.1). This is an order of magnitude estimate, since the value of ρ_U in this table is limited to the luminous portion of the universe. As will be shown later, luminous material only accounts for a few percent of the universe's total energy content, so an absolute meter's true value is closer to 0.1 mm. *The universe fits in a hypercube with an edge length of a fraction of a millimeter.*

Mathematicians perform operations on N-dimensional spaces and treat the fourth dimension as if it were simply another extension of space. Physicists puzzle over the implications of a fourth dimension and have reason to suspect that it is not fully equivalent to the other three. The physicists are correct; the fourth dimension is profoundly different from space. It is the universe's *closure dimension.* It is linear, true enough, and it has a defined metric as required by closure. But it is not space and ultimately possesses only a fleeting resemblance to it. The universe's volume has two entirely equivalent expressions: infinite three-dimensional space and finite hypervolume. *The addition of a single dimension allows us to express infinite volume as finiteness and look directly at the totality of existence.* Properties of unit hypervolume are properties of the entire universe. It contains an infinite amount of volume yet can be written as the finite quantity of $\Diamond_4$.

THREE CONSTANTS, ONE UNIVERSE

The universe contains one and only one finite constant because it has but a single finite boundary in unit hypervolume. Its effect is literally everywhere, beginning with the existence of particles and photons and extending through their quantum interactions. But unit hypervolume isn't the only universal constant; it is just the only *finite* constant. Dimensional closure contains three elements, *and they are all universally constant*:

- Infinite space, ∞^3. Governs the conservation of volume and energy. Also symbolized as V_U.
- Spatial continuity, $1^4/\infty^3$. Governs the compositional relationship between space and dimension. Originally defined as *zero's internal perspective* by (Ψ2.13), $1^4/\infty^3$ will henceforth be referred to as *netherspace*, δ_3, in order to provide more specific terminology.

- Unit hypervolume, 1^4. Governs the quantization of matter and energy. Also symbolized as $\Diamond_4$.

Dimensional closure is responsible for each and every universal constant because it is the only relationship spanning all levels of scale. Even though the speed of light is constant and universal, it is not a universal constant. It is a *ratio* of distance and time, not a quantity in and of itself. Equation (2.20) can be rewritten in terms of netherspace, universal volume, and unit hypervolume as:

$$\Diamond_4 = \delta_3 V_U \quad (2.25)$$

This is the most accurate expression of dimensional closure because it treats infinite smallness, infinite largeness, and finiteness as distinct and constant values.

Absolute meters are the proper units for all three of the terms in the dimensional closure of Equation (2.25). The volume of space is ∞^3 cubic absolute meters, and the length of netherspace is $(1^4/\infty^3)$ absolute meters. *Unit hypervolume represents totality, and as such defines the scale of all levels of size.* Indeed, infinity is only the correct number for the universe's diameter *when it is in units of absolute meters.* The meter used by the SI system is an arbitrary human invention, not a universal geometric standard, and the universe's diameter in meters will be called *metric infinity*, ∞_m. Since the absolute meter is about ten thousand times smaller than a meter, ∞ is about ten thousand times larger than ∞_m. However, for the sake of simplicity and practicality, the term ∞ will be used exclusively throughout the remainder of *Null Physics*, even in the unbounded particle field integrations of Part III whose results are evaluated using SI units. This is permissible because particle fields go to zero at infinite distance, so the difference between ∞_m and ∞ has no effect on calculations done over finite volumes such as those of atomic nuclei and black holes.

Most of this chapter's content has tended toward the abstract, but can be summarized by two key concepts. Our analysis of spatial substructure began with Equation (2.1)'s simple premise:

$$0 = (\ldots + 0 + 0 + 0 + \ldots)$$

and ended with Equation (2.25):

$$\Diamond_4 = \delta_3 V_U$$

It might be easy to lose track of how the first is related to the second, because they are two distinct spatial manifestations. Equation (2.1) has to do with space's *composition* while Equation (2.25) describes its *dimensional configuration*. Although composition ultimately governs dimensional closure, they are separate considerations, and the relationships among 0, 1^4, and $\aleph$ will be more fully explored in the next chapter.

2.9 GENERAL CONCLUSIONS

The majority of this chapter was spent tearing space down into its constituent point structure in the search for geometric closure. The conclusions arising from this process define our position within reality and provide the first meaningful step towards understanding energy:

- The universe has three compositional dimensions (space) and one closure dimension (fourth dimension). Space is the three-dimensional surface of this four-dimensional construct.
- The universe has a finite four-dimensional hypervolume. This represents its one and only universal boundary condition and is proportional to the value (hc/ρ_U) (the product of Planck's constant and the speed of light divided by average universal energy density).
- Distance has an absolute unit of measurement called the absolute meter. It is equal to the fourth root of the universe's hypervolume and has a length equal to a fraction of a millimeter.
- The linear density of space is (∞^3) points per absolute meter.

Universal constants pose quite a mystery for conventional physics. How, for instance, can Planck's constant operate the same for atoms on Earth as it does (as deduced from distant electromagnetic spectra) for atoms in the stars of galaxies in deep space, billions of light-years away? Even the idea of a *universal constant* seems incongruous when applied to submicroscopic environments. How can an operator that regulates something as small as an atom be the same throughout the universe? Herein lies the true motivation behind string theory. The hidden spatial dimensions it postulates provide storage locations for universal constants as well as a direct connection between nuclear and cosmic scales. String theory maintains that the constants that govern elementary particles are universal because the infrastructure of space itself is universal. While this is certainly a reasonable premise, it ultimately fails because it provides no assurance that this infrastructure *is* fixed throughout the universe.

The truth is far simpler and far more eloquent. Universal constants control submicroscopic processes because the four-dimensional size of the universe, unit hypervolume, has a tiny spatial extent and is *by definition* the same everywhere. It is the one and only connection between the micro and macro universe because it is the finiteness that of necessity exists at the midpoint of the universal size scale. Unlike the hidden dimensions of string theory, unit hypervolume cannot vary throughout space because it represents its overall size.

3. FINITE DIMENSIONALITY

Unit hypervolume has sweeping implications, one of which seems contrary to first principles:

> *If the entirety of the universe is nothingness, yet has a size equal to finite four-dimensional volume, does nothingness itself have a finite four-dimensional volume?*

The answer from first principles is an emphatic no. Nonexistence is the lack of quantity, whereas the four-dimensional size of the universe represents the quintessential quantity; *the quantity of quantity*. The only way it can exist within nothingness is by being voided by some internal relationship. On the one hand, the universe sums to nonexistence and many of its global properties can be deduced from this. On the other, the state of nonexistence is the absence of the universe itself, so any aspect of existence, whether it is infinite space or finite hypervolume, has a substructure allowing for complete and utter cancellation. Closure reveals the internal size of reality but the Null Axiom defines its nature. *If unit hypervolume is a form of the universe's overview it must contain the elements of its own nullification*. But how?

Let's return to first principles.

3.1 HALF OF EVERYTHING

Existence is partial nonexistence and nonexistence the sum of existence. But what, precisely, is the difference between these two states? They are both composed of the same fabric, nothing, and their duality vanishes in existential closure. *Nonexistence is the vacant truth on both ends of reality and existence is everything in between*.

Existence and nonexistence have the same underlying composition, so there is no scale where existence suddenly becomes nonexistence or vice versa, but there is a scale where the two states are completely equivalent. The midpoint of their duality occurs at the level of *half of existence*. *The nonexistence of half of the universe is equivalent to the existence of the other half*.

This is the only scope where these two states have equal magnitudes. It will be called *existential parity*:

Ψ THEOREM 3.1 - EXISTENTIAL PARITY {Ψ1.8}
THE EXISTENCE OF ANY HALF OF THE UNIVERSE IS EQUAL TO THE NONEXISTENCE OF THE OTHER HALF

Existence and nonexistence are contextual states, each defined in reference to the other. Reality collapses to a singularity at totality, but at half this scale the two existential states are fully equivalent. They can't be equal at the level of the entirety of the universe, as it would require the simultaneous presence of both, making the universe twice its size regardless of how large it happens to be. There is one and only one scale where existence and nonexistence are of equal size - half of totality.

COMPOSITIONAL SAMENESS

The ultimate distinction between the two existential states tends to blur to some extent regardless of scale. The only difference between them is how totality happens to be apportioned. This is particularly evident at the level of existential parity and will be referred to as *existential uniformity*:

Ψ THEOREM 3.2 - EXISTENTIAL UNIFORMITY {Ψ3.1}
THE TWO HALVES OF EXISTENTIAL PARITY ARE INDISTINGUISHABLE

Existential closure and uniformity are the two governing dynamics of the relationship between the existential states. In existential closure, infinite smallness and largeness are inverted perspectives of nonexistence; in existential uniformity, the states of partial existence and partial nonexistence are mirror images of each other. Like individual points in space, their only difference is physical separation. Unlike points, there are only two such regions.

The largest possible sum of existence is the sum of the two halves of the universe, with the result simply being the universe. This satisfies the completeness of nonexistence but not its voided nature. Just as nonexistence is supposed to be the counterpoint of existence, some operator has to offset the cumulative effect of the summation of nothing. This operator, naturally the converse of summation, is the *difference* operator. The largest possible difference in the universe is the one occurring between its halves.

This will be referred to as *existential dichotomy:*

Ψ THEOREM 3.3 - EXISTENTIAL DICHOTOMY {Ψ3.1}
THE LARGEST DIFFERENCE IN THE UNIVERSE OCCURS BETWEEN ITS TWO HALVES

The nonexistence of the universe is not the neutral summation of countless numbers of null images; *it is the difference between the two largest summations within it.* This is how existence is reconciled with nonexistence at the level of totality. The difference between any two halves of the universe has the internal expanse of its total size and the external expanse of an utter void.

Our initial formulations of nonexistence are hereby refined through the contextual relationship between the existential states:

Ψ THEOREM 3.4 - NONEXISTENCE {Ψ3.2, Ψ3.3}
THE STATE OF NONEXISTENCE IS THE DIFFERENCE BETWEEN ANY TWO HALVES OF THE UNIVERSE

The question now is how this universal difference manifests itself.

POLARITY

The universe is four-dimensional. Its three infinite extents are defined by summation, leaving only one degree of freedom for the difference operator and the nullification of space - the fourth dimension. *The fourth dimension has polarity.* Existence and nonexistence are related by:

$$\aleph = \left(\frac{\Diamond_4}{2}\right) - \left(\frac{\Diamond_4}{2}\right) \qquad (3.1)$$

and this difference occurs along the fourth dimension.

The magnitude of any half of the universe is equal to half of unit hypervolume ($\Diamond_4/2$). Nonexistence is the difference between two such magnitudes. Summation occurs along the three infinitely large dimensions of space whereas difference occurs along the fourth dimension. In keeping with existential uniformity the twin halves of totality are indistinguishable from each other and their difference is null.

Positive and negative directions along the fourth dimension are entirely equivalent. Their only difference is their opposition to each other:

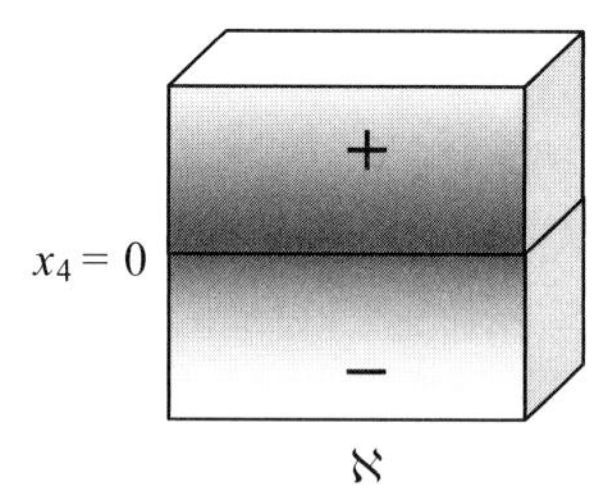

Figure (3.1) The most complete description of totality is as a self-nullifying four-dimensional finite

The designation of positive and negative axes is therefore purely a matter of convention, much like our assignment of charge for positrons and electrons.

The fourth dimension's polarity is the result of a curious, albeit inescapable, set of relationships. Existential closure requires infinite largeness and infinite smallness. The juxtaposition of these limits produces totality's finite hypercube. This hypercube must sum to nothingness in order to exist within it, and in so doing nullify the spatial summation inherent in dimensional closure. Space sums to zero by being composed of zero; its resultant hypercube sums to zero by being the difference of two compositions. This is how the fourth dimension allows the totality of space to achieve nonexistence.

Ψ THEOREM 3.5 - CLOSURE POLARITY {Ψ3.4}

THE FOURTH DIMENSION IS INTRINSICALLY POLAR

Closure polarity and the symmetry of totality require polar attributes for the universe. This is why there is an infinite wealth of positive and negative electric fields scattered across space. From Figure (3.1), look again at totality's hypercube:

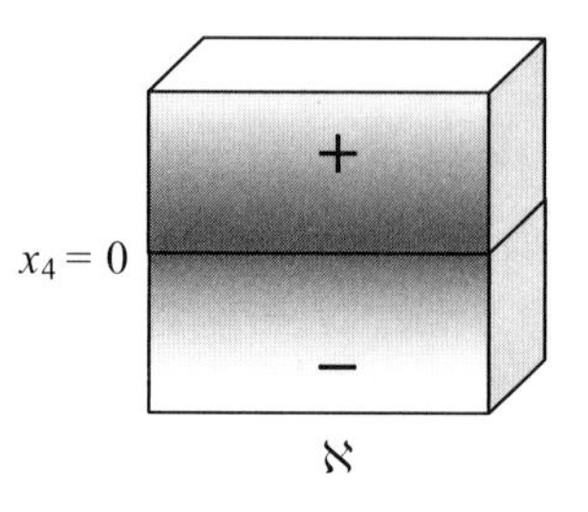

It consists of an infinite number of layers of three-dimensional space at various fourth-dimensional elevations. The fourth dimension is not composed of points because it represents their fourth-dimensional *deflection* - a difference of position. Space exists because the sum of nothing is nothing and the fourth dimension exists because the difference of nothing is also nothing.

Space has no center but fourth-dimensional deflection does, at least in reference to neutral space. The uniquely defined coordinate ($x_4 = 0$) is the hypervolumetric expression of existential parity. It is the only case where the universe has two equal and opposing halves. *This means space has to exhibit fourth-dimensional displacement in order to achieve four-dimensional symmetry.* This in turn tells us that the positive and negative fields of energy are physical deflections of space along the fourth dimension. Space sums to nonexistence *through its displacement along the fourth dimension.* The only compositional dimensions the universe has are those of space and the only way it can achieve a polar attribute is by displacement along x_4. In short, the universe is ultimately composed of space, so anything *not space* is by default spatial distortion. Just as postulated in our original derivation of the Null Axiom, reality consists exclusively of space and distortions of space. Our more refined characterization of nonexistence now reveals that space literally could not exist without the distortions connecting it to nonexistence.

But how does spatial deflection actually nullify space? If a region of space were deflected ten meters into the positive fourth dimension followed by the same distance into the negative, the result would be neutral, undeflected space, not nonexistent space. While this is true, it is not the existential perspective, because energy is associated with these deflections, and energy is conserved. Matter and antimatter do not merge to vanish into nothingness; they annihilate to produce light, a different form of positive/negative spatial deflection. Space's nullification, like every other existential operator, occurs only in totality's external perspective. The relationship between spatial deflection and energy will be fully defined in Parts II and III, but for now our premise is that the nullification of spatial magnitude is the *raison d'etre* for the curious polarity displayed by energy's elementary formulations.

Fourth-dimensional deflection changes space in a fundamental way; it has to. A spatial point and its deflection are not separate entities. Charged space is charged down to the level of individual points. If it were possible to superimpose the two halves of totality, the result would be nothingness, not space - a direct consequence of the universe's boundless energy distribution. Existence is composed of two and only two things, space and curved space. It contains the exact amount of positive and negative electric fields (spatial deflection) necessary to provide perfect four-dimensional symmetry.

ENERGY'S CRITICALITY

The universe could not exist in the absence of charged space and charged space cannot exist in the absence of energy. Space needs the electric fields of matter, antimatter, and photons in order to nullify the baseline existence of its reference frame. So while matter requires the existence of antimatter to balance within nothingness, space needs both to cancel the magnitude of its own closure.

Another way to look at energy's criticality is to view space with and without energy as two different states of existence. Existence is by first principles a case of partial nonexistence. Empty space is therefore the *nonexistence of energy*. It is half of a twin set of states. The existence and nonexistence of energy is similar to the existence and nonexistence of the universe; one cannot exist without the other. *Empty space is dependent on the existence of energy because it is by definition the nonexistence of energy*. Nothingness is the lack of all things. Space is the lack of *something*. Observation tells us this is energy.

Space and energy have a codependent existential relationship, so space could no sooner exist in the absence of energy than energy could exist in the absence of space.

Ψ THEOREM 3.6 - ENERGY'S EXISTENTIAL CRITICALITY {Ψ3.4}

SPACE COULD NOT EXIST IN THE ABSENCE OF ENERGY

Space reveals the extent of existence while energy shows us its four-dimensional symmetry.

3.2 TIME, THE VARIEGATED EXTRASPATIAL REALM

Although much has been made of the universe's infinite size, it is not necessary to actually reach totality to get a glimpse of its closure. It glimmers at us in every waking moment.

What is:

- Fourth-dimensional?
- Intrinsically polar?
- External to space?
- A metric for spatial distance?

Time. Time, like space, is an inevitable consequence of unit hypervolume. Space might constitute the composition of reality but time is the cause and effect binding it all together.

Ψ THEOREM 3.7 - TIME {Ψ2.12, Ψ3.5}

TIME IS THE DIFFERENCE OF SPACE

This embodies all of time's properties in a single statement: its polarity, association with space, and extraspatial nature. The most basic nature of anything is the way it sums to zero, the elemental reality allowing it to exist. Closure represents the additional dimension of time, and unit hypervolume is the distribution of space along this extent. Time is not a compositional dimension, it is a *difference* dimension. Within the context of three-dimensional space, it takes on two seemingly unrelated forms - the displacement caused by motion and the polarity of electric fields.

Time is not an independent dimension. If it had the degrees of freedom of four-dimensional space, the universe would have more than one future and more than one past (although supporters of the many-histories idea claim it does). Nor is time the same as spatial extent. It is an extension beyond space yet remains inextricably bound to it - the difference of space.

METRIC EQUIVALENCE

Nonexistence is symmetric, so the universe's hypervolume is fixed with equal extents in all four of its dimensions. Although time may be infinitely thin for infinite space *it has a length of unity in unit hypervolume*. Just as the absolute meter m_a is the absolute unit of distance, the *absolute second*, s_a, is the absolute unit of time. The existential significance of meters and seconds may differ, but symmetry ensures that their magnitudes are the same:

$$|1\ m_a| = |1\ s_a| \tag{3.2}$$

Much like the contextual relationship between the existential states, spatial and temporal distance are by definition equivalent because they can only be compared to each other. A distance in time *is not the same thing* as a distance in space, but their absolute metrics are identical because time is a polar *component* of space. This will be called *metric equivalence*.

Ψ THEOREM 3.8 - METRIC EQUIVALENCE {Ψ1.8}

SPATIAL AND TEMPORAL DISTANCE HAVE EQUIVALENT MAGNITUDE

Even though space and time are compositionally unique, the symmetry of nothingness, from which unit hypervolume is derived, ensures their equivalence. The speed of light quantifies this equivalence and its value is unity when units of distance and time are scaled according to their innate relationship:

$$c = \frac{1\,m_a}{1\ s_a} \equiv 1 \tag{3.3}$$

The speed of light is an expression of dimensional symmetry. It is not a physical constant, it is one of our universe's unbreachable *geometric* limitations.

THE DUALITY OF UNIT HYPERVOLUME'S UNITS

Unit hypervolume is the overview, and the overview is symmetric, so it has units of distance^4. It has the units of four-dimensional space but not the resolution, because four-dimensional resolution requires four unbounded, simultaneous, contiguous extents. The universe only has three infinite dimensions, and when these are stacked into a finite hypercube, all four dimensions are filled with spatial points at three-dimensional density, and the resulting hypercube has units of quadric distance. However, any *incomplete* representation of this hypercube, such as half of unit hypervolume, has units of time-distance^3. Totality is neutral and symmetric, whereas its internal composition is polar and asymmetric. *Time is the dimension that bounds, not extends, three-dimensional space.*

The symmetry between space and time allows a meaningful conversion between these two hypervolumetric formulations:

$$\text{distance}^4 = c\left(\text{time}\cdot\text{distance}^3\right) \tag{3.4}$$

Unit hypervolume is the absolute magnitude of the universe's internal size, and therefore the largest possible hypercube. Time, after all, only serves to nullify space's magnitude in an *external* context. The relationship between space and time is entirely contextual. Just as time is the difference of space, so too is space the difference of time.

ETERNITY

The speed of light is the dynamic reality of metric equivalence. Since the universe's diameter is the largest possible distance, the longest possible duration is the time required for light to traverse it. This will be called *eternity*. It is the cosmic balance between space and time.

Ψ THEOREM 3.9 - ETERNITY {Ψ3.8}

THE TIME REQUIRED FOR LIGHT TO TRAVERSE THE UNIVERSE IS ETERNITY, ∞/c

where c is in absolute units of $(m_a/s_a) = 1$.

Certainly something moving slower than light would ostensibly require a timeframe longer than eternity to cross space, but as shown in the next chapter, such a journey is not possible. The speed of light is the underlying dimensional relationship between time and distance. It provides the standard for unbounded duration, much as the universe's diameter provides the standard for unbounded distance.

Like infinite distance, eternal time is difficult to conceptualize. Eternity is the age of the universe; the duration of existence. The breadth of the universe has no beginning or end, nor does its perpetuity. This is consistent with the lack of an origin for nonexistence. What it also means, interestingly enough, is that every moment in time is *equivalent*. The space of our universe has no center *and neither does its history*. This will be referred to as *temporal equivalence*:

Ψ THEOREM 3.10 - TEMPORAL EQUIVALENCE {Ψ3.8}

ALL MOMENTS IN THE HISTORY OF THE UNIVERSE ARE EQUIVALENT

History has no universal bound, so there is no way to unambiguously specify a particular moment within it. In space, for instance, the relationship between any given location and the entire universe is the same. While this is impossible to visualize it is nonetheless true. Similarly, the relationship between any moment of time and eternity is the same. This is difficult to reconcile with the subjective reality known as the *present* moment of time, but it is still a necessary consequence of a boundless universal history. From the vantage of our finite existence, time is delineated into past, present, and future, *and the present is a uniquely separate state*. Temporal equivalence indicates that this is not an accurate assessment of time. There is no *now* just as there was no beginning. This curious situation will be resolved in Chapter 4.

MOMENT DENSITY

The totality of time is composed of an infinite number of separate moments. Each is delineated by the motion of material through space. Distance has a resolution of ∞^3 points per absolute meter, so one might think motion occurs at the same resolution. This is not the case. The infinitely fine resolution *of any entity* is defined as the inverse of its unbounded totality. Motion reaches its existential maximum at eternity. As such, its resolution is equal to that of one-dimensional space.

Ψ THEOREM 3.11 - MOMENT DENSITY {Ψ2.8}

THE RESOLUTION OF TRANSLATIONAL MOTION IS ∞ MOMENTS PER ABSOLUTE SECOND, (∞/s_a)

This will also be referred to equivalently as *translational density* or *motion density*.

The confluence of eternity and moment density provides the total number of moments in the history of the universe:

Ψ THEOREM 3.12 - ETERNAL VARIABILITY {Ψ3.8, Ψ3.11}

ETERNITY CONTAINS ∞^2 *SEPARATE MOMENTS*

This is the amount of temporal variability found in the universe, the total number of different states of matter and energy in space.

3.3 SPATIAL CONSERVATION

Conservation is the invariant counterpoint to change. The amount of energy in the universe must stay precisely the same from one moment to the next because each and every particle is required to properly express the unboundedness of nothingness. This is why the compositional quantities of reality such as joules of energy and cubic meters of space are preserved in any interaction. The ultimate level of spatial composition is the geometric point, and space is equal to the sum total of every point in the universe. To change the number of points in space is to violate the basic premise of its composition:

Ψ THEOREM 3.13 - POINT CONSERVATION {Ψ1.6, Ψ2.5}

GEOMETRIC POINTS ARE CONSERVED, THEY CAN NEITHER BE CREATED NOR DESTROYED

Closure provides an immutable equivalence between totality and a single geometric point. Destroying a point is tantamount to destroying a universe. It is also somewhat nonsensical. A point is the epitome of nothingness - it has no properties to remove. Creating a point is just as illusory for it has no properties to add. Conservation is an inevitable consequence of a point's inherent nature.

Point conservation is abstract but not merely an exercise in abstraction. It is evaluated in terms of the density and volume of space. Like energy density, point density varies throughout space but must balance around a universal average. This is confirmed by the existence of gravitational fields, which are derived in Part III.

Volume conservation follows from point conservation as well as first principles:

Ψ THEOREM 3.14 - SPATIAL CONSERVATION {Ψ3.13}

SPATIAL VOLUME IS CONSERVED, IT CAN NEITHER BE CREATED NOR DESTROYED

3.4 COMPOSITIONAL PARITY

Dimensional closure spans the breadth of reality from the diameter of a point to the volume of the universe. It is responsible for the layers of summation between quantity and singularity, the existence of time, the existence and quantization of energy, and the relationship between infinite linear smallness, finiteness, and infinite volumetric largeness. Yet as sweeping as its implications are, it does not provide the basis for spatial *composition* across a similar scale.

Space is a three-dimensional volume composed of infinitely small volumes and ultimately points, but it is not constructed of point diameters, as they are too dimensionally restricted. *Space is built of infinitely small volumes.* So although composition and dimensionality are inseparable regardless of scale, closure cannot govern composition. But it does provide a clue about the responsible agent. Just as closure is driven by the distance between two points, or *dimensional* infinite smallness, composition is mediated by a different standard of infinite smallness - *compositional* infinite smallness. This will be called *metavolume*. It is the N-dimensional inverse of N-dimensional space:

$$\widehat{V}_N = \left(\frac{1^2}{\infty}\right)^N = \frac{1^{2N}}{\infty^N} \tag{3.5}$$

Our universe's metavolume is $(1^6/\infty^3)$. The six dimensions of its unity term correspond to the simultaneous existence of infinitely small and large volumes in space. Both use the same three spatial dimensions and are separated *only by scale*; there is no six-dimensional compositional totality.

Metavolume is the foundational building block of spatial volume, but is still far larger than the volume of an individual point, which will be symbolized ξ_N. In N-dimensional space this is equal to the Nth power of its diameter:

$$\xi_N = \left(\frac{1^{N+1}}{\infty^N}\right)^N \tag{3.6}$$

Our universe's metavolume is $(1^6/\infty^3)$, while its point volume is $(1^{12}/\infty^9)$. Unlike the power of unity in metavolume (1^6), the power of unity in point volume (1^{12}) consists of duplications of the three dimensions of space as well as the fourth dimension. *A point has the same diameter along each of its spatial dimensions, and unit hypervolume is implicit in each.*

LAYERED COMPOSITION

From our finite perspective, metavolume might seem closer to a point than to the volume of space, but in reality it is not. It is just infinitely small and we are unable to compare infinitely small and large volumes with our instruments. *Metavolume is the compositional inverse of space.* Just as space is composed of a certain number of metavolumes, metavolume is composed of a certain number of points. A single point and the totality of space are two forms of the same thing, composed of the same fabric. Metavolume is the layer needed to isolate them from each other, a compositional mirror embedded within the dimensional mirror of closure. A point is as compositionally removed from metavolume as metavolume is from space, so there are as many points in a metavolume as there are metavolumes in space. This is the concept of *compositional parity.*

Ψ THEOREM 3.15 - COMPOSITIONAL PARITY {Ψ2.4}

THE METAVOLUME OF SPACE IS ITS NUMERICALLY COMPOSITIONAL MIDPOINT

Metavolume is precisely midway, in volume, between the volume of a point and the volume of space. Our universe has an evenly balanced spectrum of size from its entirety down to a single point because *space's total volume and the individual points within it are two forms of the same thing*:

$$\sum_{1}^{(\infty_0)^6}(\xi_3) \rightarrow \hat{V}_3 \ldots\ldots \sum_{1}^{(\infty_0)^6}\hat{V}_3 \rightarrow \left(\infty^3\right) \tag{3.7}$$

where ∞_0 is a unitless unbounded count, $\infty_0 = (\infty/1)$.

Metavolume is composed of $(\infty_0)^6$ points and space is composed of $(\infty_0)^6$ metavolumes, thereby providing a symmetric compositional balance between infinitely small and large volumes. It is the mediator of multidimensional composition, infinitely small because it lies below the level of continuousness. Metavolume governs composition much as unit hypervolume governs dimension.

Closure is inverted by scale but composition is not. It is by necessity symmetric for both ends of infinity, and this is what the term *numerically compositional midpoint* means for (Ψ3.15). Distance is continuous and so is volume. Compositional parity maintains the volume continuity of space. On one end of the scale is a single point which by definition is continuous. On the other end is space, also continuous. The system is entirely and perfectly symmetric.

Compositional parity is the relationship between the universe's three levels of composition:

$$\hat{V}_N = \sqrt{\xi_N \infty^N} \tag{3.8}$$

Metavolume is the geometric mean of the product of the volume of a point and the volume of the universe.

3.5 FINITE DIMENSIONALITY

We are now prepared to address one of the most difficult questions about space:

Why does the space of our universe have only three dimensions?

If this were purely the product of chance, the probability of three dimensions out of a range from one to infinity is infinitely small. The anthropic solution is that only a three-dimensional universe can produce humans, but this is true only because it is the statement of the patently obvious. Humans are the by-product of the universe's material. If the universe were significantly different than it is, *we would be significantly different than we are.* Using the anthropic argument for universal properties is like saying the Earth is the way it is because otherwise birds wouldn't be able to fly.

But the nontrivial problem of finite dimension remains. If all dimensions are possible, the likelihood of three-dimensional space is vanishingly small. Either we inhabit an infinitely unlikely universe floating among countless others of various dimensions, or three-dimensional space is the only viable solution to the state of nonexistence. The latter is the more reasonable assertion. Regardless of the potential existence of universes of other dimensions, *some operator defines the number of dimensions in our universe.* It is built into the structure of space just as surely as unit hypervolume. The characteristics of our universe are a certainty, fully defined by its relationship to nonexistence.

Our universe does not have three spatial dimensions because some other universe has more or less. The term *universe* signifies a self-contained distribution of nothingness. A universe is an integrated formulation of zero. Any properties evident in the solution to the zero equation, from the dimensions of space to the density of the matter within it, must be precisely the way they are in order to satisfy this equation. There is no other cosmic requirement. The space of our universe *has to be* three-dimensional. And if it is required for our universe, it is *one of the defining parameters of a universe.* This suggests that three dimensions might be the only viable configuration for a spatial solution within nothingness.

Suppose for the sake of argument there are many universes, some with three dimensions and others with seven. If they are truly independent of each other, the total number of three-dimensional universes has nothing to do with the number having seven. Hence the dimension of any particular universe is purely a matter of *chance.* However, chance presupposes a defining moment - a time where one of two paths is taken. This in turn requires a universal origin which, by first principles, is utterly contrary to the fundamental nature of existence. So if our universe's dimension follows from nonexistence (and again there is no other available source), then space *has to be* three-dimensional. *Universes similar to ours but of other dimensions are simply not viable.* Why is this?

Look again at the compositional balance between the two expressions of totality, Equation (3.8):

$$\widehat{V}_{\mathrm{N}} = \sqrt{\xi_{\mathrm{N}} \infty^{\mathrm{N}}}$$

Substitute Equation (3.5) for metavolume and Equation (3.6) for point volume:

$$\frac{1^{2\mathrm{N}}}{\infty^{\mathrm{N}}} = \sqrt{\left(\frac{1^{\mathrm{N}+1}}{\infty^{\mathrm{N}}}\right)^{\mathrm{N}} \infty^{\mathrm{N}}} \tag{3.9}$$

Square:

$$\frac{1^{4\mathrm{N}}}{\infty^{2\mathrm{N}}} = \left(\frac{1^{\mathrm{N}+1}}{\infty^{\mathrm{N}}}\right)^{\mathrm{N}} \infty^{\mathrm{N}} \tag{3.10}$$

Divide by ∞^{N} and simplify:

$$\left(\frac{1^{4}}{\infty^{3}}\right)^{\mathrm{N}} = \left(\frac{1^{\mathrm{N}+1}}{\infty^{\mathrm{N}}}\right)^{\mathrm{N}} \tag{3.11}$$

Take the Nth root:

$$\frac{1^{4}}{\infty^{3}} = \frac{1^{\mathrm{N}+1}}{\infty^{\mathrm{N}}} \tag{3.12}$$

Behold. The only N for which this is true is (N = 3), the space of our universe. *The singularity of nothingness demands existential closure. Existential closure demands compositional parity. Compositional parity demands cubic space. Our universe's dimensionality is as inevitable as its existence.*

3.6 GENERAL CONCLUSIONS

This chapter is very much an extension of the last and its conclusions provide the foundation for energy as well as the uniqueness of three-dimensional space:

- The most inclusive and accurate portrayal of nothingness is as the difference between the positive and negative halves of the universe's hypervolume.
- Time is the difference of space.
- The four-dimensional symmetry of the universe's totality requires space to have positive and negative fourth-dimensional displacements.
- Unit hypervolume is neutral and symmetric, with units of distance^4. Subcomponents of this hypercube have units of time-distance^3.
- Eternity is the time it takes for a photon to traverse infinite space.
- All moments in history are equivalent, and motion has a resolution of ∞ movements per absolute second.
- Eternity contains ∞^2 separate moments.
- Conservation extends down to the level of geometric points.
- The space of our universe is three-dimensional because this is the only dimension whose volume is compositionally consistent through all levels of infinite size while forming the surface of its own hypervolume.

4. ULTRASTASIS

Time appears to extend beyond the here and now of our daily struggles, but is this truth or appearance? Does it reflect the existence of other universes? Is our universe's history merely one of infinitely many different timelines? The best way to understand the fourth dimension's full extent is to look for other universes. As it turns out, however, *there are none*.

4.1 ONE EXISTENTIAL SINGULARITY, ONE UNIVERSE

Our universe's infinite thinness along the fourth dimension gives the initial impression that reality has more than enough room for an infinite number of parallel universes. However, if this were true, dimensional closure would fail because the space of our universe would no longer be the surface of its own totality. It would instead lie infinitely close to other spaces along the fourth dimension, eradicating the boundary of its own closure. Composition is synonymous with summation. Space-time can't be filled with space without also being composed of space. Regardless of the apparent redistribution of the universe's material over time, its entire history can be no larger than the size of any given moment. *Time does not expand the size of the universe*.

Time is a fourth-dimensional interconnection among energy forms. Although trillions of years may pass before two distant quanta eventually interact, *if it occurs in the future, the certainty of this future occurrence existed throughout the infinite past*. Every event is preordained. If it were possible to look at eternity in its entirety it would be clear that every fragment of energy in the universe ultimately has or will interact with every other. Changing, coruscating, fluctuating, translating. This inconceivably complex matrix finds its roots in *nothingness*. Where lies the possibility of differentiation? One featureless source is responsible for all of this variety. The relationship among the quanta of our universe throughout time is a *single relationship*. They are all part of a *single omniquanta*. Just as nothingness defines the number of dimensions in our universe, it also defines its present state. The distribution of energy in space at this very moment is not a statistical accident, it is an absolute certainty. *Nonexistence is not complete by virtue of statistics; it is complete by virtue of immensity*. Nothingness and the sum of our universe are indistinguishable, and both are complete. *Our universe is the only universe*.

There is but a single universe, it is the only possible universe, and we are living in it.

Ψ THEOREM 4.1 - ONE UNIVERSE {Φ1.1, Ψ1.6, Ψ2.4}
OUR UNIVERSE IS THE ONLY UNIVERSE

Our universe is not a permutation; it is not one of legions of realities that through dark and mysterious means interact to achieve nonexistence. It is a singularity and a certainty, the one and only personification of the void. But even though it constitutes the full breadth of reality, it is more extraordinarily immense than is easy to imagine, as should soon become apparent.

4.2 A SINGLE UNIVERSE OF CHANGE

Our analysis now turns to how change operates on the largest scale. Time is a dimension as well as a phenomenon. On one end of the existential realm lies nothingness, a state of complete stasis. On the other end are atoms, planets, stars, and galaxies; all sites of riotous transmutation. The relationship between the two is the ultimate description of phenomenological time.

The terms *time* and *change* will be used interchangeably throughout this chapter, although the predominant reference will simply be *time*. Our focus is the global relationship between what is perceived as time and the universal distribution of material across space. Although this is certainly related to time's fourth-dimensional nature, its net effect is motion.

Our universe constitutes all of reality, but it is as yet unclear whether or not it actually changes from one moment to the next. Time might be a nonspatial extension centered on space whose zero-sum is sequential instead of simultaneous. Perhaps there has to be such an extension to fully embody the permanent nature of existence. Nonexistence has unbounded extent as well as perpetuity. Is time the compromise between static quantities, like the amount of energy in the universe, and everything else?

Only one question needs to be answered to achieve a full understanding of time.

Does the universe change from one moment to the next, and if so in what way?

If the universe changes it needs a way to somehow nullify its variation in order to exist within nonexistence. Otherwise how is the motion that is so manifestly evident across the heavens instantaneously and continuously cancelled as it occurs? Earth is finite and has changed significantly over the past billion years. The universe is infinite and this is where change and other finite considerations coalesce into featurelessness.

The portion of the universe visible to our instruments, consisting of billions of galaxies scattered across billions of light years of space, is all part of a *single perfect equation*. Our local surroundings are defined down to the last electron in order for the entire universe to achieve its exact null summation. Existence contains all details and all information in an infinitely precise balance. Thinking of the universe as a gigantic chaotic contraption forced to obey a few general rules is incorrect. The truth, though much more difficult to comprehend, is that reality is an utterly flawless system, fully defined and determinant, and huge beyond reckoning. Energy's causality and connectivity allow for no other possibility.

In the overview there is nothing; in existence there is every possible thing and each and every one of these things is vital both in form and substance. The slightest deviation from the universe as it exists at the present time in all its current majesty would produce an imbalance in the overview and violate the relationship between the existential states. Every electron and photon have the exact momentum they are supposed to have, to infinitely fine resolution. *This is the only way they can exist within a featureless hypercube*. The best way to understand this is to consider the alternative. Suppose the universe was exactly the way it is, except with one of its photons displaced by 1 centimeter. This modified universe is not possible, because the photon has a *history* and the energy of which it is composed has an infinite history. Displacing a photon any distance in any direction breaks its connectivity to the rest of the universe. Again the question must be asked:

Does the universe change from one moment to the next, and if so in what way?

A changing universe faces two major challenges:

- Temporal centricity.
- Nonlocal mass conservation.

TEMPORAL CENTRICITY

If the universe has *N* different states, what defines its current state? Existence is a certainty down to the very last detail. This requires nonexistence to contain the data for all possible universal states as well as its *current state*. Even if time is a cycle able to circuitously cancel change starting from any given state, it is not possible to define what *zero change* means in the absence of a reference state. The present moment is necessarily a reference, creating an asymmetry comparable to the problem of a universal origin. Every universal state is equivalent, yet if the universe changes it must exist in one of *N* states at any given moment. This creates an intractable separation between the current state and all of the other states and will be referred to as *temporal centricity*. Since this is in direct violation of temporal equivalence (Ψ3.10), it represents a formidable obstacle for the premise of a changing universe. The universe cannot change if there is no present moment.

NONLOCAL MASS CONSERVATION

The mass of every star in the universe varies. They gain it during their formation and lose it as they burn, and are separated from each other by vast and in some cases infinite distances. The universe's total mass is on the order of ∞^3 kg. Either it changes over time or it does not. If it is fixed, the creation and destruction of mass throughout space has to be controlled with infinite precision over infinite distances. Conversely, if universal mass changes, how large is its variation? Does it oscillate around a mean value at one kg/s? A billion kg/s? Or an infinite number of kg/s? What is the length of its cycle? Either the universe's mass is miraculously and meticulously maintained throughout all of space or it oscillates at a certain number of kg/s. Yet if mass changes, universal states are no longer quantitatively equivalent and temporal equivalence is violated.

A changing universe is simply at odds with unbounded time.

4.3 FROZEN FOR ETERNITY

Regardless of what the whirling galaxies, exploding stars, and general chaos in space seem to indicate, *the universe, when viewed in its entirety, remains precisely the same from one moment to the next*. The universe does not change for the same reason it cannot have an origin - temporal equivalence. The perpetuity of nothingness is not maintained by an additional dimension of existence; it is maintained by its utter stasis. How blatantly unimaginable is that? The universe's mass does not vary, nor does anything else, because its entire energy distribution remains constant down to the relative position of the last electron.

This concept will be referred to as *ultrastasis*, the most comprehensive realization of temporal equivalence. All moments in time are equivalent because they are one in the same:

Ψ THEOREM 4.2 - ULTRASTASIS {Φ1.1, Ψ4.1}

THE UNIVERSE'S OVERALL MATTER/ENERGY DISTRIBUTION IS INVARIANT

The universe had no origin; there was no special moment when everything began. Nor will it ever end. There are no special universal moments at all. Ultrastasis is the perfect antidote to temporal centricity. The universe at one moment is exactly the same as at any other moment. It is governed by the utter immutability of nothingness.

Ultrastasis is at first spectacularly counterintuitive, but its underlying reasoning is inescapable and its ramifications staggering. A static universe means every moment in the universe's history, and therefore Earth's history, *is located somewhere in space at this very instant*. This would initially seem impossible since the observable universe appears to change with time. How could the universe be large enough to hold its own history?

4.4 OMNIPATTERN

The universal engine churns throughout an infinite number of galaxies with no net effect, but change is no more an illusion than existence. The only way the universe can achieve perfect stasis is if every one of its particles move in a precise way. Change vanishes in the overview because of its infinite precision, not because it is ephemeral. Think of the universe's energy as a single entity connected by causality throughout an eternal history. If this universal configuration could change from one moment to the next it would be *incomplete*. Space is complete by virtue of boundlessness. The same is true for energy, but its completeness is enormously more complex. The universal energy distribution is fixed because to change is to contain new information. New information would mean the prior state was incomplete. There is one past and one future because they are exactly the same, the ultimate expression of nothingness as well as causality. The infinite distribution of matter and energy throughout space will be referred to as the *omnipattern*.

§ DEFINITION 4.1 - OMNIPATTERN

THE UNIVERSE'S UNBOUNDED, INVARIANT MATTER/ENERGY DISTRIBUTION

How is it possible for this scintillating mosaic to remain constant over time? The first step to solving this riddle is to look more closely at the essential nature of the large-scale universe, beginning with its size.

COSMIC PROPERTIES

Theorists speculate about alternate universes because they underestimate the breadth of their own. The universe's volume is ∞^3, so it contains ∞^2 regions of ∞ volume. This in itself is curious. An object whose volume is ∞ *has no spatial boundary*. So not only does the universe lack a boundary, it is so large that it contains an infinite number of unbounded subunits. This property will be referred to as *ultraunboundedness*:

Ψ THEOREM 4.3 - ULTRAUNBOUNDEDNESS {Φ1.1, Ψ4.1}

THE OMNIPATTERN CONTAINS AN INFINITE NUMBER OF AN INFINITE NUMBER OF INFINITELY LARGE PATTERNS

Material content is the other property to consider. Matter is composed of elementary particles that exist in entirely equivalent positive and negative varieties. Any pattern composed of matter must also exist as antimatter. This will be called *ultrasymmetry*:

Ψ THEOREM 4.4 - ULTRASYMMETRY {Ψ3.8, Ψ3.10}

THE OMNIPATTERN IS ELECTRICALLY SYMMETRIC

For every moment of the Milky Way's lifespan, scattered throughout the universe, there is a complementary moment of an anti-Milky Way.

4.5 A COSMIC QUANTUM

The universe is ultrastatic, so its internal change corresponds to a variety of material/energy distributions in space. There is a one-to-one correspondence between the number of states in eternity and the number of subpatterns in the omnipattern. Eternity (∞) and the linear resolution of motion ($1^2/\infty$) require ultrastasis to be composed of a total of ∞^2 separate states (Ψ3.12). This means the omnipattern contains ∞^2 separate subpatterns. These will be referred to as *omnielements*. They represent the *infinite* quantization of space-time.

Ψ THEOREM 4.5 - OMNIELEMENT, Ξ {Ψ3.12, Ψ4.2}

THE OMNIPATTERN IS COMPOSED OF ∞^2 SUBUNITS CALLED OMNIELEMENTS JUST AS ETERNITY IS COMPOSED OF ∞^2 INFINITESIMAL CHANGES

Omnielements and their related parameters will be symbolized by Ξ. The universe has a volume of ∞^3 and contains ∞^2 omnielements, so the volume of any given moment of time is ∞ cubic absolute meters:

$$\Xi_V = \infty \tag{4.1}$$

Omnielements are infinitely large, so their lifespan is infinitely long. But more can be said than this. An omnielement's lifespan is eternity because change's resolution ($1^2/\infty$) defines both the difference among omnielements as well as their total number. The maximum difference among omnielements is the full extent of their lifespan, which is the inverse of

$(1^2/\infty)$, ∞. Omnielement lifespan and the universe's maximum duration are one in the same.

Ψ THEOREM 4.6 - OMNIELEMENT LIFESPAN {Ψ4.5}
OMNIELEMENTS HAVE ETERNAL LIFESPANS ($\Xi_\tau = \infty$)

The universe doesn't have multiple universal states, but it does contain ∞^2 *internal* states. Infinite omnielement lifespans provide enough time for the energy and matter of which they are composed to shift through every available internal state. *Eternity is long enough for an infinite amount of matter/energy to express every possible spatial distribution.* Omnielements are the endpoint of causality. They are infinitely large, yet physically interact to ensure that they never occupy the same state at the same time. Ultraunboundedness provides the sheer size to allow the omnipattern to process an infinite number of separate moments simultaneously.

Since all moments of time are equivalent (Ψ3.10), omnielements possess individual ultrasymmetry.

Ψ THEOREM 4.7 - OMNIELEMENT ULTRASYMMETRY {Ψ3.10, Ψ4.4, Ψ4.5}
OMNIELEMENTS ARE ULTRASYMMETRIC

Omnielements are composed of equal parts matter and antimatter.

4.6 A VAST AND IMMUTABLE SPACE-TIME

Consider one of the more significant ramifications of ultrastasis:

All moments in Earth's history are permanently distributed throughout infinite space.

Let τ_E represent the duration of Earth's entire lifespan in absolute seconds, from its formation to eventual dissolution. Each of its infinitesimally brief moments will be referred to as an *instance*. Since time's instance density is equal to motion's moment density (∞ instances per unit time), Earth's entire history contains a total of ($\tau_E\infty$) different instances. In general, the total number of instances of any object will be called its *instance number*, N_τ:

$$N_\tau = \tau\infty \tag{4.2}$$

where τ is its lifespan in absolute seconds. Ultrastasis means each of these distributions exist in space at all times. The longer an object's lifespan, the more instances it has scattered across the universe. The minimum number of cosmic instances any object can have is one. Similarly, the minimum lifespan any object can have is $(1^2/\infty)$ because this is the resolution of change.

An object's instances have an associated spatial volume, their *instance volume*. Think of it as the average size of the spatial region that contains a single instance of an object's history. It is given by the ratio of the universe's volume to the object's instance number:

$$V_\tau = \frac{V_U}{N_\tau} = \frac{\infty^3}{\tau\infty} = \frac{\infty^2}{\tau} \tag{4.3}$$

with units of volume/instance. The shorter an object's lifespan, the fewer omnielements its instances occupy. Omnielements are not *identical*; they are *equivalent*. They have the same size and total energy but don't contain the same patterns. Note that when lifespan is eternal (∞), instance volume is equal to omnielement volume. *The only object with a lifespan long enough to be present in every omnielement is an omnielement.* This is consistent with their role as the intermediary between change and ultrastasis.

What Equation (4.3) tells us is *moments of phenomenological time are isolated from each other by infinite distances*. Volumes containing instances of Earth's history are limitless, so their relationship to each other within space is dimensionally structured yet quantitatively unbounded. The set of all of Earth's instances constitutes its actual lifespan, scattered across space at this and every other moment of time. The universe is large enough to contain its own history because *it is tightly restricted by causality and contains an infinite number of infinite volumes*. Each of us has a finite stature, yet the smoothness of change in our day-to-day existence causes our individual identity to *span the entire universe*. We are sprinkled throughout the firmament on immeasurably distant Earths, the confluence of discreteness and totality.

Time is the difference of space. Whether Earth's nearest moment exists in its distant past or future is at present unknown, but the more sobering reality is that *reaching it from our current location would require crossing an infinite distance*. Time travel is simply not possible, at least in the physical sense. We might be able to bring a human body's internal change to a stop with a cryogenic environment or use virtual reality to simulate a bygone era, but we will never be able to actually visit a different moment of history.

4.7 HIERARCHY OF VARIABILITY

The universe's nonstatistical nature can be argued in terms of the precision it requires to sum to nonexistence, but there is more profound evidence to consider. It has to do with the way objects *vary* throughout space. The cosmically average galaxy (excluding dwarf galaxies) contains about a hundred billion stars. This means the total number of *different* galaxies in the universe is at least a hundred billion times *less* than the total number of *different* stars. This in turn means variability actually *decreases* as the scale of size is ascended, an effect that will be referred to as *variability inversion*. It is a valid result irrespective of any of our previous conclusions or null principles.

Variability inversion is the opposite of how a statistical universe should operate. The larger the object, the more complex. From a purely statistical perspective, an object's universal variability ought to increase with size. The truth is although large objects appear to express greater diversity, they are physically unable to achieve it. Their size dictates their cosmic number density. This in turn constrains their cosmic variability. The *total* number of any type of object in the universe is $\rho_N\infty^3$, where ρ_N is its average number density in space (number per cubic absolute meter) and ∞^3 is the universe's volume, V_U. The universe contains $\rho_N\infty^3$ *different* objects of any given type.

Ψ THEOREM 4.8 - VARIABILITY INVERSION {Ψ2.3}

THE UNIVERSAL VARIABILITY OF A TYPE OF OBJECT IS EQUAL TO $\rho_N\infty^3$, GENERALLY INVERSELY PROPORTIONAL TO ITS SIZE

As the scale of size extends beyond the galaxy into the supercluster, then beyond it eventually into the omnielement, cosmic variability continues to fall off. *There are fewer stellar patterns than atomic patterns. There are fewer galactic patterns than stellar patterns. There are fewer supercluster patterns than galactic patterns. This leads inexorably past the omnielement to the universe itself and the cessation of variability. There is only one universe and it has only one state.*

The ultrastatic premise of omnielements stepping through a cosmic state machine is utterly counterintuitive, and wholly unlike the regal procession of celestial objects. *Yet the logic of variability inversion leads to the same conclusion from an entirely different direction and with precious few prerequisites.* There is an unfathomable scale of size, regardless of any other dynamical consideration, that has an intrinsic universal variability of ∞^2. This is the omnielement. For additional analysis of the universe's large-scale variational structure please see Appendix D.

4.8 TEMPORAL CLOSURE

One of the many interesting aspects of living in a boundless reality is the fact that the universe is not the only thing without a first origin. There are a number of reiterations of the cosmic chicken and the egg problem - matter and light being a prima facie case. Particles and antiparticles annihilate to create high-energy photons, and the collision of high-energy photons with matter creates particles and antiparticles. *Matter is necessary to create light and light is necessary to create matter.* There can be no first origin of matter because it would require a first origin of light, which in turn would require a first origin of matter, and so on. The universe had no first origin, and neither did its fundamental energy forms.

This concept will be called *temporal closure*:

Ψ THEOREM 4.9 - TEMPORAL CLOSURE {Ψ4.2}

TEMPORALLY CLOSED OBJECTS ARE THEIR OWN PRECURSORS

Temporal closure is not restricted to elementary forms of matter. Take the case of stars. Stellar ignition requires the presence of heavy and perhaps even radioactive elements. Yet the only place where this kind of material is created is in a star's core. This is not paradoxical because there *never was a first star*. The universe is a perfect steady-state equation with no origin. It is therefore permissible, required in fact, to have finite objects within it whose existence is contingent upon an environment they themselves create. The key is balance and equilibrium. The question of where everything comes from has already been answered by the Null Axiom.

Stellar genesis requiring stellar by-products is one thing, but what about life? Does its emergence require the delivery of certain molecules by way of meteorites, molecules produced only by living things? Panspermia might be the de facto standard for the onset of evolution. If this is the case, then the connectivity Earth has with the rest of the universe has just taken a monumental step forward.

It is important not to use temporal closure as a crutch to avoid real science, but it is just as important to understand nature's interrelationships on a grand scale. Temporal closure must not become a new version of the anthropic argument where the faithful claim, for instance, that life can only come from life. While this may in fact be the case, it can only be established by removing all other possibilities and arriving at a true understanding of what life is.

4.9 ULTRACONTINUITY

The universe's energy is complete because it contains its own history and future. But how far does this completeness go? What about alternate histories? If every possible energy distribution exists in space, does the universe contain a set of Earthlike worlds with *N* degrees of continuous variation between them? An Earth where Hitler won the war? An Earth where Martin Luther King wasn't shot?

This concept will be referred to as *ultracontinuity*. Any object, whether a galaxy, planet, or person, has a prodigious number of similar examples in space varying along *every possible parameter*. Only a portion of these are causally related as a sequence of instances. The question becomes:

> *How closely can the nonhistorical instances of any given type of object resemble each other?*

Are there legions of different versions of each of us that achieved or failed, lived and died, in near infinite variety? Think of this in terms of a morphing algorithm where images of any two objects can be connected and related with a continuous series of intermediate forms. A cat becoming a tiger. A biplane becoming a jet. Even in the small sample space of Earth, it is relatively easy to find individuals with no genetic relationship who appear virtually indistinguishable. How close can they become in the immensity of the universe's sample space?

Geneticists say that the percentage of DNA separating chimpanzees from humans is less than 2 percent. Between human races this decreases to less than 0.1 percent. Two identical twins have the same DNA but not the same experiences. Two people in the same quantum state have the same molecular configuration (to include DNA) as well as the same experiences. Any person on the planet could, at least theoretically, be transformed into any other with a variation of quantum states and molecules. Ultracontinuity is the idea that intermediate states of a certain resolution exist somewhere in space at this very moment. The question is not whether or not ultracontinuity is true; it is true by virtue of the universe's permutational variability. The question is *what is its resolution?*

Every possible combination of matter and energy exists, but we currently lack the computational power to determine what is possible and what is not. The task is daunting. A change of one state in one atom requires a precursor form and a series of prior events cascading backward through infinite time and across infinite space. Even so, perhaps scientists will someday be able to estimate the universal resolution of Earth-like planets. The nearest match might be nearly unrecognizable. Or it might be virtually the same as our current home; a planet whose only differences exist at the atomic level, too small for the human eye to see. Given the universe's infinite variety, the latter seems more likely.

Nature's deep connectedness goes far beyond that so eloquently envisioned by Francis Thompson:

All things by immortal power
Near or far,
Hiddenly,
To each other linked are,
That thou canst not stir a flower
Without troubling of a star

Francis Thompson, *Mistress of Vision*

4.10 OMNIPATTERN ETERNAL

A question has followed humankind through the ages; a question near and dear to us all - *is there an afterlife?* What happens to us when we die? Are our souls immortal? Will we be reborn into some new, improved incarnation? *Null Physics* makes it possible to address these inquiries scientifically.

Every loved one we have ever lost is alive and well in the omnipattern. They are separated from us by infinite distances but are indisputably alive. Right now. This instant. The omnipattern is not a heaven in the sense that these distant souls are removed from the pains and anguish of daily existence. Nor is it a hell where they lie in torment. All of our past images move through their lives precisely the same way they did before their passing, their joys and sorrows forever etched on the universe's fabric.

Moments are never *lost*. The universe holds every touchdown pass, every footstep on the moon, and every rainbow and solar eclipse. All our laughs are still echoing and all our teardrops are still falling. Dinosaurs roam distant lands now just as they did during the Jurassic period on Earth; they're just too far away to reach. And we are not alone. Every civilization that has existed or will ever exist on any world shares a similar posterity. We are all part of the indelible community of certainty.

4.11 GENERAL CONCLUSIONS

One of the most surprising aspects of this chapter is the breathtaking symmetry between space and time. What began as the premise of existence within the void blossomed into a crisp perspective of the universe's largest structure:

- Our universe constitutes the full extent of reality. It is the only universe and the only possible universe.
- The universe's overall matter/energy distribution is utterly static. It does not change in the least between one moment and the next.
- All moments of every object's history exist permanently in space, separated by infinite distances.
- The universe's largest level of structure is constructed of infinitely large components.
- The larger an object, the less its universal variability, precisely the opposite of what would be expected in a statistical universe.

PART II: PHYSICS OF ENERGY
INTO THE LIGHT

Energy's substructure

- *Quantum neorealism*
- *Maximum potential*
- *A new interpretation of a black hole's surface*
- *Absolute space*
- *Energy's dimensionality*
- *Electrically curved space*
- *Differential photon geometry*
- *Unit hypervolume*

I do not know what I may appear to the world, but to myself I seem to have been only like a boy playing on the seashore, and diverting myself in now and then finding a smoother pebble or prettier shell than ordinary, whilst the great ocean of truth lay all undiscovered before me.

Sir Isaac Newton, 1642-1727

Figure (II) Hodge 301 stellar nursery in the Tarantula nebula (lower right)
(Courtesy NASA/Hubble Heritage Project)

Prerequisite Concepts for Part II

The following concepts are carried forward from Part I:

- The universe exists because it is a distributed form of nothingness. Its existence within nonexistence requires it to sum precisely to zero. This is why it is electrically neutral and why matter is composed of curved empty space. It is also why momentum and charge are conserved.

- The universe's volume, V_U, is ∞^3. It is unbounded because nothingness is unbounded. It is eternal because nothingness is eternal.

- The universe has four and only four dimensions, three of space and one of time. Dimensional closure relates these dimensions with a constant called *unit hypervolume*. Unit hypervolume represents the finite four-dimensional size of the unbounded three-dimensional space of our universe. It is a symmetric hypercube whose size has units of distance^4. The length of one of the edges of this hypercube is referred to as an absolute meter.

- The fourth-dimensional thickness and minimum divisibility of space is called netherspace, δ_3. It is infinitely small and has a value of $(1^4/\infty^3)$.

- The product of netherspace and the volume of the universe is unit hypervolume.

- Time is the difference of space.

5. QUANTUM NEOREALISM

Before embarking on a new physics for matter and energy, let's review some of the problem areas where contemporary physics falls apart.

5.1 ORBITALS AND QUANTUM EFFECTS

Physics has become increasingly infused with a form of mysticism called *quantum reality*. This conceptual deterioration began in the early 1900s when scientists discovered a number of peculiar, nonclassical effects as they probed more deeply into atomic structure. The bizarre antics of the submicroscopic realm turned physics from a staid and confident science into an epidemic of suppressed hysteria that remains prevalent to this day. The root of this delirium isn't energy quantization; it's the curious *wave* nature of atoms and particles. While this might be exceptionally difficult to understand from an neorealistic perspective, a frantic departure from rationality is not warranted.

Energy density is uniquely defined at every point of space. It has no intrinsic statistical nature as its very existence depends on a precise universal summation to zero. Let's tackle the quantum reality question and look closely at the phenomena that have led most physicists to believe the universe lives and dies by the roll of the cosmic dice.

THE ATOM

Soon after electrons and protons were discovered, physicists began to investigate their dynamical relationship within an atom. The electrostatic attraction between charged particles has a distance/strength relationship similar to the gravitational attraction of celestial objects, so electrons were originally thought to orbit nuclei the way planets orbit the sun (only much faster). This was called the *Rutherford atom*. On another scientific front, photon emission from the acceleration of charged particles successfully explained the energy released by radio antennas when subjected to varying electric fields. Shortly after they were conceived, these two ideas came to blows with each other.

If accelerating electrons emit photons, how can they accelerate around the positive nuclei of atoms with no attendant energy loss? Electrons in these orbits ought to radiate energy until they crash into the nuclear surface. The rate of this calamitous energy loss would be given by the Larmour formula for accelerating charges:⟨10.1⟩

$$\frac{dE}{dt} = \left(\frac{q^2}{6\pi\varepsilon_0 c^3}\right)a^2 \tag{5.1}$$

where a is acceleration and q is unit elementary charge.

The alleged energy loss associated with the motion of the Rutherford atom's orbital electrons is called the *Rutherford catastrophe*, and it heralded the birth of quantum mechanics. Heisenberg made bold, nonsensical statements like *atoms are not things*. This eventually led to the concept of the *orbital*, a statistical spatial distribution of the relative likelihood of where an atom's electron exists at any given time. Unfortunately, no one knew what an atom was anymore. Electrons didn't orbit a nucleus; they were probability clouds existing everywhere and yet nowhere. The specter of a Rutherford catastrophe:

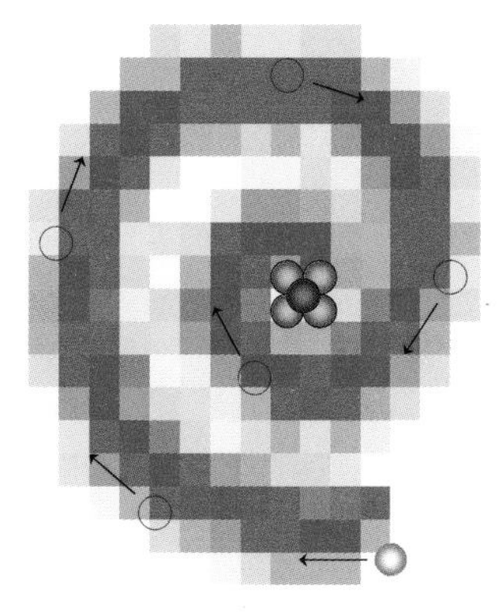

Figure (5.1) Rutherford catastrophe

eventually led to the illusory probability clouds known as *orbitals*:

Figure (5.2) Atomic orbitals

As it turns out, however, the orbital stability of an atom's electrons has nothing to do with their *realness*; it simply represents *the known fact* that a certain amount of time is required for photon emission.

Indeed, the Heisenberg relationship:

$$\Delta E \Delta t \geq \frac{\hbar}{2} \tag{5.2}$$

maintains that a change in energy is linked to and requires a corresponding interval of time.

The true solution to the Rutherford catastrophe is *charged particles emit photons when accelerated if and only if the period over which their net acceleration occurs is longer than the time necessary to emit the photon related to their net change in momentum.* This is the difference between a radio antenna and an atom. Photons are not emitted in relation to acceleration at some *instant* in time because they are not emitted at an *instant* in time. The net acceleration of a charged particle and the emission of a photon are both the result of an *interval* of time. Since the orbital period of an atom in its ground state is on the order of the amount of time required to emit a photon, and since the net acceleration over a complete orbit is zero, no photon is emitted.

Take the case of the hydrogen atom. The radii of its primary quantum states are given by:

$$r = \frac{n^2 h^2 \varepsilon_0}{\pi m_e q^2} \tag{5.3}$$

where n is the number of the state. The lowest energy state is called the *ground state* ($n = 1$). The following shows the relative sizes of hydrogen's first seven energy states:

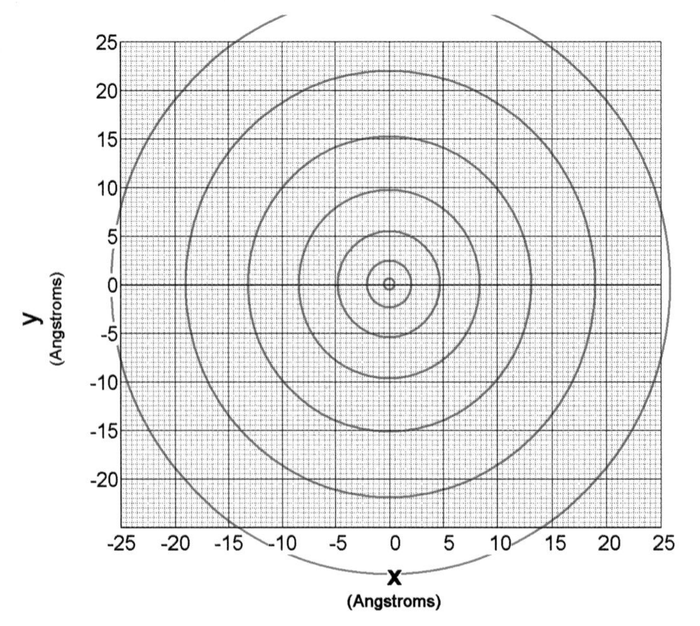

Figure (5.3) Hydrogen's atomic radii

The innermost radius at (n = 1) is referred to as the *Bohr radius* and is only 0.529 Angstroms (10^{-10} m). At (n = 3) the orbital radius is nine times larger. An electron's speed in a particular state is given by:

$$v = \frac{q^2}{2nh\varepsilon_0} \tag{5.4}$$

The time it requires to complete an orbit ($2\pi r$) of the nucleus (proton) is given by combining Equations (5.3) and (5.4) for radius and velocity:

$$t = \frac{4n^3h^3\varepsilon_0^2}{m_e q^4} \tag{5.5}$$

or $1.5(10)^{-16}$ s for the ground state.

The energy a photon can emit in this interval is constrained by Planck's constant:

$$E_{mr} = \frac{h}{t} = \frac{m_e q^4}{4n^3h^2\varepsilon_0^2} \tag{5.6}$$

and will be called the *minimum release energy. It is the minimum amount of energy that can be released within the timeframe of an electron's single nuclear orbit.* The time necessary to release a photon is inversely proportional to its energy. As the available time decreases, energy can only be transferred in *larger* packages.

The kinetic energy an electron has in a given state is:

$$E_K = \frac{m_e q^4}{8n^2h^2\varepsilon_0^2} \tag{5.7}$$

This decreases at higher energy levels, but more slowly (by one factor of n) than the minimum release energy of Equation (5.6). The ratio of kinetic energy to minimum release energy is given by:

$$f_A = \frac{E_K}{E_{mr}} = \frac{\left(\dfrac{m_e q^4}{8n^2h^2\varepsilon_0^2}\right)}{\left(\dfrac{m_e q^4}{4n^3h^2\varepsilon_0^2}\right)} = \frac{1}{2}n \tag{5.8}$$

The only state where kinetic energy is less than minimum release energy is the ground state. At (n = 2) the two energy levels are equal. This is why high energy states can decay into lower states while it is not possible for the ground state to decay into a lower energy level. The ground state cannot decay because the minimum energy photon that could be emitted in its orbital time exceeds its available kinetic energy by a factor of two. Equation (5.8) will be called the *atomic stability ratio*, f_A.

Although quantum mechanics is able to effectively describe many aspects of a hydrogen atom's energy states using integer variables and differential equations, the underlying structure of even this simplest of atoms is vastly more complex than quantum expressions would suggest. But comprehension should come before description and the bottom line is electrons simply orbit atomic nuclei. An abrupt departure from reality is unnecessary.

Atoms and antennas do not differ in realness, but are significantly dissimilar when it comes to their size and dynamics:

Ψ THEOREM 5.1 - ATOMIC QUANTUM STABILITY {Planck Relation}

GROUND-STATE ELECTRONS ORBIT ATOMIC NUCLEI IN LESS TIME THAN IT TAKES TO EMIT A PHOTON WITH ENERGY EQUAL TO THEIR AVAILABLE KINETIC ENERGY

This is why atoms have the size they do - their electromagnetic stability is purely a matter of scale. Orbital electrons are just as real as the planets in our solar system. They are simply much smaller, more discrete, and move a great deal faster. Experiments are beginning to bear this out, as physicists in Germany and Austria were recently able to use extreme ultraviolet light to calculate hydrogen's ground-state orbital period at $1.5(10)^{-16}$ s, the value expected from Equation (5.5).[35] Note that (Ψ5.1) only tells us why atoms have stable ground states, even though their electrons move in normal, circulating orbits. It doesn't define the actual size of these orbits, as this is a function of the quantization of the electrons' orbital momenta.

ELEMENTARY MAGNETIC MOMENTS

The Rutherford atom did not go quietly into oblivion. When excited hydrogen atoms are exposed to a strong magnetic field, their spectra exhibit a fine structure consistent with the tiny magnetic moments their electrons ought to have as they circulate about the atomic nucleus. In 1913, Niels Bohr proposed an atomic model that postulated the electromagnetic stability of classical electron motion, but unlike (Ψ5.1), it had no supporting justification. Physicists were so eager to *describe* reality that they eventually created another (in addition to orbitals) entirely fictitious phenomenon to account for fine spectral structure - *intrinsic* elementary magnetic moments. Let's calculate the moment of hydrogen's ground state.

The magnetic dipole moment of N turns of wire carrying i amperes of current around an area A is given by:

$$\mu = NiA \tag{5.9}$$

The current due to an orbiting charge is given by the product of its charge and orbital frequency:

$$i = q\nu = q\left(\frac{\mathrm{v}}{2\pi R}\right) \tag{5.10}$$

where v is its velocity and R the radius of its orbit.

Substitute this into Equation (5.9) and let ($N = 1$) for the case of a single circulating particle:

$$\mu = \frac{q\mathrm{v}}{2\pi R}\pi R^2 = \frac{q\mathrm{v}\mathrm{R}}{2} \tag{5.11}$$

This is the general expression for a circulating charge's magnetic moment. Now substitute Equations (5.3) and (5.4) for the velocity and Bohr radius of hydrogen's ground state:

$$\mu = \frac{q}{2}\left(\frac{q^2}{2h\varepsilon_0}\right)\left(\frac{h^2\varepsilon_0}{\pi m_e q^2}\right) = \frac{qh}{4\pi m_e} = \frac{q\hbar}{2m_e} = \mu_B \tag{5.12}$$

The result is called the *Bohr magneton. Its value is within 0.1159% of an electron's actual ground-state orbital moment, as deduced from the spectral signature of hydrogen exposed to a strong magnetic field.*

If electrons don't orbit atomic nuclei, why is the magnetic moment of hydrogen's ground-state electron so astonishingly close to the value an orbiting particle would have? This should have posed quite a dilemma for the statistical wave functions of the orbital concept, but it did not. Physicists simply chose to disregard the most straightforward interpretation of these measurements - that electrons actually circulate around atomic nuclei. Instead they postulated intrinsic magnetic moments, where electrons spin about their own tiny axis. More egregious still, *these moments have never been observed for individual particles.*

Intrinsic moments are said to be undetectable in a beam of free electrons or protons because their effect is weak and easily swamped by the beam's momentum distribution and the Heisenberg Uncertainty Principle. But this is not the true reason they are only visible in combinations of particles. Intrinsic moments can't be found in free particles *because they are not an intrinsic property*. They are solely the product of quantized orbital motion.

The counter argument provided by quantum realists is if an electron's moment were due to the alignment of its atomic orbit with an external magnetic field, an optically strong double refraction would be present. No such refraction has been observed, so it follows that these moments are intrinsic. *This view is currently accepted doctrine even though the observed moment is within 0.1% of the expected value.* Here again is the selective and erroneous application of macroscopic considerations to the atomic scale. Double refraction assumes that magnetized valance electrons significantly alter their trajectory, when the actual interaction might simply be a compensatory adjustment of their orbital radius. The mark that quantum reality has left on physics is indelible and pervasive, leaving in its wake a blatant ignorance of the basic dynamics of atomic electrons.

The original problem, the Rutherford catastrophe, was resolved by simply denying the orbital motion of atomic electrons. This demonstrates no small theoretical hubris given their known electrostatic attraction to positively charged nuclei. Subsequent experiments with magnetic fields screamed for reconsideration of this interpretation, but instead another level of abstraction was layered over the problem and electrons were interpreted as spinning on some tiny internal axis like submicroscopic magnets. This is particularly incongruous since electrons are also considered point particles with no size. Finally, when the newly created elementary spin couldn't be demonstrated in beams of free particles, obscure, untestable justifications were postulated and quantum reality plodded onward.

None of this is necessary, and the truth is far more eloquent:

- Electrons orbit atomic nuclei in accordance with their mutual electrostatic attraction and quantized energy levels.

- The ground state of these orbits occurs at a size too small to allow sufficient time for the emission of electromagnetic radiation.

The orbital dynamics of atomic electrons is perhaps the easiest quantum misinterpretation to correct, which is why it was presented first. The interference and diffraction of these same particles are far more difficult to explain. These are the demons that physicists confronted at the turn of the twentieth century. As will soon become apparent, the behavior of light and elementary particles pushes even the most logical minds to the brink of denying nature's rationality.

5.2 WAVE-PARTICLE DUALITY

There were plenty of experiments causing physicists to run for cover a hundred years ago, but some of the most striking and inexplicable were the diffraction and interference of elementary particles. An experiment often considered a prima facie case for quantum reality goes as follows:

> Take a thin piece of metal and cut two submicroscopic patterns through it. The first is a single vertical slit 0.2 microns wide and a millimeter long. The second is two slits similar to the first, parallel and closely spaced at 1 micron apart:

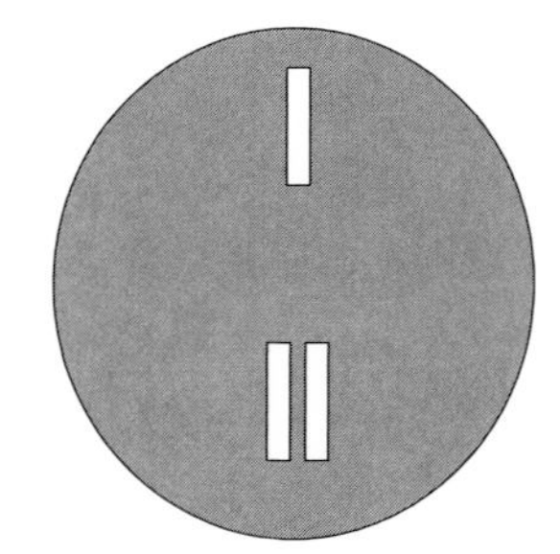

Figure (5.4) Single and double-slit apertures in a test plate (not to proportion or scale)

> Place this thin metal plate in a hard vacuum between an electron gun and a phosphor screen as follows:

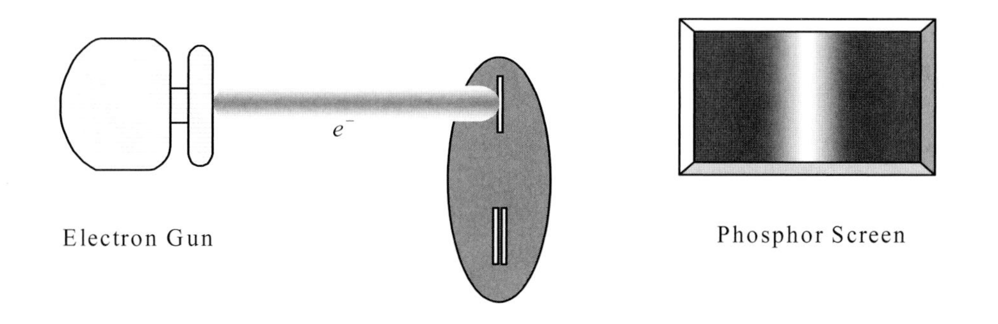

Figure (5.5) Interference test apparatus

> Fire a beam of electrons with 25 EV of energy *at the single slit*. They are moving at about 1% of the speed of light at this energy, just sufficient to excite phosphorous atoms.

A diffuse *diffraction pattern* will appear on the phosphor screen:

Figure (5.6) Single-slit diffraction pattern

Now direct the same beam toward the double-slit aperture. A diffuse *interference-diffraction pattern* will appear:

Figure (5.7) Double-slit interference pattern

The bright regions indicate where the electrons strike the phosphor and are referred to as *maxima*; the dark regions between them are zones the electrons avoid called *minima*. If one of the slits in the double-slit aperture is covered, the interference pattern of Figure (5.7) immediately vanishes and is replaced with the single-slit diffraction pattern of Figure (5.6).

Moving particles possess wave-like characteristics that in many cases make them indistinguishable from light. They appear to have a *de Broglie wavelength* given by:

$$\lambda = \frac{h}{m\mathrm{v}} \tag{5.13}$$

When electrons carry enough energy to be easily detectable, their wavelength is over a thousand times smaller than visible light. This is why the slits in the experimental test plate shown in Figure (5.4) are so small.

The idea that electrons have an intrinsic wavelength is not particularly troubling until *interference* appears. If electrons are real physical entities, how can they superimpose to deposit greater numbers in the bright maxima and fewer in the dark minima of a double-slit pattern? The initial explanation would be that electrons interact prior to or just after passing through the slits, altering their distribution into a series of maxima/minima. Measurement of the amperage leaving the electron gun matches the current collected on the phosphor plate, so it is clear that no electrons are being created or destroyed in this experiment. Or at least it is clear that if electrons are being created and destroyed a balance is somehow achieved so the net change is zero. However, the idea of particles being created or destroyed at these meager energies is nonsensical.

In order to isolate the presumed electron-electron interaction necessary for a reasonable explanation of the interference effect, reduce the beam current to see if a lower electron density changes the pattern. The pattern remains the same, albeit dimmer. Now reduce the beam so that only one electron per second passes through the slits and document the individual flashes occurring on the phosphor screen over an extended period of time (perhaps days). Lo and behold, *the pattern remains the same*! *The interference pattern persists even though the electrons could not have possibly interfered with each other.* Exit rationality, at least for modern physics.

Quantum reality's explanation is *electrons do not exist until they are observed at the phosphor screen*. The interference pattern is thought to be caused by the superposition of all possible paths the electrons *might* take to the screen. This is patently ridiculous but is nonetheless the current stance of contemporary physics. Heisenberg's famous quotation, *"atoms are not things,"* echoed through the physics halls of universities everywhere nearly a hundred years ago and continues to do so. Even Einstein, with his penetrating insight, was unable to put physics back on a firm causal footing.

The wave-particle nature of light is just as bizarre as that of matter. If light is indeed composed of photons and energy and photon number are in fact conserved, how is it possible for two light sources to *interfere* with each other? Take the ultimate case of a stellar interferometer. This device can be used to measure the diameter of distant stars and more recently has been used to search for planets beyond our solar system. Photons are collected at two locations, perhaps 100 meters or even 100 kilometers apart, and electronic collation of both sources will form an *interference pattern* that accurately corresponds to the source's diameter.

How can electrons interfere with each other if they pass through an aperture one at a time? How can two photons arriving at detectors hundreds of kilometers apart *interfere* with each other? Moreover, the curious behavior of light and elementary particles *is precisely the same at any comparable wavelength.* These simple observations are responsible for unhinging modern physics from reality.

There is a straightforward, albeit surprising, explanation for the interference patterns described in this section. The answer begins with the resolution of the most desperate extrapolation of quantum reality, Bell's Theorem.

5.3 BELL'S THEOREM AND THE EPR PARADOX

In 1935 Einstein, Podolsky, and Rosen used momentum-correlated electrons to demonstrate that quantum mechanics is an inherently incomplete description of reality. David Bohm later revised and simplified their approach by using polarization-correlated photons, but the general argument is still referred to as the EPR paradox in recognition of the original theorists.

Its premise is the following:

> *In contemporary quantum mechanics, photons are viewed as having no intrinsic properties other than their wavelength/energy. Yet certain light sources (such as mercury) produce twin photons with the same polarization relative to each other. This is referred to as pair-polarization. How can twin photons have the same polarization at distant detectors if neither is able to carry this information?*

Suppose, as shown in the following figure, a source **S** of pair-polarized photons is set up between two distant polarization detectors **A** and **B**:

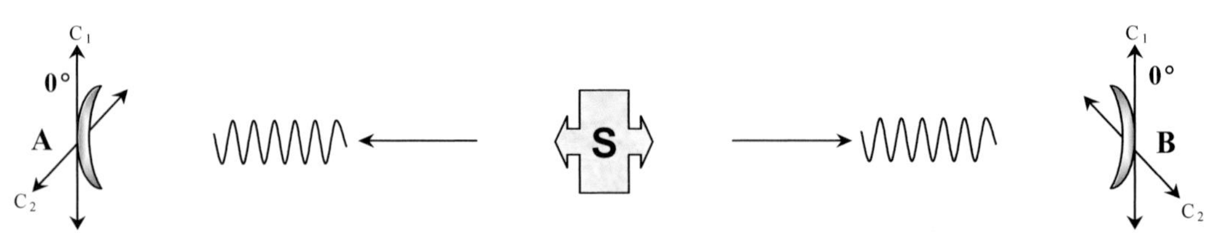

Figure (5.8) Pair-polarized photon detection in the EPR paradox, both detectors set to 0°

The source is constructed with an electric current so feeble it emits only a single pair of photons at a time, and the detectors are so sensitive they can detect a single photon. Each detector has two collectors, C_1 and C_2, set at right angles to each other. Any photon either detector receives will register in one or the other of its collectors. Since source polarization is random, each collector at each detector captures ~50% of the incoming photons. The angle between the C_1 collectors of **A** and **B** is initially zero degrees as shown in the figure. Every time detector **A** scores a hit on C_1 so does detector **B**. The same is true of C_2. If photons do not have intrinsic polarization, why are the detectors' results the same 100% of the time?

The EPR paradox didn't slow the headlong rush to develop quantum mechanics because it is viewed (on the rare occasions it is considered at all) as more of a philosophical than scientific problem. It deals with the underlying basis of quantum mechanics, which in its own estimation is nonexistent. The salient issue is whether or not physical theories should have an underlying basis. Quantum realists not only claim they do not, *they also maintain they need not*. This is why modern physical theories are littered with a never-ending succession of dead end concepts.

Nearly 30 years after the discovery of the EPR paradox, a physicist named John Bell devised what is considered by many a proof that not only is there no underlying basis of quantum phenomenon, *all physical interactions have superluminal (faster than light) connectivity*.

The facts thought to support this concept are listed below. Referring to the Figure (5.8):

- The distances between the source and each detector are immense and equal, and source **S** has been emitting pair-polarized photons at regular intervals for a long time.

- For either detector, a photon logged in C_1 is a hit while one logged in C_2 is counted as a miss. Each of the detectors at **A** and **B** produce a photon *detection sequence*, a code of hits and misses in a series of the form: H M H M H H H (H-hit, M-miss).

- A detector's angle is defined to be the angle of its C_1 collector. If the angles of detectors **A** and **B** are both at 0°, the correlation between their detection sequences is 100%. They are perfectly *correlated* (in practice there are occasional errors due to detector imperfection, but for the purposes of this discussion these can be ignored).

- When one detector is at 0° and the other at 90°, the correlation between their two detection sequences is 0%. Every hit one detector gets is scored a miss at the other detector. They are perfectly *anti-correlated*.

Now for the experiment:

- Move detector **A** clockwise to 30°. Comparison of the two detection sequences contains a 25% error rate.

- Move detector **A** back to 0° and detector **B** counterclockwise to −30°. Comparison of the two detection sequences will again contain a 25% error rate.

- Now move detector **A** clockwise to 30° and detector **B** counterclockwise to −30°. Comparison of the two detection sequences contains a 75% error rate.

The practical application of Bell's premise is this:

> *The detection sequence is purely a function of the photons already in flight. Ergo, changing the orientation of one detector should have no effect on the results at the other detector. The maximum combined error rate should never be greater than the sum of the individual error rates:*
>
> $$\delta(30°\mathbf{A}) + \delta(-30°\mathbf{B}) \leq \delta(30°\mathbf{A} + -30°\mathbf{B})$$
>
> *where δ is error rate.*

This is the Bell Inequality. It should hold if and only if the events at the detectors are truly independent of each other. *Since the combined error rate is three times the individual rates, a number of quantum realists believe that some form of superluminal connectivity exists between the detectors, juxtaposed on their local existence.* In the most general sense, this result is taken to mean reality itself is *nonlocal*, where causality between events is not isolated in time and space by the speed of light. Not only don't photons carry their own polarization, this information is communicated to both detectors at superluminal speed.

This interpretation makes quantum realism seem rational by comparison. The idea that nature is interconnected by some mysterious governing super-reality is at best disingenuous. The universe is an infinite whole that sums to zero, but this is accomplished by the known laws of physics, not by some omniscient controlling entity. While superluminal signals might exist, they are not specifically designed to distort polarization measurements. Let's take a closer look at the Bell Inequality to find the turn where the wheels came off the wagon.

When one detector is at 30° and the other at 0°, the two sequences the detectors produce, based on the above experiment, might look like the following:

A(0) H M H $\underline{\text{H}}$ M M $\underline{\text{M}}$ H M H $\underline{\text{H}}$ H

B(+30) H M H $\underline{\text{M}}$ M M $\underline{\text{H}}$ H M H $\underline{\text{M}}$ H where they differ precisely once out of every four measurements (underlined). In reality they differ more than this, but the experimental setup and the concept of the Bell Inequality are both limited to binary accuracy.

If each hit and miss is scored *with the intrinsic photon polarization angle*, it would take the form:

A(0) $H^{+12}\ M^{+85}\ H^{+15}\ \underline{H}^{-20}\ M^{+80}\ M^{-50}\ \underline{M}^{+60}\ H^{+8}\ M^{+82}\ H^{-10}\ \underline{H}^{-35}\ H^{-5}$

B(+30) $H^{+12}\ M^{+85}\ H^{+15}\ \underline{M}^{-20}\ M^{+80}\ M^{-50}\ \underline{H}^{+60}\ H^{+8}\ M^{+82}\ H^{-10}\ \underline{M}^{-35}\ H^{-5}$

Any angle greater than ±45° between the photon and detector is a miss and any angle less is a hit. The sequences only appear to correlate by 75% because this is the full extent of the experimental resolution. Even at 30°, none of the events *match perfectly* because the detectors have different orientations to the incident photons. As a rare case, a photon might have approximately the same magnitude of angle relative to both detectors, such as +15 and −15, but with sufficient angular resolution a difference will exist such as +15.02 versus −15.05.

Polarization detectors capture a fraction of incident photons in accordance with:

$$f_\gamma = 1 - \cos^2\theta \tag{5.14}$$

where θ is their relative angle of polarization. Photons whose relative angle is 30° are a third as likely to be captured as those at 60°. *Polarization is not linear.* Thus:

$$f_\gamma(30°) + f_\gamma(30°) = \left(\frac{2}{3}\right) f_\gamma(60°) \tag{5.15}$$

is consistent with the observed results. *Reality's much touted nonlocality is nothing more than a round-off error.* The reason why Bell's inequality fails is because photons have intrinsic polarization, not because reality is nonlocal.

Quantum reality is an incomplete perspective of energy. But we don't need paired polarization to tell us this; we just need a more honest evaluation of our ability to describe atomic structure. Quantum physics' deficiency is painfully obvious whenever it is used to try to fully describe even the simplest atomic and molecular systems. Physical chemists try to overcome this with a qualitative mixing of states, often describing a molecular interaction as ~3% of one state and ~97% of another, but they know this theory's limitations.

The evidence of quantum physics' incompleteness is not subtle; it is overwhelming. If this theory were complete, it could define the angle between the hydrogen atoms in a water molecule (~104°) from first principles. It cannot - it does not even come close to this level of descriptive power. There are more than 112 known elements with more than 1500 different isotopes. Quantum mechanics is able to quantitatively describe one of these, the hydrogen atom. This is a success rate of 0.07%. To call it *complete* is ludicrous. Quantum mechanics is a powerful tool that is useful for relatively simple systems and vast collections of similar particles. It is simply not qualified to make claims about the limits of reality. Chairs are real, planets are real, and the atoms of which they are composed are just as real as they are.

5.4 WAVE NATURE

Wave nature is the most enigmatic aspect of particles and photons. Like electrons, photons form interference patterns even when they pass through a double-slit aperture *one photon at a time*. This effect will be referred to as *sequential interference*. It must be resolved before a physical description of quantum effects is possible. Our investigation will now focus on photons because their wave nature is more accessible than particles and is due to the same underlying cause.

When individual photons pass through a double-slit, even at widely spaced time intervals, they interfere with each other, somehow eliminating energy from minima and transferring it to maxima:

Figure (5.9) Double-slit interference pattern

This pattern is irrespective of the material of which the slits are composed - the only requirement is that it must be sufficiently opaque to prevent the passage of light.

In the initial pandemonium following the discovery of quantum effects, electromagnetic waves became statistical carrier envelopes for photons, much like the waves at the beach are composed of water molecules. This provides no lasting respite from the baffling sequential interference effect, however, because sooner or later the photons need to be connected to the waves. The bottom line is something causes interference patterns when photons pass through a pair of closely spaced slits one at a time, and there is a short list of potential candidates to choose from:

- Aperture's wave nature. Not viable. Although matter is surrounded by standing electromagnetic waves, as evidenced by the Casimir force, they are not the source of interference. The pattern a single slit produces is the same as the pattern generated by a double slit when one if its slits is masked. These two cases have different material distributions, so they also have different standing wave distributions, yet their effect on light is identical.

- Photon beam. Not viable. Photons superimpose in space, but this has nothing to do with the observed interference pattern. The reason why there is a notable reduction of energy density in an interference pattern's minima is *because there is also a corresponding increase in the energy density of its maxima.* Photons have not been nullified; they have been *redirected.*

The combined output of twin slits is specifically not the superposition of each slit's output. Regardless of the information content in the incident beam, the mathematical interaction of said information cannot produce the observed interference pattern. If the second slit did not exist, photons destined to pass through it would be reflected or absorbed, not removed from the beam passing through the first slit. Opening a second slit steals photons from the first slit's output *with no change in the incident beam*, and it does so in an amazingly precise way. The lingering question is how this can possibly be reconciled with the discrete energy packages of which light is composed. Sequential interference is a deeply perplexing phenomenon. No wonder Bohr and Heisenberg lost their faith in subatomic reality.

5.5 QUANTUM HYSTERESIS

Sequential interference doesn't mean the universe is irrational or that atoms are not real. Its resolution hinges on a single crucial observation: *adding a second slit to a single-slit aperture changes the first slit's output*. So the question becomes, how could photons (or energy) passing through one slit conceivably affect the way photons pass through the other? Since an interference pattern shows up even when photons move through the slits one at a time, it means that one photon affects the motion of the next even though they have a wide physical separation between each other. Although this seems patently impossible, not only is it possible, *it is an absolute necessity*.

The universe is causal and deterministic. Each photon *leaving* the double-slit aperture does so with a particular momentum and energy. This is not a random event; *it is the direct result of the aperture's state and the incident photon's momentum*. The material of which the aperture is composed is a system with an existing quantum state that interacts with an incoming photon. There is a sequence of at least three different states the aperture must occupy during this process:

1. Photon is incident.

2. Photon merges with aperture.

3. Photon is transmitted.

If incident photons arrive by way of a laser, their momenta and energy are nearly identical. Transmitted photons, however, have an angular variance caused by the width and spacing of the slits. Energy and momentum are always conserved. What this means, and this is the pivotal realization, is *the aperture's state changes depending on the direction the transmitted photon takes*. A transmitted photon released at an angular deflection of 5° corresponds to one aperture state; 10° is a different state. The aperture is a macroscopic object. Its quantum states are virtually continuous and it has an extremely fine state resolution, *but it still exists in only one state at a time*.

One photon sets the aperture's state based on its exit trajectory; the next to follow encounters an aperture whose state was set by the previous photon. *This provides a causal relationship between the two photons even though they might be widely separated in time*. Each is deflected based on the aperture's state, which in turn is driven by the deflection of previous photons. This will be called *quantum hysteresis*, a system's quantum memory.

Ψ THEOREM 5.2 - QUANTUM HYSTERESIS {Ψ5.1}

AN OBJECT'S QUANTUM STATE IS THE STORED RESULT OF ITS ACCUMULATED ENERGY INPUT/OUTPUT

Quantum hysteresis is why a system's complex diffraction/interference output can be calculated as the integral of its emitting surface at a given wavelength. A double-slit aperture's high-resolution quantum state stores the transmitted momentum of any photon passing through it, then uses this information (and its current state based on all prior transmissions) to chart a path for the next incident photon. Quantum hysteresis is particularly evident with a laser source because all of its photons have virtually the same momentum/energy.

Photons (or electrons) interfere with each other as they pass singly through an aperture because the result of their passage must, by the conservation of momentum, be preserved in the aperture's current state. Photons have momentum scaled by their individual energy:

$$p = \frac{E}{c} \tag{5.16}$$

The aperture that deflects them has to store the complementary response to this. Their deflection is an *elastic* collision. This preserves their energy/wavelength, but changes their momentum as well as that of the aperture they strike.

A double-slit interferometer is a quantum switch. It directs photons onto various paths as a function of its previous switch setting combined with the state of the incoming photon. Its size is defined by the physical extent across which comparative measurements are made. This is the full scope of the so-called observer-created reality. If two photons are logged from apertures a millimeter apart, the quantum switch is for all intents and purposes a millimeter wide because this is the span where momentum has been sampled and must balance. If the same photons are collected a thousand kilometers apart, the quantum switch consists of all of the material between the detector locations, to include mountains, lakes, and cities. Momentum and energy are conserved to infinite resolution. A lone photon impacting a planet's surface leaves an indelible mark on its quantum memory.

There is relentless and tumultuous activity between the widely spaced detectors of the interferometers currently being used in the search for extra-solar planets. Cars drive between them; electricity surges through high-power transmission lines; the sun heats the ground. But none of this commotion is relevant to the photons in the detectors. The only interference it is measuring, indeed the only conceivable interference that might exist, *is between the photons being compared*. Regardless of what the ambient noise level might be in an interferometer spanning a continent, a photon striking one detector will have an effect on the one to follow at another detector because momentum is conserved to infinitely fine resolution. Isolating the few photons incident on widely spaced mirrors effectively nullifies the entire countryside because it shows how photons from one detector, and only those photons, *change the momentum of the entire system relative to the photons arriving at the other detector*.

UNIFORM AND UNCONDITIONAL REALITY

Bohr and Heisenberg were spectacularly wrong about reality. Atoms are things, and so are particles and photons.

This concept will be referred to as *quantum physicality*:

Ψ THEOREM 5.3 - QUANTUM PHYSICALITY {Ψ5.2}

PARTICLES, PHOTONS, AND NEUTRINOS ARE REAL, PHYSICAL OBJECTS

Null Physics has a basic theme - the equivalent and unconditional reality of all things, from space to photons (Ψ1.5). Contrary to the quantum reality rampant in modern physics, the universe is wonderfully deterministic and rational. *If an experimental result makes no sense, then look at it from another rational perspective.* Instead, many theorists are quick to assume the universe itself is nonintuitive or *beyond common sense.* Rationality is like the conservation of energy; we are not free to choose where and when it applies.

Many philosophers and theorists look down on common sense as some base, puny manifestation of reason because it lacks the structure and formalism of mathematics and logic. This could not be further from the truth. Common sense is the most encompassing test of any theory or idea. It is the sum of our experience of what exists and what does not. This kind of rationality is not easy to formalize because the universe cannot be reduced without the loss of some of the attributes essential for its universal nature. Anything less is ... less.

5.6 FOCUS ON THE BUILDING BLOCKS

The material (nonspatial) universe is primarily composed of particles and photons. Our usage of the term *photon* has the same meaning as in conventional physics: a quantum of electromagnetic radiation. The term *particle*, however, will differ markedly from common usage. Unless otherwise specified, it will refer exclusively to elementary particles that are unconditionally stable in free space. This is limited to electrons and protons and their antiparticles.

§ DEFINITION 5.1 - PARTICLE

ONE OF FOUR STABLE ELEMENTARY COMPONENTS OF MATTER:
ELECTRONS, POSITRONS, PROTONS, AND ANTIPROTONS

None of matter's neutral elementary forms are stable in free space. Although neutrons appear stable within atomic nuclei, they decay in approximately 10.3 minutes outside of this environment. Unstable particles can become increasingly stable at high speed, due to the relativistic dilation of their half-lives, but can never move fast enough to become stable. A muon, for example, has a rest energy ~200 times greater than an electron and decays into an electron and neutrinos in about a millionth of a second.

It might seem reductionist to limit our investigation to only two masses of particles, particularly to anyone familiar with current doctrine. Particle physics views the universe in terms of families of particles, the overwhelming majority of which are unstable. Some are so transient their actual existence is contested. In actuality, the characteristics of this veritable elementary zoo are largely irrelevant in the search for matter's immutable essence. Contrary to what particle physicists currently believe, mesons, pions, and their cousins reveal little to nothing about matter's inner structure ... yet. Since there was no Big Bang and therefore no moment in the universe's ancient past where fundamental forces sprang into being, the use of unstable particles as a probe for the first few moments of cosmic history is entirely misdirected. In the future, when electrons and protons are fully understood, we can move on to mesons, but at our present state of knowledge they offer no insight whatsoever (as should become abundantly clear after the introduction of the matter field in Part III). We need to do our particle arithmetic before trying calculus.

Our theoretical development shares virtually no common ground with particle physics' standard model, and we will forego even a tenuous connection to the existing vernacular by taking the liberty of redefining a few of their terms, as follows:

- Unless specified otherwise, such as in reference to antimatter, the term *electron* will be used to refer to either an electron or positron, and protons and antiprotons will collectively be referred to as *protons*.

- In contemporary particle physics, neutrinos are considered leptons - allegedly related to electrons. Since no one has the faintest clue what a neutrino actually is, it is premature to attempt to classify it at this stage of our investigation. The term *neutrino* will be used to signify either a neutrino or antineutrino. No distinction will be made between the two, for reasons that will become evident later. The more global term *radiant energy* will be used to refer to any form of quantized energy that propagates at the speed of light - photons and neutrinos.

- In particle physics' standard model, neutrons are classified as baryons, in the same category as protons. This is a misnomer, as a neutron isn't even an elementary particle, but is instead a compound system consisting of a tightly bound electron and proton. A neutron is actually a baryon-lepton, not a baryon.

Particles and photons will collectively be referred to as *quanta* or singly as a *quantum*. This is the most general expression for discrete bundles of energy governed by unit hypervolume. Attempts will be made to avoid confusion between quanta and the quantum energy levels present in atoms and molecules. Although it is not as yet clear exactly how the energy in particles or photons is packaged by unit hypervolume, the geometry of Part I tells us that it is the only finite boundary condition capable of providing universal energy segmentation.

5.7 ACTION AT A DISTANCE

Quantum reality's grotesquely erroneous interpretation of nature isn't limited to the electrons orbiting atoms or striking the screens of our television sets. It also lends itself to the distortion of another important concept: the field. The phenomenon of "action at a distance" has puzzled physicists for centuries. The Coulomb force between charged particles, for example, is currently thought to be exerted through the exchange of intermediate virtual photons. This is consistent with its ability to transfer energy across space at the speed of light, but creates more questions than answers because the physical agent that controls the emission and absorption of virtual photons is utterly inscrutable. If particles are surrounded by clouds of virtual photons, what are they emitted from and what are they absorbed into? Space? If so, what is the difference between the space near a charged particle and the space without the particle? The concept of virtual photons, like quantum reality, is nonphysical, inconsistent, and obscures a more eloquent truth.

When an electron and proton combine to form a hydrogen atom, they emit a photon with a few electron volts of energy - 13.6 EV for the ground state. Their interaction is of the form:

$$e^{-} + p^{+} \leftrightarrow H + \gamma \tag{5.17}$$

Energy is always conserved, so the only energy going into this reaction is what was present in the electron and proton in the first place - their rest energy. There is nothing else. Two particles react and give off energy. Where did this energy come from? Any physics student will say it comes from a difference in *electrostatic potential energy*. When a proton and electron approach each other their potential energy is converted into motion and released as electromagnetic radiation. Although this might be a helpful way of acknowledging the system's energy change, it does not provide a legitimate source. The concept of potential energy was created to account for the energy differences produced by field interactions. It describes changes in energy without the slightest reference to an underlying substance. There is no tangible *cause* for the Coulomb force in contemporary physics (particles are charged, no reason) and ergo there is no *cause* for the energy change associated with this force. The reality, however, cannot be circumvented. The only possible source for the Coulomb potential in the reaction shown in Equation (5.17) is the rest energy of the interacting particles. This is why a hydrogen atom's mass is slightly less than the sum of the masses of an isolated electron and proton.

Like the Coulomb force, gravity is the archetypical *action at a distance* - in many cases enormous distances. Many physicists believe it is carried by gravitons: hypothetical particles so elusive that quarks are easy to isolate by comparison. This avoids the action at a distance problem but is entirely inconsistent with Einstein's profoundly successful depiction of a gravitational field as a space-time distortion. Yet Einstein was reluctant to embrace the full implication of action at a distance. In General Relativity, no reason is given for why space is curved near a mass; it just is. This effectively decouples gravitational fields from rest energy and in so doing sacrifices their physicality.

When molecules, atoms, or particles approach a massive object and cannot maintain their increasing kinetic energy, they typically emit radiant energy into space (such as in the case of an accretion disk around a black hole). *The ultimate source for this radiation is the rest mass of the incident matter.* Like the Coulomb potential in the hydrogen atom of Equation (5.17), there is no other source. *Matter is not distinct from the fields it generates - matter/field is a continuum.* The gravitational field an object produces is as physically real as it is.

The modern interpretation of a particle's elementary nature varies depending on the interaction under investigation:

- High-energy collisions. Here electrons are thought to be point particles while protons are composed of point-like subunits called quarks. This generates two unwieldy inconsistencies. First, if all of a particle's rest energy is stored in infinitely small volume, it has infinite energy density and ought to be a gravitational singularity. Second, why should the atomic or nuclear proximity of an electron and proton incur a change in the energy stored in their infinitely small regions?

- Low-energy interaction. Here electrons and protons are probability waves propagating through space whose very existence is dependent on interactions with other particles or photons.

Neither of these concepts is compelling or enlightening.

Contemporary physics has created and nurtured phantasmagorical explanations to try to circumvent the simplest explanation for action at a distance:

> *The energy of a particle's rest mass is distributed through space. When two particles interact at a distance it is because bits of the energy density of their respective rest masses coexist there.*

The photon emitted from the formation of a hydrogen atom is composed of energy from the fields of the electron and proton within it. *A portion of their rest masses has been ejected from the system*. There is no alternative. Regardless of how the energy density or wavelike phenomena of matter might be interpreted, the rest masses of electrons and protons are *composed of bits of energy, these bits are removable, and the magnitude of their ejection is dependent on the spatial separation between the particles.* This is not to say rest energy maintains the same structure when released as a photon. Certainly a transformation takes place, but the ultimate source for the radiating energy is the interacting particles' fields. Again, this is the only available supply.

Our investigation of matter's fields will provide unshakable evidence that the wave functions used to describe atomic energy levels in quantum physics are a glimpse of energy's distributed composition. Nature is a continuum. Physicists reduce particles into points with separate compartmentalized properties and then try to recombine the pieces in the search for a unified field theory. A particle is not a set of the form (mass, charge, spin); nor is it a superstring. It is a certain amount of energy with a single distributed substructure that generates all of its properties, and is far simpler, yet at the same time more complex, than most particle physicists would care to believe.

5.8 MAXFIELD

A wealth of observational evidence supports the following:

- Matter is composed of fields of distributed energy.
- When more than one field of matter is present, changing their proximity changes their constituent energy.
- Matter and its fields are inseparable.

Various enigmatic processes might be championed by modern physicists to account for the action at a distance phenomenon, but they cannot escape the conservation of energy. The energy of any interaction is linked to the compositional energy of its interacting particles because they are one in the same thing:

Ψ THEOREM 5.4 - FIELD/MATTER DUALITY {Ψ5.3}

ENERGY RELEASED BY THE NEGATIVE POTENTIAL OF ANY INTERACTION ORIGINATES FROM THE REST ENERGY OF THE REACTANTS

This follows directly and inexorably from the observed properties of fields, and it has sweeping repercussions for matter's most elemental nature.

Field/matter duality dictates that a system's maximum negative field potential is equal to its available rest energy. This will be referred to as the *Maxfield* constraint.

Ψ THEOREM 5.5 - MAXFIELD {Ψ5.4}

ENERGY RELEASED BY THE NEGATIVE POTENTIAL OF ANY INTERACTION CANNOT EXCEED THE AVAILABLE REST ENERGY OF THE REACTANTS

This is an inevitable consequence of energy conservation once all of its forms are assigned a uniform level of realness.

5.9 GRAVITATIONAL POTENTIAL LIMIT

Maxfield reasoning can be applied directly to the gravitational force. The potential energy of a rest mass m_0 in the weak nonrotating field of another mass M is given by:

$$\Phi_g = -\frac{Gm_0 M}{r} \tag{5.18}$$

This is a special case of the relativistic formula:⟨4.2⟩

$$\Phi_g = m_0 c^2 \left(e^{\Phi} - 1\right) = m_0 c^2 \left(\sqrt{1 - \frac{2GM}{c^2 r}} - 1\right) = m_0 c^2 \left(\sqrt{1 - \frac{R_S}{r}} - 1\right) \tag{5.19}$$

where R_S is the *Schwarzschild radius*:

$$R_S = \frac{2GM}{c^2} \tag{5.20}$$

Equation (5.19) applies to a spherically symmetric nonrotating field of any strength, up to and including that of a black hole.

The Schwarzschild radius is the limit where the magnitude of *gravitational potential is negative unity*. This corresponds to an energy loss equal to the rest mass of the captured object:

$$\Phi_g(R_S) = -m_0 c^2 \tag{5.21}$$

The Schwarzschild radius is the gravitational confirmation of Maxfield.

One of the caveats of Equation (5.19) is that within the Schwarzschild radius, at $r < R_S$, gravitational potential becomes a complex number - *it is imaginary*. Theorists speculate about what happens in this region. It is often said that gravitational force and potential go to infinity along with energy density. A number of different concepts have been proposed to describe a black hole's interior,[4.1] but none are correct, as they all lack the most pertinent governing dynamic - the Maxfield constraint. The gravitational potential of any particle will never exceed its own rest mass because the ultimate source for this energy *is its own rest mass.*

5.10 BLACK HOLES IN AN ETERNAL UNIVERSE

One of the requirements of an eternal universe is the renewability of all matter and energy. *This means a black hole's material capture is a reversible process.* For every particle falling into its deep, gravitational abyss, another particle escapes. How is this possible? It is possible because *gravitational potential is independent of kinetic energy.* Why this is the case will become evident in Part III, once the source of the gravitational field is identified, but for now all that is necessary is a brief inspection of Equation (5.19). Gravitational potential is related to a particle's *rest energy* (rest mass), not total energy (mass). A proton with a kinetic energy of $2.99m_pc^2$ at a gravitational potential of -0.99 has, upon reaching free space, a remaining kinetic energy of $2.0m_pc^2$. Not only has it escaped a near-unity gravitational potential, but it is still moving close to the speed of light. *The maximum energy loss caused by a particle's departure from the surface of a black hole is one rest mass.* Energetic particles such as cosmic rays have orders of magnitude more energy than this. Any material whose kinetic energy exceeds its rest energy can escape from a black hole.

The relationship between a particle's rest mass m_0 and total energy (as relativistic mass m) is given by the expression:

$$m = \frac{m_0}{\sqrt{1 - \frac{\mathrm{v}^2}{c^2}}} \tag{5.22}$$

When a particle's kinetic energy equals its rest energy, its total energy m is $2m_0$. Substitute this into Equation (5.22) and solve for a black hole's escape velocity:

$$\mathrm{v} = \frac{\sqrt{3}}{2}c \tag{5.23}$$

or 86.6% of the speed of light. The electrons emitted from the source of a typical hospital X-ray device have about 20% of this energy level.

Unlike matter, light cannot carry kinetic energy, so it loses all or nearly all of its intrinsic energy to a black hole's surface potential. A photon emitted from a gravitational potential of –0.99, for instance, loses 99% of its energy. *Black holes separate matter from light*:

Ψ THEOREM 5.6 - GRAVITATIONAL FILTER {Ψ5.5}

MATTER WITH A KINETIC ENERGY GREATER THAN ITS REST MASS CAN ESCAPE FROM A BLACK HOLE; LIGHT IS EITHER TRAPPED OR REDSHIFTED TO NEGLIGIBLE ENERGY

The Schwarzschild radius of Equation (5.20):

$$R_S = \frac{2GM}{c^2}$$

is not a place where reality ends. It is simply a region specifying a negative unity gravitational potential energy. This has a one-to-one correspondence with any other form of energy and represents the full extent of its effect on matter. If particles couldn't carry more than their own rest mass of kinetic energy, or if gravitational potential could collapse to negative infinity, the material in a black hole would be trapped forever. However, neither of these situations is even remotely similar to a black hole's actual dynamics.

The time dilation caused by gravitational potential is often cited with the bold claim that objects never really fall into a black hole - they are forever frozen at the threshold of its event horizon. This couldn't be further from the truth. Gravitational potential dilates the rate of internal change of a system; it has nothing to do with its progress through space. Moreover, maximal gravitational potential only exists near a black hole's surface, which is relatively thin. Beneath it the mass enclosed by some radius r decreases and so does gravitational potential. *Potential is zero at a black hole's center.* The lifespans of unstable elementary particles, for instance, are maximized on a black hole's surface. In its interior, they decay at a rate consistent with the prevailing gravitational potential they encounter.

A black hole is certainly an unusual environment, but not a mystical one. Its physics is more accurately understood in terms of gravitational potential than as spatial/temporal distortion. The physical location of a black hole's maximum potential will be referred to as its *gravitational veneer*, or more simply its *veneer*:

Ψ THEOREM 5.7 - GRAVITATIONAL VENEER {Ψ5.5}

BLACK HOLES ARE BOUNDED BY A VENEER WITH A GRAVITATIONAL POTENTIAL EQUAL TO OR APPROACHING NEGATIVE UNITY AND RADIUS EQUAL TO OR APPROACHING $R_S = (2GM/c^2)$

Black holes have been grossly misinterpreted. They are not gravitational cul-de-sacs from which Hawking radiation is the only escape. They are vital, dynamic components of a steady-state universe. Their cosmic function as presented in Part IV is dramatically different from contemporary dogma, and far more interesting. Theorists get so caught up in the mathematics of curved space that they lose track of its far more pragmatic physics.

5.11 GENERAL CONCLUSIONS

In Part II, *Null Physics* moves away from the null and into the physics:

- Electrons orbit atoms in much the same way as planets orbit stars, albeit significantly faster. The reason why they don't emit photons in a Rutherford catastrophe is because, in their ground state, they orbit nuclei in less time than it takes to release a photon corresponding to their available kinetic energy.

- The Bell Inequality doesn't prove the nonlocality of material interaction; it demonstrates the incompleteness of quantum theory. Our reality's substructure is deterministic. Photons and particles are real objects whose energy is distributed within spatial volume. Light's wave nature is due to the quantum hysteresis of matter, not to its intrinsic statistical nature.

- The ultimate source of the energy released by the negative potential fields of matter: Coulomb, gravitational, and Strong, is the rest energy of the matter generating the field.

- The limit of a black hole's gravitational potential is negative unity. It has no *event horizon* - its maximum escape velocity is $0.86c$ for elementary particles and c for photons. Particles lose one rest mass of kinetic energy when they escape a black hole. Photons' energy loss is governed by the gravitational potential of the black hole's veneer. At a potential of –0.99, for example, photons lose 99% of their energy.

Quantum phenomenon and action at a distance are both the result of matter and energy's spatially distributed nature. Contemporary physics has attempted to separate a particle from its effects on other particles, and in the process lost the ability to explain either. The time for a new physics is upon us and mysticism should be summarily banished to the dark caves in which it originated. Welcome to the universe we actually live in: infinite, perfect, and deterministic. Everything it contains shares an equal level of realness, irrespective of size or composition.

6. ABSOLUTE SPACE

6.1 AETHER WIND

Photons are real, particles are real, *and space is real.* The reason contemporary physics has stagnated is because it has abandoned the reality of the universe's key constituents. Like matter, space used to be considered a material substance. Toward the end of the nineteenth century, most physicists, Maxwell in particular, believed space was an elastic solid through which matter and energy propagated. This view fell into disfavor when attempts to detect *absolute space* (also called the *aether*) failed miserably. The first and certainly most famous aether experiment was performed in 1887 by Albert Michelson and Edward Morley. The reasoning behind their experiment was if light propagates through the aether and the Earth is also moving in some direction through this continuum, then the speed of light should vary along differing directions in space. This effect was termed the *aether wind.*

To test this hypothesis, Michelson and Morley set up two mirrored paths in perpendicular orientations. They split a beam of light and sent it along these paths, recombining the separate beams in an interferometer. No interference patterns were detected when the entire apparatus was rotated through 360°. This meant the beams returned to the collection site at the same time regardless of the path they took. Repeated experiments failed to reveal interference patterns - no aether wind was found. A number of comparable measurements have been done since then with greater accuracy and consistently negative results.

But absolute space didn't go down without a fight. Shortly after Michelson and Morley performed their original work, a few theorists argued that their experimental setup had contracted and expanded relative to the direction of motion in such a way as to prevent the aether's detection. Subsequent experiments served to invalidate this premise.

While it is certainly true that the aether wind is undetectable, this doesn't mean absolute space does not exist.

6.2 RELATIVISTIC CONUNDRUM

Although Michelson and Morley were looking for a subtle effect of the aether wind near Earth, its consequences would be anything but mild. If such a wind existed:

- Terrestrial telescopic images would vary over time. At the current rate of Earth's motion around the Milky Way, a light source only 10 km distant would be blown several meters off course by the aether wind during the course of a few hours of the Earth's daily rotation. This would cause telescopic images of fixed objects to rise and fall during the day, an effect that would have been obvious hundreds of years ago.

- Our GPS system would not work. Its operation is dependent on precisely timed signals arriving from space. An aether wind would introduce complex variations on the order of milliseconds, quickly disabling the network.

The question now is, *if absolute space exists and photons are real objects moving at constant speed through it, where is the aether wind?*

GRADATIONAL DISTANCE

Consider a baseline of two points **A** and **B** in absolute space separated by a distance *d*:

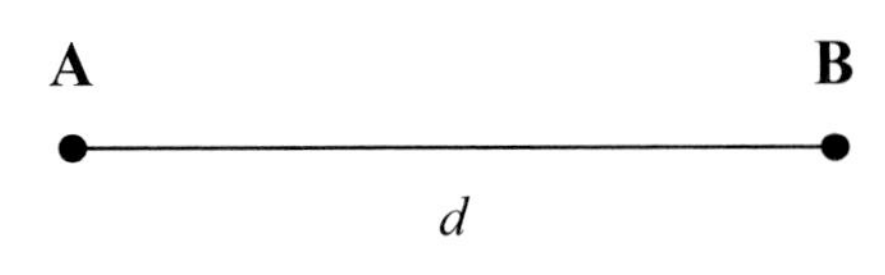

Figure (6.1) Distance in absolute space

This distance is simultaneously composed of two *directional* paths, **AB** and **BA**. Space is inherently neutral, so the only way to define the directional difference between the paths is through the use of an additional dimension, time. *Time is the difference of space* (Ψ3.7). It is easy to claim that a vector lies in a given direction, but in truth a vector is merely the relationship of a point in space to an arbitrary coordinate system. Direction as a physical reality requires a substantive difference in space, which in turn requires time.

Just as matter and antimatter are intrinsic components of the universe, balanced spatial differences are intrinsic components of distance:

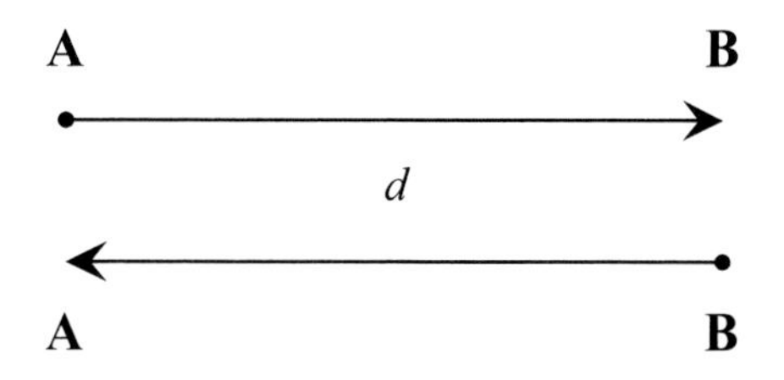

Figure (6.2) Distance's intrinsic composition

This will be called *path dichotomy*:

Ψ THEOREM 6.1 - PATH DICHOTOMY {Ψ3.7}

SPATIAL DISTANCE IS COMPOSED OF TWO DIRECTIONAL PATHS OF OPPOSITE SIGN

Since the paths **AB** and **BA** are simultaneous components of *d*, the distance between two points in the third dimension is the average of their lengths:

$$d = |\mathrm{AB}| = \frac{|\overrightarrow{\mathrm{AB}}| + |\overrightarrow{\mathrm{BA}}|}{2} = \frac{d_a^b + d_b^a}{2} \tag{6.1}$$

where the vector origin is subscripted. Component path lengths such as d_a^b and d_b^a will be referred to as *gradational distance*. The term d_a^b, for instance, is the gradational distance from **A** to **B**.

In the absolute reference frame, time is time and distance is distance; there is a pure dimensional separation between the two. The same is not true for a moving reference frame because it is a blend of distance and velocity. Let's set a baseline **AB** in motion through absolute space toward **B** at some velocity $\mathbf{v}_{\mathrm{AB}}$, inducing a complementary *spatial velocity*, $\mathbf{v}_\mathbf{s}$:

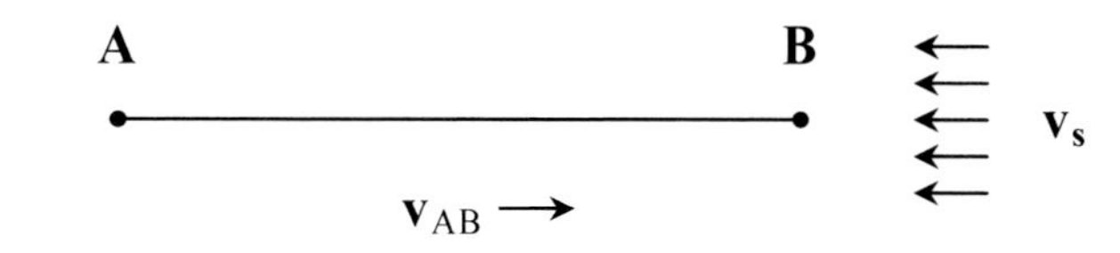

Figure (6.3) A baseline moving through absolute space at $\mathbf{v}_{\mathrm{AB}}$, inducing $\mathbf{v}_\mathbf{s}$

Velocity, like gradational distance, is a fourth-dimensional operator. It is a certain amount of distance in a certain amount of time, but more importantly represents *a difference in space*. So on one hand gradational distance is an intrinsic vector spatial difference, and on the other velocity is the physical reality of this difference.

Metric equivalence (Ψ3.8) allows us to scale any velocity in terms of c to allow a direct comparison to gradational distance. The conversion between spatial difference and spatial velocity has the form:

$$\Delta d = \frac{\mathrm{v_S}}{c} d \tag{6.2}$$

The baseline in Figure (6.3) moves in **B**'s direction, so the net motion of absolute space is directly opposite, oriented in **A**'s direction. Space's velocity is therefore positive (along the same direction) for path **BA** and negative for path **AB**. Applying Equation (6.2) to (6.1) yields component path distances in terms of spatial velocity $\mathbf{v_S}$:

$$d = \frac{\left(d_a^b - \frac{\mathrm{v_S}}{c} d\right) + \left(d_b^a + \frac{\mathrm{v_S}}{c} d\right)}{2} = \frac{d_a^{b*} + d_b^{a*}}{2} \tag{6.3}$$

where the paths modulated by spatial velocity are:

$$d_a^{b*} = d_a^b - \frac{\mathrm{v_S}}{c} d \tag{6.4}$$

$$d_b^{a*} = d_b^a + \frac{\mathrm{v_S}}{c} d \tag{6.5}$$

Spatial velocity will also be referred to as the *aether wind*. Note that the presence of this wind has no effect on the three-dimensional distance d between points **A** and **B**, as defined by Equation (6.1), because the velocity terms cancel when the paths are combined.

Since the static path lengths are equal to d:

$$\left|d_a^b\right| = \left|d_b^a\right| = d \tag{6.6}$$

the ratio of the modulated path lengths can be solved by substituting Equation (6.6) into Equations (6.4) and (6.5) and dividing:

$$\frac{d_a^{b*}}{d_b^{a*}} = \frac{\left(d_a^b - \frac{\mathrm{v_S}}{c} d\right)}{\left(d_b^a + \frac{\mathrm{v_S}}{c} d\right)} = \frac{\left(1 - \frac{\mathrm{v_S}}{c}\right)}{\left(1 + \frac{\mathrm{v_S}}{c}\right)} = \frac{(c - \mathrm{v_S})}{(c + \mathrm{v_S})} \tag{6.7}$$

Or:

$$d_a^{b*} = \left(\frac{c - \mathrm{v_s}}{c + \mathrm{v_s}}\right) d_b^{a*} \tag{6.8}$$

Gradational distance is shortened in the direction of the baseline's motion. Distance is invariant with respect to motion, but not the gradational path lengths within it. This will be referred to as *gradational asymmetry*:

Ψ THEOREM 6.2 - GRADATIONAL ASYMMETRY {Ψ6.1}

GRADATIONAL PATH LENGTH CONTRACTS IN THE DIRECTION OF AN OBJECT'S MOTION AND EXPANDS IN THE DIRECTION OPPOSITE ITS MOTION

It is not possible to measure gradational distance because it is intrinsically fourth-dimensional. A ruler applied to **AB** will always indicate the same distance as **BA**. If **AB** is at rest relative to absolute space, the two paths do in fact have the same length. In any other case they do not.

6.3 MOVING PATHS

Let's now look at a moving reference frame from an empirical perspective. Suppose Earth is moving through absolute space at $\mathbf{v}_{AB}$, and a baseline is again set up from points **A** to **B** in $\mathbf{v}_{AB}$'s direction, as shown earlier in Figure (6.3). Atomic clocks are then used to measure the time it takes photons to travel from **A** to **B** and from **B** to **A**. In accordance with the relativistic equivalence of reference frames, the travel times for both directions are the same. Why is this?

Photons move at c in the absolute reference frame, so the distance a photon moving from **A** to **B** must traverse is:

$$\mathrm{d}_{AB} = \left(\frac{c}{c - \mathrm{v_s}}\right) d_a^{b*} \tag{6.9}$$

where d_a^{b*} is the motion-modulated gradational distance from **A** to **B**, d_{AB} is the actual distance a photon travels in absolute space, and $\mathbf{v_s}$ is the aether wind ($\mathbf{v_s} = -\mathbf{v}_{AB}$). In the case of Equation (6.9), absolute distance is longer than d_a^{b*} because **B** is receding as the photon moves toward it against the aether wind. The absolute distance in the opposite direction (with the wind) is shorter than d_b^{a*}:

$$\mathrm{d}_{BA} = \left(\frac{c}{c + \mathrm{v_s}}\right) d_b^{a*} \tag{6.10}$$

The ratio between the two absolute distances is:

$$\frac{\mathrm{d}_{\mathrm{AB}}}{\mathrm{d}_{\mathrm{BA}}}=\left(\frac{c+\mathrm{v}_{\mathrm{s}}}{c-\mathrm{v}_{\mathrm{s}}}\right)\left(\frac{d_a^{b*}}{d_b^{a*}}\right) \tag{6.11}$$

Substitute Equation (6.7) for the ratio of gradational distance:

$$\frac{\mathrm{d}_{\mathrm{AB}}}{\mathrm{d}_{\mathrm{BA}}}=\left(\frac{c+\mathrm{v}_{\mathrm{s}}}{c-\mathrm{v}_{\mathrm{s}}}\right)\left(\frac{c-\mathrm{v}_{\mathrm{s}}}{c+\mathrm{v}_{\mathrm{s}}}\right)=1 \tag{6.12}$$

The time required to complete either path is the same because photons move through absolute space at a constant speed of c and the actual distance they travel is the same in either direction.

In contemporary physics, the distance between two points in space is not considered a vector operator unless some form of field is present. Yet the only way to define moving distance is to tag it with a vector identifier. *Moving distance is not the same as stationary distance.* Relativity treats a moving reference frame as if it were a scalar field. This is a reasonable simplification since it is not possible to differentiate between moving distance and stationary distance without access to absolute space, but it is not an accurate portrayal of the underlying reality. Two reference frames in uniform motion relative to each other may be observationally equivalent, but they are not dimensionally equivalent. Motion is a fourth-dimensional process. The only way **AB** can differ from **BA** is in terms of this dimension. Moreover, *if there were no difference between a stationary **AB** and a moving **AB**, motion would not exist.* Please see Appendix E for further information about the relationship between relativistic effects, such as time dilation, and absolute space.

6.4 SPACE IS REAL

The universe's material has no net velocity, just as nothingness has no net velocity. This defines the *absolute reference frame* - the aether. Matter and energy are real distributions in this medium and it is motionless with respect to the universe as a whole. Contrary to some contemporary interpretations of Einstein's work, *the aether is not in any way at odds with relativity.* Relativity's basic premise is that uniformly moving reference frames are mathematically equivalent. It makes no claim as to the existence or nonexistence of absolute space. As it turns out, however, moving frames are not *entirely* equivalent.

Consider a reference frame where a particle approaches a planet with a uniform speed of 0.99999*c*. According to relativity, this is indistinguishable from another reference frame where the particle is at rest and the planet is approaching it at 0.99999*c*. If these two systems were *physically* equivalent, the amount of energy would be equal in both, *and it is not*. There is far more energy in the frame of the fast-moving planet than the fast-moving particle. It is not possible to determine the *absolute* amount of kinetic energy in a system any more than it is possible to detect absolute space. But they are in fact related and relativity provides an incomplete perspective.

Suppose two objects of mass m and speed $\mathbf{v}$ move in opposite directions of the same axis so that their net momentum is zero. Now apply an axial motion $\mathbf{v}_n$ to the entire system.

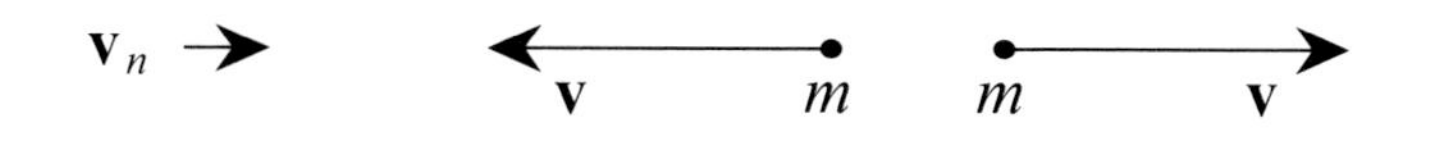

Figure (6.4) Applying a net motion to a zero-momentum system

$\mathbf{v}_n$ doesn't change the objects' motion relative to each other, but the system's total kinetic energy increases according to:

$$E = \left(\frac{m}{2}\right)\left((\mathrm{v}+\mathrm{v}_n)^2 + (\mathrm{v}-\mathrm{v}_n)^2\right) = m\left(\mathrm{v}^2 + \mathrm{v}_n^2\right) \tag{6.13}$$

For any set of relative motions, a system's minimum kinetic energy is achieved with a velocity field whose net motion is zero. Net motion increases a system's energy because it takes more energy to increase velocity from a given magnitude than is saved by decreasing it from the same magnitude. Since the universe is motionless, *it exists at the minimum kinetic energy its relative internal motions can have.*

The universe's total energy is fixed, yet varies with the frame of reference it is measured in, so the absolute reference frame is the state corresponding to *zero momentum and minimum total energy*. It is the one and only coordinate system that satisfies both. All reference frames might be mathematically equivalent to a certain extent, but their equivalence ends there, and doesn't tell us what we need to know about the nature of space.

Absolute space exists physically by virtue of its volume and dynamically as a zero point of net universal momentum:

Ψ THEOREM 6.3 - ABSOLUTE SPACE {Ψ1.5}

ABSOLUTE SPACE IS THE UNIVERSAL REFERENCE FRAME OF ZERO NET MOMENTUM AND MINIMUM TOTAL ENERGY CONTENT

Michelson and Morley demonstrated that absolute space cannot be directly measured, but (Ψ6.3) shows us that it is nonetheless real.

Finding the universe's minimum energy state from its celestial dynamics is a daunting proposition. It can't be found by evaluating the velocity distribution of any astronomically small region because even on the scale of galaxies and superclusters there are massive net material flows, such as those generated by the Great Wall and Great Attractor. The proper and only way to detect the universe's net momentum is when the sum within a cosmologically large sample set is insignificantly small in relation to the magnitude of the sample's average momentum. As sample size increases from this, an accurate tally of net momentum for the universe's absolute frame will approach zero. *Earth-based measurements ought to reflect Earth's motion through absolute space as a net motion of the cosmic material distribution.*

As it turns out, however, we can find absolute space without plotting the trajectories of millions of galaxies. NASA has conducted two comprehensive surveys of the universe's Cosmic Microwave Background (CMB) radiation intensity, an early effort called the Cosmic Background Explorer (COBE) followed by the higher-resolution Wilkinson Microwave Anisotropy Probe (WMAP). Both show Earth's unequivocal motion artifact through this field. Microwaves are blueshifted in the direction of our planet's movement and redshifted in the opposite direction:

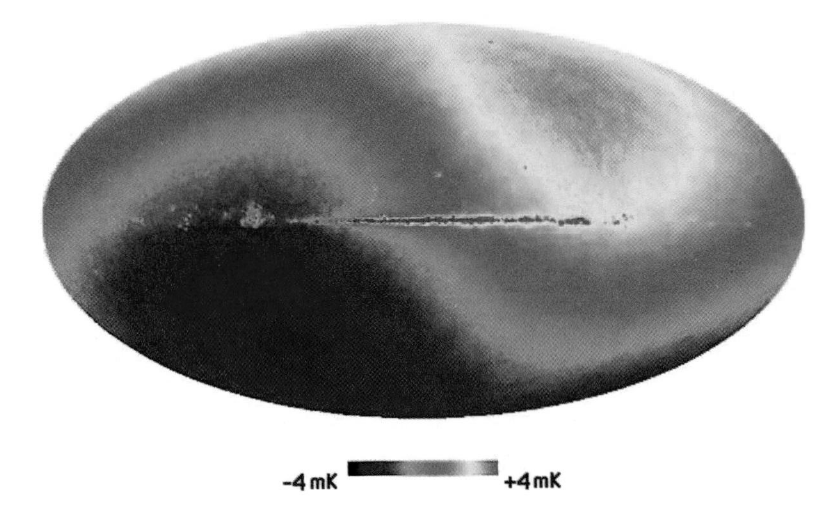

Figure (6.5) Earth's motion through the CMB
(Courtesy NASA/WMAP Science Team, shown previously as Figure (1.1))

This is called the *CMB dipole anisotropy*. Earth's motion through the CMB has been calculated[36] to be 370 ± 3 km/s toward the galactic coordinates (l^{II}, b^{II}) = (264.4 ± 0.3°, 48.4 ± 0.3°). The right ascension-declination (1950 epoch) for these coordinates is (RA, DEC) = (168°, −6.7°). Earth is moving toward the constellation Crater.

When Earth's dipole anisotropy is subtracted, the CMB is isotropic to better than one part in 10^5:

Figure (6.6) WMAP without Earth's motion artifact
(Courtesy NASA/WMAP Science Team)

The red horizontal region is our galaxy's microwave emission. The subtracted image of Figure (6.6) is far grainier then Figure (6.5) because its temperature resolution is significantly higher once the Earth's large motion dipole has been removed.

When the Milky Way's galactic emissions are also eliminated, the result is a clear view of the universe's microwave background:

Figure (6.7) WMAP without galactic emissions or Earth's motion dipole
(Courtesy NASA/WMAP Science Team)

The CMB, as derived in Part IV, is an equilibrium concentration of the oldest photons in the universe. Their net momentum, like that of the universe's material content, is zero. *Since the net momentum of the CMB rest frame shown in Figure (6.7) is so close to zero (isotropic), it constitutes a superbly accurate measurement of absolute space.*

A careful galactic redshift survey ought to reveal a dipole anisotropy comparable to the one seen in the WMAP. *Absolute space is the universe's rest frame.* Earth is moving through it toward the constellation Crater at a speed of ~370 km/s.[6.1]

ABSOLUTE LIGHTSPEED

Absolute space means absolute motion, which in turn means an absolute value for the speed of light:

Ψ THEOREM 6.4 - ABSOLUTE LIGHTSPEED {Ψ6.3}

ABSOLUTE LIGHTSPEED IS LIGHT'S SPEED THROUGH PERFECTLY EMPTY SPACE

Since this requires the absence of all fields, to include gravity, it is a theoretical limit that will always be slightly faster than a photon can move anywhere in the universe.

6.5 GENERAL CONCLUSIONS

Space is the foundation of matter and energy; it is not possible to understand one without understanding the other:

- Photons move at the speed of light relative to absolute space at all times and in all directions. The reason why a moving reference frame preserves this relationship is because the superposition of motion and distance is a fourth-dimensional vector field scaled by the speed of light.

- Absolute space exists. It is the universal state of zero momentum and minimum kinetic energy. Earth is moving through it at a speed of ~370 km/s toward the constellation Crater.

Our resistance to the concept of real space probably originates from a primordial need for solidity. Falling from a tree, having nothing to claw at but air, can instill a profound respect for substance. We have to let go of primitive concepts. Matter and energy will never make sense until they are firmly and unilaterally connected to the medium they inhabit.

7. ENERGY'S GEOMETRY

7.1 THE OTHER UNBOUNDED CUBIC

Netherspace is the infinitely small distance of dimensional closure, as described earlier by Equation (2.25):

$$\lozenge_4 = \delta_3 V_U$$

The dimensional symmetry inherent in $\lozenge_4$ is lost when expressed as the asymmetric product of V_U and δ_3, but δ_3 still has units of distance, not time, because a single point can have no polarity. Netherspace's most literal interpretation in this context *is the universe's fourth-dimensional thickness, not its spatial continuity.* It constitutes distance, since a point's width is by definition composed of points.

Unit hypervolume has to contain polar components to allow for its own self-nullification, so dimensional closure has an equivalent expression of the form:

$$\lozenge_4 = \delta_3 \left(\frac{V_U}{2} - \frac{V_U}{2} \right) \tag{7.1}$$

In this internal case, unit hypervolume has units of time-distance^3, as time is the difference of space. The only way that δ_3 can be *distance* in this polar representation is if the infinite volume of Equation (7.1) has units of time-distance^2. But we don't need linear continuity to tell us that this infinite volume is clearly present or what it represents.

Just as unit hypervolume contains an infinite amount of spatial volume, it also holds an equally large amount of the cubic time-distance^2. Its symmetric structure provides two and only two dimensionally distinct, infinitely large cubics. Not coincidentally, the universe contains infinite amounts of two and only two phenomenologically distinct substances. One is distance^3, *space*. The other is *energy*.

Ψ THEOREM 7.1 - ENERGY'S DIMENSIONALITY {Ψ2.12}

ENERGY IS A THREE-DIMENSIONAL QUANTITY OF TIME-DISTANCE^2

In summary:

- Hypervolume contains two and only two dimensionally unique, three-dimensional quantities: distance^3 and time-distance^2.
- There are two fundamental units of measurement: distance and time.
- The universe is composed of two substances: energy and space.
- Energy moves through space.

The only reasonable conclusion is time is an intrinsic dimensional component of energy. Joule is an arbitrary unit chosen without reference to reality's substructure. Energy's true units are time-distance^2.

ENERGY'S EXPLICIT DIMENSIONALITY

Energy's dimensional structure follows directly from unit hypervolume's symmetry, but the same result can be derived from an entirely different line of reasoning.

If:

- There are only four dimensions in the universe, three of distance and one of time.
- Energy is distributed into space at a certain finite average energy density. This requires energy and space to possess a comparable number of dimensions.
- Any physical quantity has a dimensional structure limited to some combination of the universe's available dimensions. Space's three dimensions are equivalent, so the universe only has two uniquely different three-dimensional combinations of its four possible dimensions - distance^3 and time-distance^2.
- Energy is different from space.

The only viable conclusion is that energy has units of time-distance^2.

The relationship between energy and space within unit hypervolume is shown below:

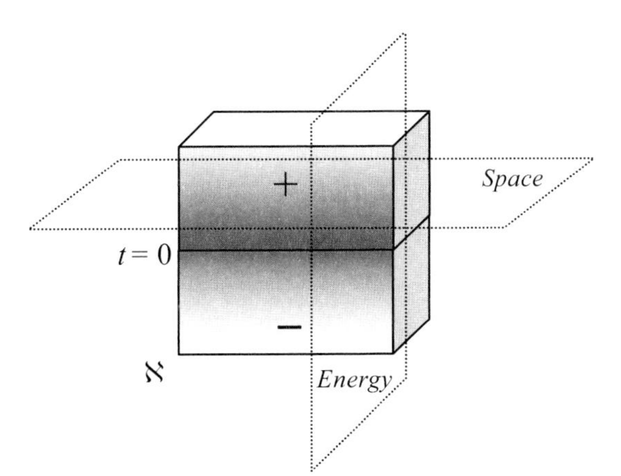

Figure (7.1) Energy is unit hypervolume's other dimensionally unique, three-dimensional cross-section

Space is associated with neutral extent while energy is associated with fourth-dimensional displacement and polarity. *The existential balance between space and time corresponds to an infinite quantity of energy.*

Unit hypervolume contains the internal product of time and space, but also the product of energy and distance:

$$\Diamond_4 = td^3 - td^3 = td^2(d) - td^2(d) \qquad (7.2)$$

Its symmetry requires space and energy to exist in equal universal quantities:

Ψ THEOREM 7.2 - SPACE/ENERGY DUALITY {Ψ2.12}

$\Sigma|\text{Energy}| = \Sigma|\text{Space}| = V_U = \infty^3$

This in turn means that the universe's *average energy density is precisely unity.*

Ψ THEOREM 7.3 - UNIVERSAL ENERGY DENSITY {Ψ7.2}

$(\Sigma|\text{Energy}|/\Sigma|\text{Space}|) = 1$

In existentially correct units, where space is (absolute meter^3) and energy is (absolute second-absolute meter^2), the universe's average energy density is 1. This is another example of the amount of nothing in nothing, except now it is in terms of quantities closer to home - energy and volume.

Time is a spatial difference. It has two remarkably divergent manifestations:

- As the positive and negative fields of photons and particles.
- As the motion of these fields through space.

Both are *spatial differences* that reflect energy's dimensional structure. *Energy moves and has polar fields because time is an essential component of its geometry.*

The magnitude of infinite space is unit hypervolume, a net value that must somehow sum to zero in order to be a component of nothingness. This is what energy accomplishes, and why universal energy density is unity. The presence of fourth-dimensional deflection adds the polarity that space needs for its quantified nullification. There is no such thing as completely empty space. Every cubic meter of the universe is filled with the fields of particles and photons. *Space is charged with fields, converting one of its dimensions to time. The resultant polarity allows it to cancel its net hypervolumetric magnitude into nonexistence.*

Since the universe's size is invariant and represents the volume of space and energy within it, energy's conservation follows directly.

Ψ THEOREM 7.4 - ENERGY CONSERVATION {Ψ2.12, Ψ4.2}
ENERGY IS CONSERVED: $\Delta|\text{Space}| = \Delta|\text{Energy}| = 0$

Interactions have no effect on energy's *quantity* because it is governed by the universe's invariant size.

MOTION

Energy density is the three-dimensional relationship between energy and space, existentially and by definition. But this is not their only relationship. Since their sub-structural difference is limited to a single dimension, distance replaced by time, they also have a *linear* relationship of the form:

$$\frac{d^3}{sd^2} = \frac{d}{s} = c \tag{7.3}$$

This is what the speed of light represents, and absolute lightspeed is its purest expression. *Motion is the one-dimensional relationship between energy and space.*

Consider for a moment how remarkably simple and intuitive these results are. Energy density is the three-dimensional relationship between energy and space. Of course it is - this is how energy is distributed into space. The concept of energy density makes no sense if energy isn't a three-dimensional substance. It is three-dimensional, yet is somehow *different* from space. *In a universe of four dimensions, the only possible dimensional structure available for energy is time-distance^2.* If any doubt exists about this structure, one needs to look no further than light's motion through space, an utterly explicit demonstration of the dimension energy has and space lacks.

7.2 DISTRIBUTED ENERGY

Energy density is the ratio of energy to volume as well as the ratio of time to distance:

$$\rho_E = \frac{E}{V} = \frac{td^2}{d^3} = \frac{t}{d} \tag{7.4}$$

This is the most precise expression of the three-dimensional relationship between space and energy. Every speck of the universe's energy, whether distributed in the fields of particles, neutrinos, or photons, has a volumetric relationship to space. Energy density in atomic nuclei is phenomenal; its presence in deep space is sparse, but there is energy, at some density, everywhere.

The speed of light is the ratio of space to energy and is fixed at unity. Energy density is the ratio of energy to space and is variable. There is only a single *dimensional relationship* between space and energy, expressed unambiguously by the motion of light. *Energy density is the distributional, not dimensional, relationship between energy and space.* This allows it to vary over a wide, albeit finite, range, and is the key difference between the ratio of space to energy and the ratio of energy to space. The amount of space in energy is existential whereas the amount of energy in space is circumstantial.

The difference between electrically charged space and empty space is *energy density*, and their relationship is defined from two governing factors:

- Energy, like everything else, is composed of space.
- Energy has a dimensional structure of time-distance^2.

It follows that energy density is *the fourth-dimensional slope of space.* This is the only ratio of time to distance that exists all the way down to space's compositional level.

Ψ THEOREM 7.5 - ENERGY DENSITY {Ψ7.1}

ENERGY DENSITY IS THE MAGNITUDE OF SPACE'S FOURTH-DIMENSIONAL SLOPE, $|t'|$

This is a logical extension of Equation (7.4):

$$\rho_E = \frac{E}{V} = \frac{td^2}{d^3} = \frac{t}{d} = |t'| \tag{7.5}$$

Fourth-dimensional spatial *slope*, in turn, requires fourth-dimensional spatial *deflection*.

SPATIAL DEFLECTION

The two long-range forces of nature, electromagnetism and gravitation, are both distributed spatial displacements, but are uniquely different. This is because there are two and only two ways space can be distorted: *externally* and *internally*.

External deflection. Electromagnetic fields are external spatial displacements into the fourth dimension. There are only two possible directions along this axis, positive and negative. This is why electrostatic fields are either positive or negative. *It is what their polarity represents.*

Ψ THEOREM 7.6 - EXTERNAL DEFLECTION, t {Ψ7.1, Ψ7.2}

EXTERNAL DEFLECTION IS THE DISPLACEMENT OF SPACE NORMAL TO THE THIRD DIMENSION; IT IS EVIDENT IN ELECTROMAGNETIC FIELDS AND IS MEASURED IN UNITS OF TIME

Internal deflection. Gravitation is the displacement of space *along space*, stretching it *within the confines of the third dimension*. This is why gravitational fields lack a discrete polar character. Unlike external deflection, internal deflection is a vector quantity that displaces space along itself in any of the infinite number of directions available in the third dimension.

Ψ THEOREM 7.7 - INTERNAL DEFLECTION, $t_{\|}$ {Ψ7.1, Ψ7.2}

INTERNAL DEFLECTION IS THE DISPLACEMENT OF SPACE ALONG THE THIRD DIMENSION; IT IS EVIDENT IN GRAVITATIONAL FIELDS AND IS MEASURED IN UNITS OF TIME

External deflections will be symbolized as t and internal deflections will be subscripted as $t_{\|}$ to denote their orientation parallel to space. Although internal deflection is a vector quantity with a magnitude and direction, it will be symbolized as $t_{\|}$ because its directional aspect is seldom relevant to the calculations done in *Null Physics*. Any deflection of space, whether internal or external, is a fourth-dimensional entity because it is noncompositional and extraspatial. It represents a spatial difference and therefore always has units of time. Internal deflection, like its external analog, has slope, and the product of slope and spatial volume is energy:

$$\rho_{E_{\|}} = \frac{E_{\|}}{V} = \frac{t_{\|}d^2}{d^3} = \frac{t_{\|}}{d} = \left|t'_{\|}\right| \tag{7.6}$$

Although this form of energy has the same units as any other form, it will be called *internal energy* and symbolized $E_{\|}$ to identify its unique relationship to space.

Spatial deflection is the final resolution to action at a distance. *The reason why fields exert influence over distance is because they are physical spatial distortions.* Our investigation will begin with external deflection because the overwhelming majority of the universe's energy is distributed in the electromagnetic fields of matter and light.

EXTERNAL DEFLECTION

Externally deflected space is the confluence of energy and space. It projects into the energy plane in the direction of maximum slope and into space in the direction of the fourth dimension:

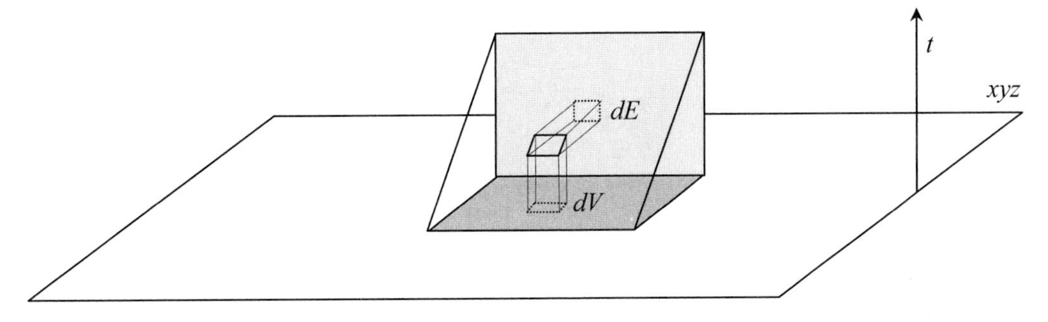

Figure (7.2) Deflected space is the combination of differential elements of energy and space.

External *slope* is energy's compositional aspect, and a spatial deflection's four-dimensional shape defines its total energy. External *displacement* is not compositional, but generates *hypervolume* as the product of spatial volume and the distance this volume is displaced into the fourth dimension:

$$d\lozenge = t\,dV \tag{7.7}$$

An unsubscripted $\Diamond$ will be used to represent the distributed hypervolume of spatial deflections while the universe's unit hypervolume remains $\Diamond_4$. The fields of particles and photons are composed of an infinite number of differential hypervolumetric elements. The hypervolume/hyperspace associated with a positive or negative spatial displacement will also be called *polarvolume* when in specific reference to its electrically charged nature. Polarvolume has the same units as hypervolume (except in the case of the totality of unit hypervolume).

Electrostatic fields are external deflection's physical embodiment. The stronger a particle's electrostatic field, the greater its external deflection. Particles have a radially symmetric charge distribution and therefore a radially symmetric field of spatial deflection:

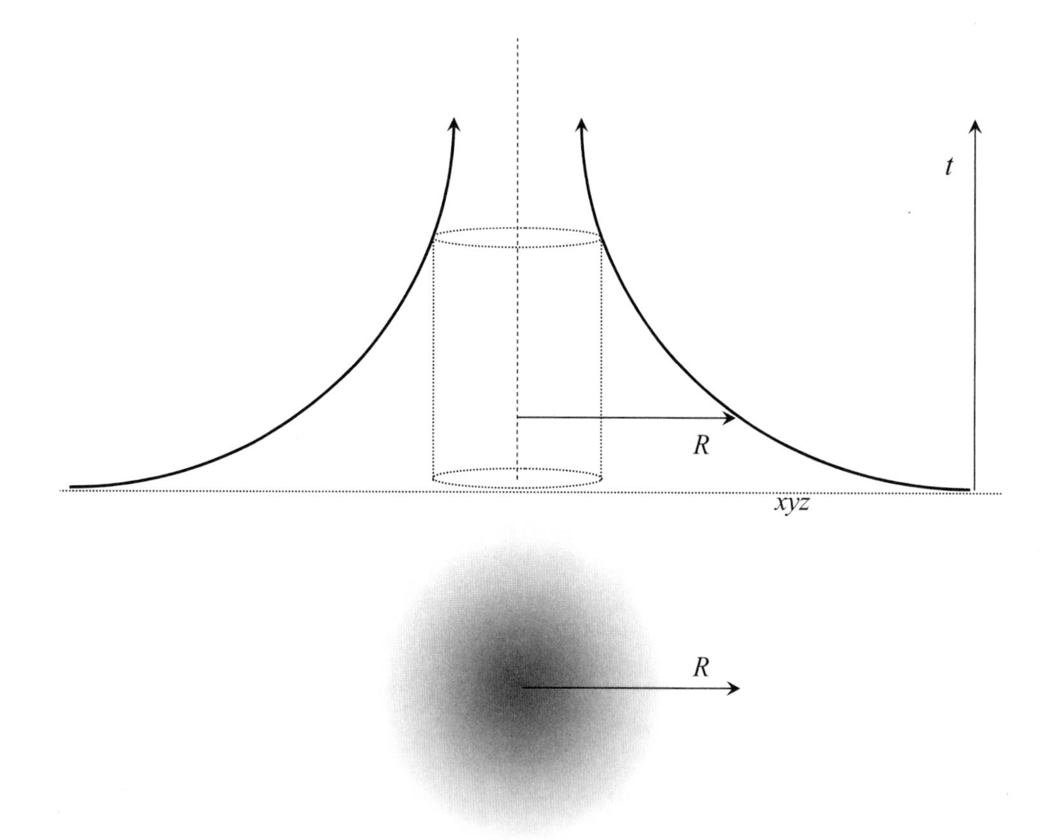

Figure (7.3) Particles have radially symmetric spatial deflection

Figure (7.3) shows a particle's field both from the perspective of the fourth dimension (top) and as viewed from within space (bottom). The polarity of its electrical *charge* is the positive or negative fourth-dimensional axis along which its deflection occurs. Its rest energy (mass) is composed of energy differentials distributed across its three-dimensional surface - a surface curved into the fourth dimension by its deflection. The symbol $\wedge$ will be used to refer to particles or their attributes.

Unlike particles, photons are electrically neutral and their equal and opposite external deflections are limited to a discrete region of space. A photon's two lobes are separate yet form a contiguous structure. Whereas particles are symmetric in space, photons are symmetric in time:

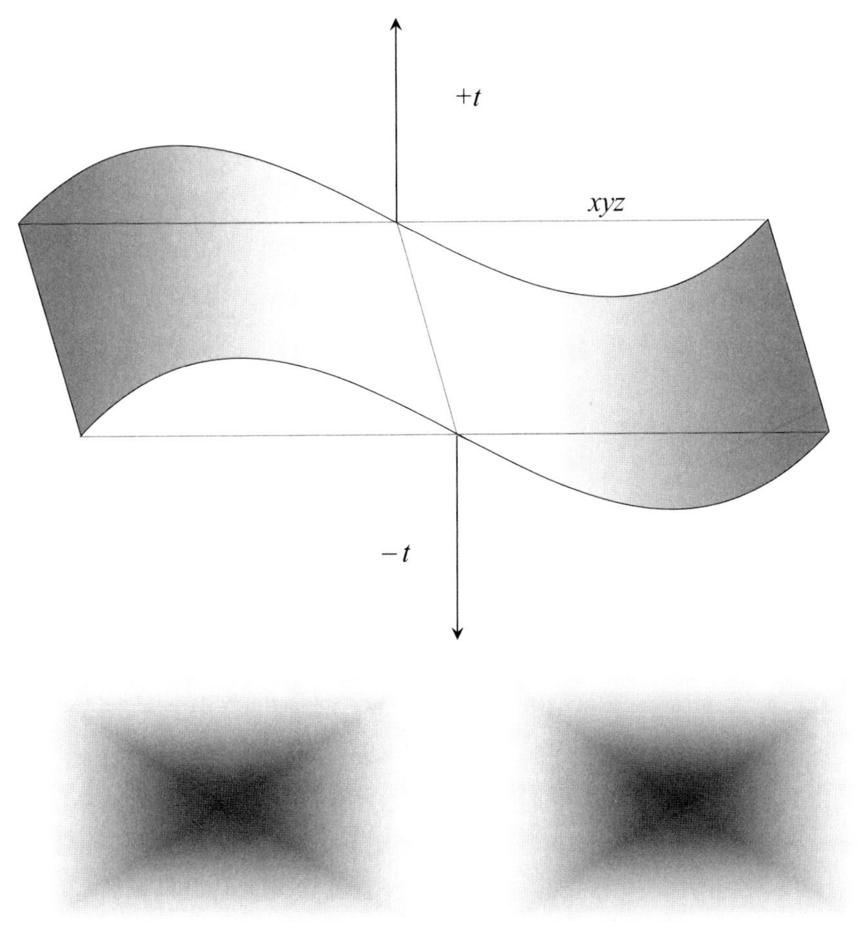

Figure (7.4) Photons have externally symmetric spatial deflection

The photon's field is shown from the perspective of the fourth dimension (top) as well as how it might appear from within space (bottom). The symbol γ will be used to refer to photons or their attributes. The orientation of a photon's positive and negative lobes relative to its motion through space is an interesting question, along with the size and shape of its actual spatial footprint. Evidence for both will surface as our analysis proceeds.

Even though particles and photons are physical spatial deflections, space is not an elastic solid as Maxwell once believed, because it contains no energy in and of itself. Space occupies a different dimensional state than energy and no interaction can occur between them.

7.3 FIELD CLARITY

Our derivation of energy density contains an implicit assumption: *The external spatial slope at any given point of space is single valued.* Is there incontrovertible evidence of this? What if quantization created entities separate in time as well as space?

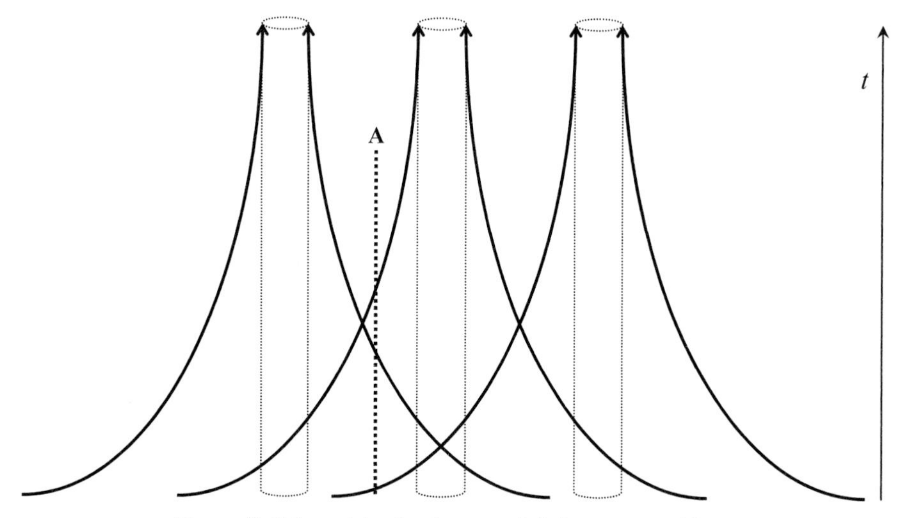

Figure (7.5) Is multi-valued external deflection possible?

The basic problem with this is energy density. The point **A** in Figure (7.5) has three different external slopes and therefore three different energy densities. What is **A**'s true energy density? The sum of the slopes? The average? Energy, like space, is three-dimensional, conserved, and compositional. If the universe's quanta were able to split spatial volume along the fourth dimension, their hypervolume would contain an infinite number of spatial instances, all at different external altitudes. This would diffuse space throughout the fourth dimension, blurring the essence of external deflection.

Space is three-dimensional, not four. The only way simultaneous occurrences of the point (x_1, x_2, x_3) could exist at t_1, t_2, and t_3 is if the universe were constructed of four-dimensional space, not four-dimensional space-time. This is not the case. *Time is the difference of space; not the space of difference.* A point can have only one external displacement. Multiple values would violate its conservation as well as continuity. This concept will be called *field clarity*:

Ψ THEOREM 7.8 - FIELD CLARITY {Ψ3.7, Ψ7.2}

EXTERNAL DEFLECTION AND ENERGY DENSITY ARE SINGLE-VALUED FOR ALL SPATIAL LOCATIONS IN ANY FIELD OR COMBINATION OF FIELDS

All of the universe's particles interact, so the deflection and energy density at any point is the net result of $\sim\infty^3$ fields. Yet it is still a single-valued result. Energy and space are no less inseparable on a global level than on a local one.

From a practical standpoint, if different particles occupied different space at different external elevations, there would no longer be an avenue for physical interaction. *Field clarity is the basis of all interactions between elementary energy forms.*

Since:

- Particles and photons are volumetric energy distributions.
- Two different energy density values cannot occupy the same spatial location at the same time.

It follows that:

- Particles and photons incur changes in their energy distributions when in proximity to other particles and photons.

All of the phenomena evident in atomic and nuclear physics and the four universal forces of nature originate from the way that the surfaces of particles and photons satisfy field clarity.

7.4 ENERGY'S INTRINSIC NEUTRALITY

Whereas external deflection has unambiguous polarity, energy density has no polarity at all. It exists as a *difference* of time per unit distance. *External slope is intrinsically neutral.* Even though energy contains a dimension of time, its compositional aspect neutralizes it:

Ψ THEOREM 7.9 - ENERGY'S NEUTRALITY {Ψ7.5}

ENERGY IS INTRINSICALLY NEUTRAL

The pivotal aspect of energy's neutrality is *the amount of energy at any location has no relationship to the actual external deflection there.* Although external deflection is the inevitable result of the accumulation of external slope, deflection and slope are geometrically separate entities. A quantity of energy is specifically not a quantity of external deflection. Deflection has polarity, whereas energy is its neutral by-product. *The universe's compositional bedrock has no polarity; polarity is the operator by which it is nullified.*

7.5 EXTRASPATIAL VOLUME

Space deflected into the fourth dimension contains additional volume in its sloped surface:

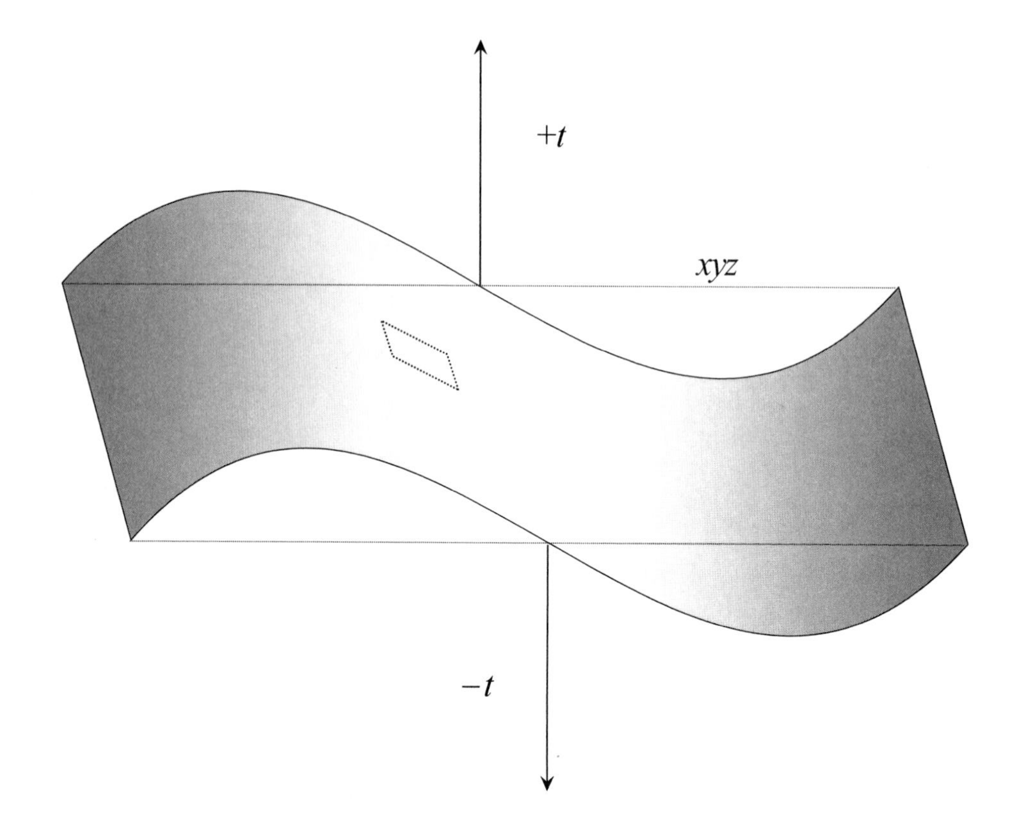

Figure (7.6) Sloped space has more volume than neutral space

The difference between empty space and space with energy density is the amount of additional volume per unit volume it has. Wood has more volume per unit volume than air. Steel has more volume per unit volume than wood. All of the solid objects around us are solid because and only because they contain more volume than the space they occupy.

Space's volumetric density, φ_V, increases in the presence of energy as:

$$\varphi_V = \frac{V_S + V_E}{V_S} = 1 + \left(\frac{V_E}{V_S}\right) = 1 + \left(\frac{\rho_E}{\rho_U}\right) \tag{7.8}$$

where V_S is a volume of space and V_E is the volume of energy within it. The energy density of Equation (7.8) is scaled by the average energy density of the universe, ρ_U, as it is measured in absolute units of volume per unit volume, not joules or second-meter^2.

Energy's presence doesn't reduce spatial volume, as space's underlying magnitude is invariant. Nor does it change the number of points in the volume it occupies. Energy's net effect on space is to increase its volumetric density with additional *extraspatial* volume. Empty space has a volumetric density of unity ($\varphi_V = 1$), so Equation (7.8) can be written:

$$\Delta\varphi_V = \frac{\rho_E}{\rho_U} \tag{7.9}$$

Energy density is the change in spatial volume density. This is inevitable. Energy and space are both conserved, volumetric quantities. Since their three-dimensional magnitudes are preserved regardless of their distribution, and energy is ultimately composed of space, *energy expands spatial volume within a spatial context*. As it turns out, there is direct and widespread evidence of this.

REFRACTION

A material's index of refraction varies with wavelength and is generally described in terms of Snell's Law. This law governs the relationship between incident and refracted angle for a photon moving freely (without scattering) across a boundary between two materials of differing refractive index. In the case of light passing from free space ($n = 1$) into a medium of refractive index n, Snell's law has the form:

$$\sin\theta_r = \left(\frac{1}{n}\right)\sin\theta_i \tag{7.10}$$

where θ_i is the angle of incidence and θ_r is the angle of refraction.

Equation (7.10) is derived from the conservation of photon momentum with the understanding that their speed through a material of refractive index n is given by:

$$\mathrm{v} = \frac{c}{n} \tag{7.11}$$

A number of physicists have recently reported *negative* refractive indexes, but these don't represent the simple transmission of light through translucent material. As such, our investigation of refraction will be limited to normal materials whose indexes have the form ($n > 1$).

The photon-level analysis of refraction's speed degradation is a topic conventional physics avoids like the plague. Although informal and generally undocumented, the currently accepted explanation of light's diminished speed through transparent substances is that its path incurs a high rate of absorption and reemission due to the strong electric fields within atoms and molecules. This process causes an accumulation of infinitesimal delays along a photon's trajectory, effectively reducing its observed speed:

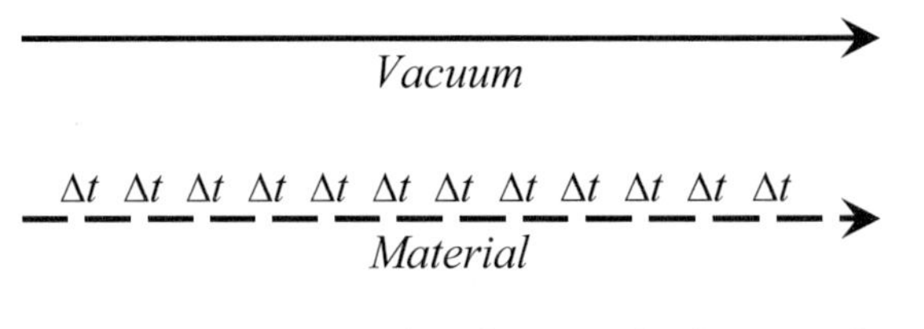

Figure (7.7) Conventional explanation of refractive index

The reason there is no scattering associated with this alleged mechanism is because the atoms or molecules absorbing the photon remain in exactly the same state until its reemission.

Even under the most cursory inspection, this is an unlikely scenario. The particles ultimately responsible for absorbing and reemitting photons in an atomic environment are valance electrons. Contrary to current doctrine, these have frenetic orbits about atomic nuclei - orbits requiring continuous changes of momentum at extremely high rates of speed. Hydrogen's ground state electron, for example, moves at two million meters a second about a tiny radius of 0.053 nm. Its entire orbit is completed in a mere $1.5(10)^{-16}$ s. By comparison, the absorption and reemission of red (600 nm) photons requires at the very least $2.0(10)^{-15}$ s ($t = hc/\lambda$), an order of magnitude longer. *A valance electron that absorbs a photon cannot possibly have the same momentum by the time such photon is reemitted.* Moreover, the addition of absorbed kinetic energy would change the shape and radius of the electron's orbit. This is why the light most objects emit isn't even coherent, let alone perfectly preserved.

The refractive index phenomenon is not due to photon absorption-reemission; it is the result of the extra volume that the energy of which matter is composed adds to space. Photons passing through regions filled with matter move at the velocity of light in straight lines, but must move further because these lines have been lengthened by extraspatial volume. As such, photons have the appearance of moving slower. But how much slower?

There are three key considerations:

- Isotropic. Space is isotropic, so the addition of extraspatial volume has the same effect in any direction.

- Distributed. Energy density is distributed through space, so its effect at one location has no relationship to its effect at another.

- Extraspatial. Extraspatial volume is a four-dimensional extension of space whose only difference from space is a single dimensional of time. This means that within space it must appear as differential distance, and differential distance is *time*.

Energy has dimensions of time-distance^2, and space has dimensions of distance^3. *They share two of the same dimensions*. The only way energy can exist within space as additional volume is by a certain density of time per distance - the units of energy density. Thus the net effect of energy density is an increase in time per distance, and therefore any distance through space is extended in direct proportion to energy density. In the absence of energy, metric equivalence tells us the ratio of time to distance is unity. This is the speed of light in free space. Adding energy literally adds time per distance, effectively slowing light.

The relationship between extraspatial volume, absolute energy density, and refractive index follows as:

$$\Delta\varphi_V = \frac{\rho_E}{\rho_U} = n-1 \tag{7.12}$$

The unity portion of refractive index is the existential amount, by metric equivalence, of time in distance. Anything above unity is energy density's contribution.

One of the simplest forms of refraction occurs when light passes through a thin gas, since few intermolecular forces exist to complicate the process. It has the form:⟨5.1⟩

$$n-1 = \frac{\rho}{\rho_0}(n_0-1) \quad \{\rho \to 0,\ \rho_0 \to 0,\ \rho > \rho_0\} \tag{7.13}$$

where n_0 is the refractive index of the gas at density ρ_0. This is contrary to the idea of the continuous absorption-reemission cycle shown in Figure (7.7) because the number of particles along a photon's path through a thin gas is not proportional to density; it increases as its cube root. Equation (7.13) is more in keeping with the idea that energy increases space's volumetric density. This is shown by the substitution of Equation (7.12) into (7.13) with a slight rearrangement:

$$\frac{\rho_E}{\rho_{E_0}} = \frac{\rho}{\rho_0} = \frac{(n-1)}{(n_0-1)} \tag{7.14}$$

As expected from Equation (7.12), *light's slowing is proportional to the energy density through which it passes*.

Refraction is a complex process that varies with a photon's wavelength. This is because photons can only move freely through energy densities less than their own. Red light, for instance, will not pass through an atomic nucleus and therefore cannot reveal its incredible volumetric density. In general, however, the smaller a photon's wavelength, the greater its refractive index in a given substance because high-energy photons have enhanced access to high-energy matter fields, and therefore greater ratios of time to distance. Absorption and other effects obscure this process somewhat, but the study of refraction in low-density gases should provide compelling evidence of the physical geometry of matter described in Part III.

7.6 MASS AND MOMENTUM

The universe is built of space and energy. Its dimensional structure allows for no additional components. Although *mass* is a convenient way to express the astounding amount of energy that is present in matter, its units, as derived from $F = ma$ or $E = mc^2$, are time^3. This cubic has no existential significance. Mass in kilograms, like energy in joules, is a redundant and dimensionally inaccurate expression for energy. Force, in the existential sense, is the acceleration of energy, not mass, where acceleration is measured directly in units of curvature, 1/distance. It is certainly much easier, for instance, to describe the gravitational acceleration at Earth's surface as 9.8 m/s^2 rather than $1.1(10)^{-16}$ m^{-1}. This is why the kilogram is useful. But this doesn't make it dimensionally precise. Eventually, even the fact that meters and seconds are arbitrary units must be taken into account, so the purest measurement of 9.8 m/s^2 is $\sim 1.1(10)^{-12}$ m_a^{-1}, in absolute meters. However, transparency must occasionally be traded for utility, and meters, seconds, joules, and kilograms are excellent tools for describing most of the universe's contents. Mass, regardless of its existential shortcomings, is particularly well suited for astrophysical calculations. It is used extensively throughout our investigation of cosmology in Part IV.

MOMENTUM

Mass is certainly not the only metric anachronism. Momentum is a conserved vector quantity, and has units of distance-time^2. Since the universe only has a single dimension of time and mass has no dimensional relevance, momentum is more accurately expressed as the motion of energy, not mass. This leads directly back to space, as the product of energy (time-distance^2), and motion (distance/time), is distance^3, *volume*. The law of the conservation of momentum is in fact an expression of spatial conservation. Distance has an implicit component of time, so momentum is the *existential* amount of space in energy. When momentum components cancel, space is not destroyed; this merely represents a conversion between energy's static (matter) and dynamic (light) representations. Both contain the same existential amount of space per energy (1/1).

All of these redundant physical units arise from treating the many manifestations of time separately. Time as spatial deflection in electromagnetic fields appears entirely distinct from time as motion through space, and neither bear much of a resemblance to time as a gravitational spatial distortion. But they are all the same, and this is the essence of the unification that unit hypervolume represents.

The vector nature of momentum represents the vector (gradational) composition of spatial distance. Any distance **AB** is composed of two temporal differences, **AB** and **BA**. When energy moves through space, it is the confluence of two different expressions of the same thing - time. The simultaneous reality of these two expressions is *momentum*. It has units of distance^3, but it is not static:

Ψ THEOREM 7.10 - MOMENTUM {Ψ7.1}

MOMENTUM IS THE PRODUCT OF ENERGY AND VELOCITY

Energy's motion is an internal perspective of space, with units of volume. This is why momentum has a three-dimensional, conserved magnitude. A photon's momentum is scaled directly by its energy, redefined in accordance with (Ψ7.10):

$$\mathrm{p}_{\gamma} = E_{\gamma} c \tag{7.15}$$

Light is the most pristine expression of energy's motion.

In conventional physics, a photon's quantization is invariably expressed in terms of its energy:

$$E = h\nu \tag{7.16}$$

In actuality, however, a photon's *momentum* is quantized, not its energy, because (a) the product of its energy's motion and wavelength is quadric distance, unit hypervolume's proper dimensional composition; and (b) a photon's quantization could not exist in the absence of either its motion or its energy. *Neutral energy forms require a neutral boundary condition - unit hypervolume.* This is also why the movement of elementary particles is quantized in terms of their momentum:

$$\lambda = \frac{h}{m\mathrm{v}} \tag{7.17}$$

where λ is a particle's de Broglie wavelength. A particle's motion is quantized by its momentum, not its kinetic energy, because momentum is a more complete representation of space. Rewriting Equation (7.17) in terms of energy yields the dimensionally correct relationship for the quantization of a particle's motion:

$$\lambda = \frac{hc^2}{E\text{v}} \tag{7.18}$$

7.7 GENERAL CONCLUSIONS

Energy's substructure slowly comes into focus:

- Energy is three-dimensional with units of time-distance^2.

- The universe's average energy density, when measured in units of absolute time and distance, is unity.

- Energy density is the fourth-dimensional slope of space.

- All quanta are four-dimensional spatial deflections whose energy is distributed within their three-dimensional surfaces.

- Energy density expands the volume per volume of space with the infusion of its fourth-dimensional component.

- Refraction is a direct demonstration of energy's extraspatial volume.

- Photons are quantized by their momentum, not their energy.

Dimension is perhaps the simplest expression of the universe's uniform and unconditional reality. Space and energy are both three-dimensional quantities, which is why their universal ratio is finite. The factor setting them apart is a single dimension of time. It is the sole distinction between empty space and space with energy. Energy's dimensional composition is one of the most important discoveries of Part II, because it allows direct access to its governing framework. In a larger sense, energy as time-distance^2 could well be one of the most important discoveries in *Null Physics*.

8. UNIT HYPERVOLUME

8.1 UNIT HYPERVOLUME CALCULATION

The time has come to revisit our place on the cosmic size scale:

Quantal size is defined by the universe's four-dimensional size.

Atoms behave differently than solar systems because the universe is granular at a certain scale midway between infinite smallness and infinite largeness. Let's calculate this threshold.

The product of a photon's energy and wavelength is proportional to Planck's constant. This in turn is defined by the universe's four-dimensional size, first introduced by Equation (2.22) as *unit hypervolume*, $\Diamond_4$:

$$\Diamond_4 = \Lambda_\gamma \left(\frac{hc}{\rho_U} \right) = \Lambda_\gamma \left(\frac{E_\gamma \lambda}{\rho_U} \right)$$

where Λ_γ is the constant of proportionality identified earlier as the *photon hyperscaling factor*. Subsequent analysis served to limit the full scope of nonexistence to our solitary, majestic universe, so the more comprehensive definition of unit hypervolume is the *absolute magnitude of nothingness*, $|\aleph|$. We also discovered, as a result of energy's dimensional context, that momentum is the motion of energy, not mass, and that unit hypervolume quantizes a photon's momentum, not its energy.

Energy is a three-dimensional substance, so unit hypervolume can be expressed in units of a photon's momentum-distance as:

$$\Diamond_4 = \Lambda_\gamma hc^2 = \Lambda_\gamma E_\gamma c\lambda = \Lambda_\gamma \mathrm{p}_\gamma \lambda \qquad (8.1)$$

The universe's four-dimensional size can be equivalently described in units of quadric distance, as in Equation (2.22), or momentum-distance, as in Equation (8.1). Both are interchangeable forms of the same cosmic boundary condition.

Unit hypervolume can be calculated by deriving the photon hyperscaling factor, Λ_γ. This is a nontrivial consideration, as individual photons are elusive. Their external deflections and spatial extents cannot be measured directly. When they interact with matter they are absorbed, refracted, reflected, scattered, or in some cases unaffected, but in all cases their interaction is quantized. The only information available from electromagnetic experiments is the flash of individual photons or the statistical effects of legions. Nor is the current literature of much use. The widespread quantum reality of the last few decades has led physics far afield from the simple concept that a photon has a physical form with energy, volume, and therefore energy density distributed across said volume.

The photon hyperscaling factor of Equation (8.1) contains two intrinsic components:

- Universe/photon scaling
  A photon's hypervolume is governed by that of the universe, but it is not entirely clear that the two are equal. Photons consist of positive and negative spatial deflections - a pair of separate and equal hypervolumes. Is unit hypervolume representative of the entire photon or its individual lobes?

- Hypervolume/(momentum-wavelength) scaling
  The product of a photon's wavelength and momentum is related to the four-dimensional volume of its spatial deflection, which is in turn related to unit hypervolume. There is no guarantee that a photon's wavelength corresponds to its actual physical extent. Most physicists would claim it does not. Our derivation of quantum physicality (Ψ5.3) suggests otherwise, however.

Let's try to determine these two relationships, beginning with the ratio between a photon's hypervolume and $\lozenge_4$.

8.2 UNIVERSE/PHOTON SCALING

The best way to deduce a photon's hypervolume is to look at other examples of energy quantization. This is a brief search since unit hypervolume's only other application is subatomic matter. Like photons, particles are quantified by the universe's four-dimensional size. Unfortunately, the similarity ends there. Particles are either positive or negative, so their field's boundary condition is a relationship between space and *one axis of time*. The fourth dimension's positive and negative axes are immutably separate as is necessary for the neutrality of nonexistence, so an individual particle expresses only one-half of the true relationship between time and space.

Conversely, light's quantization defines an electrically neutral particle. How this is related to the universe's hypervolume depends on the meaning of a photon's *antiparticle*. The particle/antiparticle relationship is straightforward in matter and antimatter, because the formation of one has never been observed in the absence of the other. They are mirror images in the fourth dimension - two halves of a greater whole. The same is not true for photons. Single photons are routinely emitted by any number of sources, from radio antennas to hydrogen atoms. Similarly, they are also absorbed one at a time. *Neither photon creation nor destruction requires the presence of an antiparticle.* Individual photons, at least in terms of their genesis and obliteration, are complete entities. This is consistent with how they have been traditionally treated in contemporary physics, where they are considered to be *their own antiparticle*. It is, however, somewhat at odds with the Big Bang cosmology.

Photon number is conserved in the absence of emission/absorption, making it difficult to explain the presence of all of the photons in the pure electromagnetic energy thought to have been released by the Big Bang. Theoretical physicists have "resolved" this difficulty by using the concept of supersymmetry to give the photon a hypothetical *superpartner antiparticle* known as a *photino*, and space is allegedly inundated with them. Photinos have no actual physical or philosophical basis; they were just created to balance the Big Bang's initial photon production.

In reality, the only conceivable antiparticle a photon has is its own mirror image along the fourth dimension, but this has little similarity to the antiparticles of matter. A photon's antiparticle has to cancel momentum as well as field, so *it is only an antiparticle to a photon moving in a specific direction and orientation*. This is a far more precise relationship than that which exists between particles and antiparticles. The probability of annihilation when a proton and antiproton combine is virtually 100%, assuming that they don't have enough kinetic energy to create a host of other particles. The probability that a collision between two gamma rays will produce a particle and antiparticle is virtually nonexistent. Moreover, temporal symmetry means there can be no basic difference between a photon moving forward or backward through space. This in turn means a photon is inherently complete. It has no *structural* antiparticle.

In the final analysis, photons have temporal symmetry, and this is nonexistence's defining attribute. Photon hypervolume and unit hypervolume are one and the same. In many ways this is inevitable, because regardless of how a photon is formed, its continued existence represents an electrically neutral space-time boundary. There is only one such constant, and it is $\lozenge_4$.

The hypervolumetric boundary associated with photons is unit hypervolume, whereas elementary particles, because of their intrinsic charge, are bound by its polar half, *unit polarvolume*, $\lozenge_q$:

Ψ THEOREM 8.1 - HYPERVOLUMETRIC QUANTIZATION {Ψ3.3, Ψ7.1}
PHOTONS ARE QUANTIZED BY UNIT HYPERVOLUME AND PARTICLES ARE QUANTIZED BY UNIT POLARVOLUME, $\lozenge_q = \lozenge_4/2$

Unit polarvolume is the *totality of polarity*, half of the *totality of neutrality*. Unit hypervolume, photon hypervolume, and unit polarvolume are related as:

$$\lozenge_4 = \lozenge_\gamma = 2c\lozenge_q \tag{8.2}$$

where a factor of c has been added to relate quadric distance and time-distance[3]. Applying Equation (8.1) to Equation (8.2) yields:

$$\lozenge_\gamma = \Lambda_\gamma E_\gamma c\lambda = \Lambda_\gamma \mathrm{p}_\gamma \lambda \tag{8.3}$$

where the hyperscaling factor Λ_γ is the ratio between a photon's hypervolume and the product of its momentum and wavelength:

$$\Lambda_\gamma = \frac{\lozenge_\gamma}{E_\gamma c\lambda} = \frac{\lozenge_\gamma}{\mathrm{p}_\gamma \lambda} \tag{8.4}$$

Let's derive this factor from a photon's known and inferred properties.

8.3 HYPERVOLUME/(MOMENTUM-WAVELENGTH) SCALING

Finding the ratio between a photon's hypervolume and momentum-wavelength requires a general assessment of its four-dimensional shape. The assumptions used in this calculation, along with their justification, are as follows:

1) Photons are solitons. Solitons are waves that preserve their energy density over extended propagation distances. They avoid dissipation by maintaining a fixed shape during motion. In macroscopic applications, this typically occurs within a propagation medium of fixed cross-section, such as longitudinal sound waves moving down steel bars of constant diameter. These waves approximate solitons because steel is virtually elastic and no energy density is lost by the wave's transverse expansion since it is constrained by the bar's fixed diameter. Similarly, a photon can

travel light years in a given direction while maintaining constant energy density. They retain 100% of their structure during motion. *No attenuation occurs with distance so no radial curvature exists in a photon's substructure.* Thus:

a) In the absence of radial curvature, the fourth-dimensional slope on a photon's deflected surface is limited to only one of its dimensions. It has no curvature in any dimension normal to this because curvature along more than one direction in space denotes a curvature along *space*, which would represent attenuation into space. Since a photon experiences no loss of energy density over (cosmologically small but astronomically large) distances, no curvature exists perpendicular to its trajectory.

b) A photon's spatial footprint is a rectangular volume with edges parallel to its direction of motion. Its external deflection varies along its wavelength while its other two spatial extents are unknown but similar.

2) The properties of electromagnetic waves and macroscopic waves such as sound are eerily similar. In general, macroscopic waves are the result of the superposition of energy and matter. They in all likelihood bear more than a symbolic resemblance to their constituent material, suggesting that a photon's external deflection is sinusoidal. This profile is virtually universal to any number of different propagating waves, and is a mathematical representation of rotation. Since rotating systems (atoms, particles) emit photons over finite time intervals, this information is undoubtedly preserved in the form of a photon's sinusoidal field.

Both of the above assertions will collectively be referred to as the *photon hyperscaling hypothesis*:

Ω HYPOTHESIS 8.1 - PHOTON HYPERSCALING {Ψ7.5}

A PHOTON IS A SINUSOIDALLY DEFLECTED RECTANGULAR SPATIAL VOLUME WHOSE DEFLECTION ONLY VARIES ALONG ITS WAVELENGTH

In keeping with the treatment of hypotheses in our investigation, (Ω8.1) will not be referenced by future theorems or hypotheses. It will, however, be used for the purpose of the calculations in this section, resulting in a value of unit hypervolume with widespread application. If future analysis should prove the value derived in this section incorrect, (Ω8.1) is the source of the error. We are now prepared to calculate the ratio of a photon's hypervolume to the product of its momentum and wavelength.

PHOTON HYPERSCALING CALCULATION

A photon's hypervolume is the integral of its external deflection over volume:

$$\Diamond_\gamma = c \int_{V_\gamma} |t| dV_\gamma \tag{8.5}$$

where absolute magnitude is used because photons have equal and opposite deflections and Equation (8.5) represents their total hypervolume. The speed of light is used to convert from time-distance[3] to distance[4].

The volume of a photon's spatial footprint has the form:

$$V_\gamma = w_\gamma h_\gamma \lambda \tag{8.6}$$

where w_γ and h_γ are its width and height normal to its wavelength. Deflection is constant along both because all of a photon's deflectional variation (curvature) is confined to λ.

Let $t_{\max}$ be a photon's maximum external deflection. Since its deflection only varies along λ and is sinusoidal, Equation (8.5) can be evaluated using Equation (8.6):

$$\Diamond_\gamma = c \int_{V_\gamma} |t| dV_\gamma = t_{\max} w_\gamma h_\gamma c \int_{\lambda=0}^{2\pi} |\sin(\lambda)| d\lambda = 2t_{\max} w_\gamma h_\gamma c \int_{\lambda=0}^{\pi} \sin(\lambda) d\lambda = 4t_{\max} w_\gamma h_\gamma c \tag{8.7}$$

where the constant term ($t_{\max} w_\gamma h_\gamma$) can be pulled outside of the integral. Equation (8.7) is a photon's hypervolume as a function of its maximum deflection, width, and height. Now let's calculate the product of its momentum and wavelength in similar terms.

A photon's momentum is the product of the speed of light and the volume integral of the absolute magnitude of the fourth-dimensional slope distributed across its three-dimensional surface:

$$\mathrm{p}_\lambda = E_\gamma c = c \int_{V_\gamma} |t'| dV_\gamma = t_{\max} w_\gamma h_\gamma c \int_{\lambda=0}^{2\pi} |t'| d\lambda \tag{8.8}$$

where the term ($t_{\max} w_\gamma h_\gamma$) can again be pulled outside of the integral. The absolute magnitude function is used because energy is neutral and external slope changes polarity three times along a single sinusoidal cycle. Equation (8.8) can be evaluated by scaling the integral by an interval where fourth-dimensional slope is everywhere positive.

This is done by segmenting the interval of integration into four equal sections as follows:

$$\mathrm{p}_\gamma = t_{\max} w_\gamma h_\gamma c \int_{\lambda=0}^{2\pi} |t'| d\lambda = 4t_{\max} w_\gamma h_\gamma c \int_{\lambda=0}^{\frac{\pi}{2}} \cos(\lambda) d\lambda = 4t_{\max} w_\gamma h_\gamma c \qquad (8.9)$$

Not unexpectedly, it yields the same result as Equation (8.7), although now in three dimensions instead of four. The product of a photon's momentum and wavelength follows directly:

$$\mathrm{p}_\gamma \lambda = (4t_{\max} w_\gamma h_\gamma c)(2\pi) = 8\pi\, t_{\max} w_\gamma h_\gamma c \qquad (8.10)$$

The ratio of this product to a photon's hypervolume is Equation (8.10) divided by Equation (8.7):

$$\frac{\mathrm{p}_\gamma \lambda}{\lozenge_\gamma} = \frac{8\pi\, t_{\max} w_\gamma h_\gamma c}{4t_{\max} w_\gamma h_\gamma c} = 2\pi \qquad (8.11)$$

A slight rearrangement yields a photon's hypervolume directly in terms of its momentum and wavelength:

$$\lozenge_\gamma = \frac{\mathrm{p}_\gamma \lambda}{2\pi} \qquad (8.12)$$

This is the boundary condition of light's quantization.

The hyperscaling factor introduced in Chapter 2, Λ_γ, is obtained by substituting Equation (8.12) into Equation (8.4):

$$\Lambda_\gamma = \frac{1}{2\pi} \qquad (8.13)$$

This factor is shape-dependent. Performing the same calculation, treating a photon as a triangular deflection, results in a photon hyperscaling factor of $(1/8)$ instead of $(1/2\pi)$:

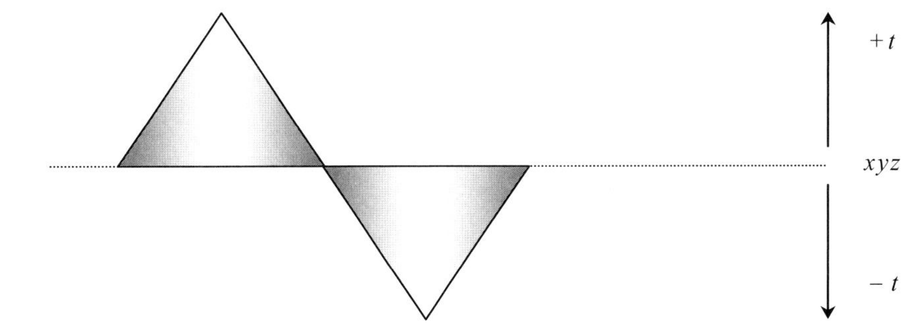

Figure (8.1) Non-sinusoidal photon topology yields a different photon hyperscaling factor

A photon's four-dimensional shape defines its ratio of hypervolume to (momentum-wavelength). The following shows a numerical verification of the hyperscaling factor with a comparison to the triangular case:

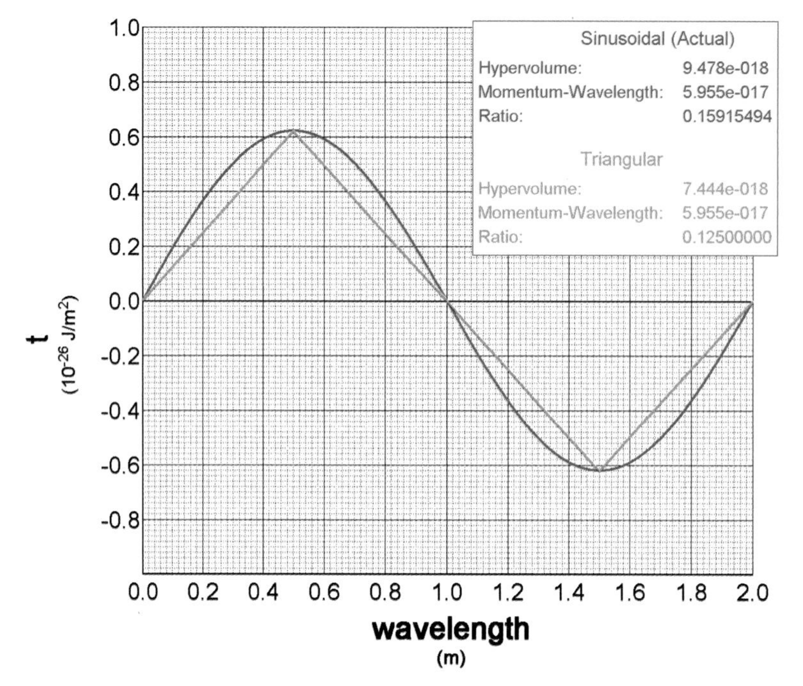

Figure (8.2) Numerical verification of the photon hyperscaling factor

Unit hypervolume can now be written directly in terms of Planck's constant:

$$\Diamond_4 = \Diamond_\gamma = \frac{p_\gamma \lambda}{2\pi} = \frac{E_\gamma c \lambda}{2\pi} = \frac{hc^2}{2\pi} = \hbar c^2 \tag{8.14}$$

in units of J-m^2/s, using the Planck relation $E_\gamma \lambda = hc$. When $\Diamond_4$ is expressed in units of energy, (J-m), instead of momentum, it is equal to the product of the speed of light and the most common representation of Planck's constant in quantum physics:

$$\Diamond_4 = \frac{hc}{2\pi} = \hbar c \tag{8.15}$$

This is no coincidence.

The photon hyperscaling calculation is not dependent on the relationship between a photon's width or height; indeed it is not even dependent on its actual physical wavelength. The only criterion is it varies sinusoidally along one dimension. If a photon were physically long, its energy density would decrease but its total energy would remain constant. The same is true of its hypervolume.

Unit polarvolume, the constant that defines matter's quantization, follows from Equations (8.2) and (8.14):

$$\Diamond_q = \frac{hc}{4\pi} \tag{8.16}$$

in units of J-m.

8.4 REALITY'S FOUR-DIMENSIONAL SIZE

Equation (8.14) yields the universe's four-dimensional volume as $9.478017(10)^{-18}$ J-m^2/s. This value has the accuracy of Planck's constant, which is far better than 1 ppm. It can be converted directly into quadric meters by dividing by universal energy density and the speed of light:

$$\Diamond_4 = \frac{\hbar c}{\rho_U} \tag{8.17}$$

Unfortunately, universal energy density is not known to great precision, but using an estimate of $4(10)^{-10}$ J/m^3 (as derived later in Part IV) results in a unit hypervolume of ~$7.9(10)^{-17}$ m^4 and an absolute meter of ~0.1 mm.

The SI units of distance, time, and energy all have absolute values defined directly by the universe's four-dimensional size, a size that can be expressed in a number of equivalent ways:

- Using absolute units of $s_a m_a^3$: 1 s_a-m_a^3.
- Using SI units of meter^4: ~$7.9(10)^{-17}$ m^4.
- Using SI units of second-meter^3: ~$2.6(10)^{-25}$ s-m^3.
- Using SI units of joule-meter^2/second: $9.478017(10)^{-18}$ J-m^2/s.
- Using SI units of joule-meter: $3.161526(10)^{-26}$ J-m.

As our development proceeds, the most commonly used form will be the last, in J-m, owing to its simplicity and the generally low accuracy of universal energy density measurements.

8.5 QUANTUM SUPERPOSITION

It might seem odd that individual quanta have sizes comparable to that of the universe. There are millions of photons and at least a few particles in every cubic meter of space. This amounts to $\sim\infty^3$ quanta. If photons have the four-dimensional volume of the universe and particles half of this, how can they all possibly fit within space? *They fit because they all contain finite amounts of energy.* If the universe were compressed into a hypercube, it would consist of solid quadric distance. Quanta have tiny amounts of three-dimensional energy distributed *on the surface of four-dimensional spatial deflections.* The exterior of each of these structures has a size equal to the universe (or half for particles), but is no more than an empty shell because the energy within it is finite, not infinite, and there is no energy or space in the four-dimensional interior of its deflection.

Just as the points of space are copies of totality, so too are the individual quanta scattered across it. The only difference is quanta represent a finite formulation of nothingness and points represent the infinite version. If all of space's points were superimposed on a single point, the result would be nothingness. Similarly, if all of the universe's quanta were superimposed into the same space-time, the result would be a single four-dimensional finite containing an infinite amount of space and energy: totality. Each and every quantum is an empty snapshot of totality's finite perspective:

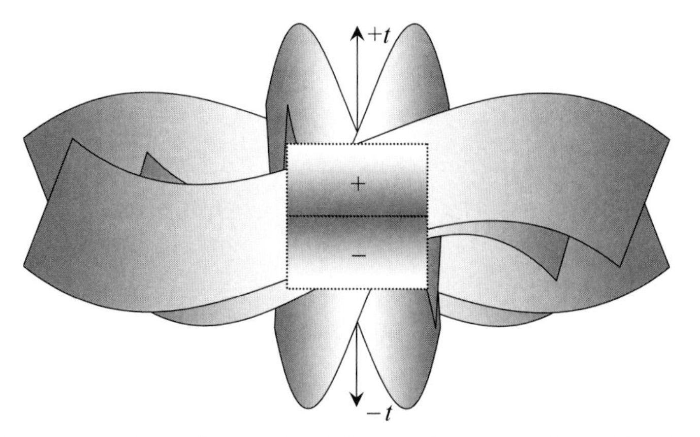

Figure (8.3) Infinite superposition of quanta is unit hypervolume, complete with infinite energy and space

Although quantal fields combine in space, the sum of their finite boundaries does not and can not exceed unit hypervolume. This is why it is not possible for particles to occupy the same space at the same time, regardless of the applied energy. To do so would exceed the four-dimensional size of the universe. Before moving on to matter, however, let's take one more look at a photon's topology.

8.6 POLARIZATION

Our derivation of a photon's hypervolume did not require a complete knowledge of the sizes of the three extents of its spatial footprint, but it is important to note that *they all exist.*

One is the photon's wavelength, and this defines its energy. This leaves two other basic photon parameters - width and height. The existence of intrinsic photon polarization suggests that the size of these two extents may not be the same. In fact, evidence suggests that they are quite different from each other. There are only two possible candidates for the basis of a photon's polarization:

- Electrical. **Premise**: A photon's positive and negative lobes are distributed perpendicular to its trajectory. **Discussion**: This would create an electrical polarity normal to a photon's motion, somewhat consistent with how polarization is currently interpreted in modern physics. The problem with this scenario is that the observed phenomenon only describes a perpendicular *axis*, not direction. Rotating a polarizer 180° produces no change in the amount of light energy it transmits. **Determination**: Unviable.

- Spatial. **Premise**: A photon's width and height normal to its trajectory are significantly different. **Discussion**: A photon's energy is distributed through space, but absorbed en masse, and refractive indexes suggest that absorption depends at least to some extent on energy density. If a photon's spatial distribution were constrained along one axis, it could align (or misalign) itself with a potential absorption agent, and this is concordant with the axial relationship that polarizers exhibit. What makes a polarizer a polarizer is the spatial distribution of its absorptive agents (molecules). A photon can only complement this by the spatial distribution of its own energy field. **Determination**: Viable. Moreover, polarizers' axial symmetry suggests that a photon's positive and negative lobes are in fact aligned *along* its trajectory. Any other orientation ought to leave a signature on its intrinsic polarization.

A photon's polarization demonstrates the asymmetrical shape of its spatial footprint perpendicular to its trajectory. Photons are fast and elusive, but not illusory. Their polarization only serves to provide further confirmation of their inescapable *realness*:

Ω HYPOTHESIS 8.2 - PHOTON POLARIZATION {Ψ7.5}

A PHOTON'S POLARIZATION IS THE RESULT OF AN ASYMMETRY BETWEEN THE SIZES OF ITS TWO SPATIAL EXTENTS PERPENDICULAR TO ITS TRAJECTORY

The ratio of a photon's spatial extents normal to its motion will be referred to as its *profile*, α_γ.

Axial symmetry is not the only clue to polarization's underlying nature. Just as significantly, it is *independent of wavelength*. The transmission of light through polarizing agents induces no chromatic aberration in the signal - all wavelengths pass with the same intensity reduction. *This tells us that a photon's profile is part of its essential character.* It is difficult to imagine that radio waves and gamma rays have the same profile, given the enormously different processes from which they arise, but in the final analysis all photons originate from the fields of matter, and all of these fields are defined by the single constant of unit polarvolume. Perhaps the more puzzling case would be if photons' profiles varied with their energy.

Randomly polarized light passing through two polarizers aligned in parallel at 0° loses about 50% of its intensity. If these same polarizers are oriented 90° to each other, a situation referred to as *crossed polarizers*, they block virtually all light. The ratio between the light intensity passing through polarizers oriented at 90° and 0° is called their *extinction factor*, and it can exceed one part in 10^5 in precision microscopy equipment. This provides some insight about the magnitude of light's spatial asymmetry. A photon's absorption by or transmission through a polarizer depends predominantly on its spatial orientation. The greatest variation this orientation can achieve is the ratio of its minimum to maximum extents. For example, if photons had profiles of 100:1, then the energy they have in their worst orientation with a polarizer is 1, and in the best case, 100. This provides a total transmission range of 100 - two orders of magnitude. Thus the maximum extinction factor, χ_γ, of crossed polarizers ought to be on the order of a photon's profile:

$$\chi_{\gamma_{\max}} \approx \alpha_\gamma = \frac{h_\gamma}{w_\gamma} \tag{8.18}$$

where a photon's height, h_γ, will be defined as its maximum extent normal to its trajectory. *If photons had profiles of 1:1, polarization would not exist.*

Extinction factors in excess of 10^6 have been demonstrated, so a photon's profile is at least $\sim 10^6$, and perhaps a great deal higher. Even so, what can be said with a great deal of certainty is that it is *bounded*. A photon cannot have a finite spatial volume as a result of infinitely small width, infinitely large height, and finite wavelength. Even a single infinite spatial extent would make it *a photon without a center*, and this is not physically viable. Photons transfer energy between particles and groups of particles, one location to the next - a process that is discrete in all three spatial dimensions.

BLACKBODY GEOMETRY

Polarization isn't the only clue to a photon's spatial geometry. Thermal radiation is a precise balance between photon energy and population density, and as such it also provides a glimpse of their three-dimensional footprints. The number of photons that can be squeezed into a given spatial region is limited by the extent to which they superimpose, which is in turn governed by their individual volumetric energy distributions. As shown in Appendix L, the relationship between average photon volume and *any* characteristic thermal wavelength, such as the location of peak energy density, has the form:

$$\overline{V}_{\gamma} \propto \lambda^{3} \tag{8.19}$$

This demonstrates that all of a photon's spatial extents scale to its wavelength, and is consistent with the invariance of a photon's profile. With this in mind, the constant ratio between a photon's wavelength and height will be called its *scale*:

$$\beta_{\gamma} = \frac{h_{\gamma}}{\lambda} \tag{8.20}$$

Scale is a useful way of characterizing the relationship between a photon's energy and maximum spatial extent, and has cosmological implications that are derived in Part IV.

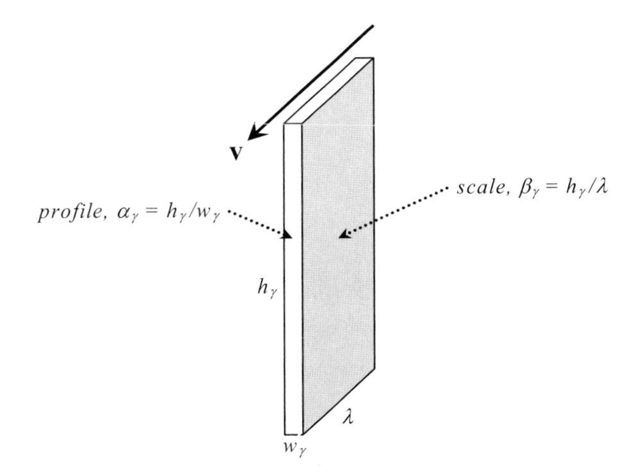

Figure (8.4) Photonic extents and proportions

8.7 GENERAL CONCLUSIONS

Energy's foundation begins to emerge:

- Light is quantized by unit hypervolume, expressed as the product of its momentum and wavelength.

- Matter is quantized by unit polarvolume, half of unit hypervolume.

- The universe's four-dimensional size is equal to $3.161526(10)^{-26}$ J-m.

- A photon's three-dimensional footprint is highly asymmetric and is responsible for its intrinsic polarization.

No experiment can demonstrate that the quantization of light originates from the universe's four-dimensional size. Nor is it possible to observe energy's true dimensional units. There have always been places that our instruments cannot reach, but *Null Physics* gives us the first chance to gain a deep understanding of the places they *can* reach. And it doesn't stop there. Its approach provides a logical connection throughout all levels of scale - a connection unfettered by observational limitations of any kind. Scientists were working with energy long before Galileo; *Null Physics finally shows us what it actually is and explains why it behaves the way it does.*

The good news is the discovery of unit hypervolume and energy's dimensional structure are transcendent advances for physical science. The bad news is in the process the lowly electron has gone from a quark-less nonentity of small mass and no volume to a four-dimensional topology problem. It is no wonder particle physicists need dozens of constants in matter's standard model. The universe's geometric fabric is, at the smallest level, *dynamic, data intensive, and four-dimensional*. It is rational; it is perfect; but it contains torrents of information in every bit of space.

Physicists have dreamed of a final theory for over a century. Most feel it is out of reach, but for the wrong reasons. It is not the enigma of existence standing in our way - far from it. The Null Axiom provides an immediate and comprehensive solution. Nor is it the complexity of its underlying fabric. It could not be simpler. The problem is even the simplest system can have an infinite amount of data. There is no magic number; there is no set of equations connecting all the dots. Those notions are hopelessly naïve. The complexity plaguing math's description of nature is not a nuance of combinatorial interaction; it reflects the amount of information within the tiniest specks of the stuff everything is made out of: *energy*. Its discrete and ordered quantization is just a cruel joke on theorists in search of analytical solutions. Planck's constant is a perfectly sharp boundary drawn in an otherwise seamless flow of twisting, extraspatial geometry.

With supercomputer simulations and the development of better four-dimensional mathematics, there will come a day when the workings of individual particles and atoms can be calculated purely from their topology to arbitrarily fine resolution. Physics will never be entirely finished, but at least now there is hope of comprehension. The dream of living in a rational universe has arrived. With it, the dream of living in a quantifiable universe has taken the next giant step forward.

PART III: PHYSICS OF MATTER
DOWN TO THE CORE

Matter's structure and function

- *Unit polarvolume distribution*
- *Differential geometry of elementary particles*
- *Particle core regions*
- *Particle core recession and the nuclear environment*
- *Bound electrons*
- *Electron and neutron degeneracy*
- *Coulomb, Strong, Weak, and gravitational forces*
- *Particle core expansion*
- *The unification of the Strong and gravitational force in black holes*
- *Gravitational veneer*

All our science, measured against reality, is primitive and childlike - and yet it is the most precious thing we have.

Albert Einstein, 1879-1955

Figure (III) Supernova 1987A
(Courtesy NASA/Hubble Heritage Project)

Prerequisite Concepts for Part III

The following concepts are carried forward from Parts I and II:

- Unit hypervolume represents the finite four-dimensional size of the universe's unbounded three-dimensional space. It is a symmetric hypercube whose size has units of distance^4. The length of one of its edges is referred to as an absolute meter and is on the order of a fraction of a millimeter.

- The product of the geometric linear resolution of space (netherspace, δ_3) and its total spatial volume (V_U) is unit hypervolume.

- Light is quantized by unit hypervolume; matter is quantized by half of unit hypervolume - unit polarvolume.

- Time is the difference of space. The speed of light is not a universal constant; it is the dimensional equivalence between space and time.

- Time has two physical manifestations, motion and electrical charge.

- Space is as real as energy or anything else.

- The ultimate source of a particle's interaction potential is its own rest energy.

- The fundamental unit of energy is time-distance^2.

- Energy density is the fourth-dimensional slope of space.

- Particles, photons, and neutrinos are four-dimensional deflections of space whose intrinsic energy is the integrated sum of the fourth-dimensional slope of their curved, three-dimensional surfaces.

9. MATTER'S QUANTIZATION

9.1 CHARGED FIELDS

The universe's size governs the quantization of all energy forms, as no other universal boundary condition exists. In light, it takes the form of Planck's constant. In matter, it is the reason *all elementary particles have the same magnitude of unit charge.*

COULOMB STRENGTH

Charge is responsible for the Coulomb force:

$$F_q = \frac{q^2}{4\pi\varepsilon_0 r^2} \tag{9.1}$$

where ε_0 is a constant known as the permittivity of free space. The Coulomb strength term:

$$\|F_q\| = \frac{q^2}{\varepsilon_0} \tag{9.2}$$

is fixed, four-dimensional, and equal to $2.9(10)^{-27}$ J-m. This is ~18% of the size of unit polarvolume *and has the same units.* Just as Planck's constant is the four-dimensional quantization of photons, elementary charge is the four-dimensional quantization of particles.

In contemporary physics, the ratio of Coulomb strength to Planck's constant is called the *fine structure constant*:

$$\alpha = \frac{\left(\dfrac{q^2}{4\pi\varepsilon_0}\right)}{\hbar c} = \left(\frac{q^2}{2\varepsilon_0 hc}\right) \cong \frac{1}{137} \tag{9.3}$$

Since its numerator originates from unit polarvolume and its denominator *is* unit hypervolume (in units of J-m), one might think its value should be 1/2. The reason this is not the case is because unlike Planck's constant, the Coulomb force's strength is not a direct assessment of unit polarvolume. It is a *by-product* of unit polarvolume just as electrostatic potential energy is a by-product of rest energy.

DISTRIBUTED POLARVOLUME

Unit hypervolume is quadric space, the difference of space-time:

$$\Diamond_4 = \Diamond_q - \Diamond_q = Vt - Vt \quad (9.4)$$

Unit polarvolume lacks unit hypervolume's *completeness*. Whereas a photon's hypervolume is geometrically *closed*, like totality, a particle's unit polarvolume is *open*. Their net polarity prevents the spatial encapsulation of their deflection. Unit polarvolume provides a *distributional* boundary condition - a family of relationships of the form:

$$t = \frac{\Diamond_q}{V} \quad (9.5)$$

The magnitude of matter's external deflection is inversely proportional to the volume it deflects.

Particles and photons have radically different properties because they express their hypervolumetric boundaries in two entirely different ways. Photons are the encapsulation of time by space; particle fields are the encapsulation of space by time:

Ψ THEOREM 9.1 - PARTICLE FIELD {Ψ8.1}

A PARTICLE IS A RADIALLY SYMMETRIC DISTRIBUTION OF EXTERNAL DEFLECTION GOVERNED BY UNIT POLARVOLUME, time = $\Diamond_q / volume$

Space is continuous and causal. The reason why a particle's central deflection diminishes with distance is because external deflection is governed by a dimensional relationship between time and space and this relationship is *constant*. An increase in distance (volume) requires a decrease in deflection. Spatial deflection *attenuates* due to increasing volume because spatial volume *causes* this attenuation.

Herein lies the meaning of *charge*. The Coulomb force increases with the amount of external deflection, but it is not simply a case of a maximum deflection somewhere in space. Charge is the *combination* of deflection and volume, and this determines the deflection at any given range.

If it were possible to see a particle's actual four-dimensional topology, it would look like an inverted funnel with spatial deflection increasing rapidly toward its center:

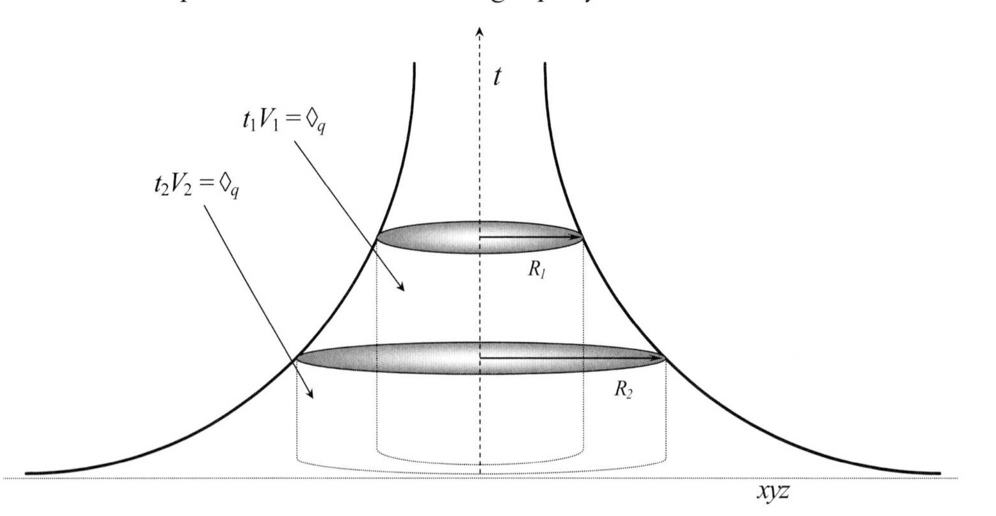

Figure (9.1) A particle is a family of concentric spherical volumes with differing fourth-dimensional elevations

The particle distribution consists of an infinite number of concentric spherical surfaces with fourth-dimensional elevations scaled by their volume and unit polarvolume. These surfaces will be called *isoexternal*, as their external deflection has a uniform magnitude over their entire spatial area. The difference between charged space and neutral space is fourth-dimensional elevation.

9.2 EXTREME DEFLECTION

Particles are radially symmetrical, so the volume element of their field distribution is a sphere. From Equation (9.5):

$$t = \frac{3\Diamond_q}{4\pi r^3} \tag{9.6}$$

External deflection decreases as the cube of the distance from a particle's center. Whereas three-dimensional deflections decrease as the square of distance, attenuated by increasing area, fourth-dimensional deflections decrease as the cube of distance, attenuated by increasing volume.

Unit polarvolume was derived in the previous chapter in Equation (8.16):

$$\Diamond_q = \frac{hc}{4\pi}$$

Substituting this into Equation (9.6) yields a particle field's external deflection expressed directly in terms of Planck's constant:

$$t = \frac{3hc}{16\pi^2 r^3} \tag{9.7}$$

with units of energy/distance^2. This can be converted to distance by applying the universe's average energy density:

$$t = \frac{3hc}{16\rho_U \pi^2 r^3} \tag{9.8}$$

As calculated later in Part IV, $\rho_U \cong 4.0(10)^{-10}$ J/m^3. Substituting this and the other constants into Equation (9.8) results in a particle field's actual physical displacement:

$$\Delta_4 \cong \frac{9.4(10)^{-18}}{r^3} \tag{9.9}$$

in meters, accurate to the resolution of the universe's average energy density. A proton's electrostatic field is active at a radius as close as 1 Fermi from its center (a Fermi is a measurement of nuclear distance equal to 10^{-15} m, about 100,000 times smaller than an atom). Substituting this into Equation (9.9) yields an external deflection of $9.4(10)^{27}$ m. *This is a distance of a trillion light years, far beyond the edge of the observable universe.* Deflections of this magnitude are possible because although c is energy's speed limit through space, there is no limit to space's speed through time. The difference between matter and empty space is matter is space that is deflected into time - *far into time.* Einstein described matter as curved, empty space, in reference to its gravitational field. This is absolutely true, but its gravitational field constitutes an insignificant fraction of its total curvature. Matter is slightly curved in space but *spectacularly curved in time.* For more detailed information about matter's *cubic distribution* ($1/r^3$), please refer to Appendix F.

9.3 FIELD ENERGY DENSITY

In accordance with Equation (7.5), a particle field's energy density is defined by the fourth-dimensional slope of its deflected surface. This is given by the first derivative of the external deflection of Equation (9.6):

$$\rho_E(r) = |t'| = \frac{9\Diamond_q}{4\pi r^4} \tag{9.10}$$

The external deflection produced by unit polarvolume can be enormous, and so can its associated energy density. Evaluating Equation (9.10) at a range of one Fermi yields an energy density of $1.13(10)^{34}$ J/m^3, comparable to but less than the energy density of atomic nuclei ($\sim 10^{35}$ J/m^3).⟨18.2⟩

Unit polarvolume is responsible for electrostatic fields, so there ought to be at least a rough correlation between Equation (9.10)'s energy density and the energy density of space charge. There is. The strength of the electric field generated by a unit elementary charge in free space is given by:

$$|\vec{\mathrm{E}}| = \frac{q}{4\pi\varepsilon_0 r^2} \tag{9.11}$$

This field's energy density is proportional to the square of its strength:

$$\rho_{E_q} = \left(\frac{\varepsilon_0}{2}\right)|\vec{\mathrm{E}}|^2 = \left(\frac{\varepsilon_0}{2}\right)\left(\frac{q}{4\pi\varepsilon_0 r^2}\right)^2 = \frac{q^2}{32\pi^2\varepsilon_0 r^4} \tag{9.12}$$

The ratio between this and a particle field's intrinsic energy density is constant and is equal to Equation (9.12) divided by Equation (9.10):

$$f_{\rho_E} = \frac{\rho_{E_q}}{\rho_E} = \frac{q^2}{72\pi\varepsilon_0 \Diamond_q} \cong \frac{1}{1200} \tag{9.13}$$

What this shows is the *effect of external deflection is proportional to its magnitude*, and the energy density of its effect is only a small fraction of the energy density of the underlying field. This is to be expected since an electric field is the consequence, not composition, of a particle's topology. It is the difference between electrostatic energy and *rest energy*. This difference was noted earlier in Equation (9.3), but it can now be quantified using Equation (9.13).

9.4 REST ENERGY AND THE PARTICLE CORE

A particle's rest energy, $E_\wedge$, is the integral of the energy density distributed across its deflected surface:

$$E_\wedge = \int_V \rho_E \, dV = \int_V |t'| \, dV \tag{9.14}$$

where energy density is the magnitude of external slope.

External slope is given by Equation (9.10) and a spherically symmetrical volume differential dV is $(4\pi r^2 dr)$. Applying these provisions to Equation (9.14) yields a particle field's total energy from some radius R to infinity:

$$E_{\wedge}(R) = \int_V |t'| dV = \int_{r=R}^{\infty} \left(\frac{9\Diamond_q}{4\pi r^4} \right) 4\pi r^2 dr = \int_{r=R}^{\infty} \left(\frac{9\Diamond_q}{r^2} \right) dr = \frac{9\Diamond_q}{R} \qquad (9.15)$$

Since particles have:

- Finite energy.
- Radially symmetric electrostatic fields of infinite range.
- External deflection decreasing as the cube of distance from their centers, defined by unit polarvolume.

They also have *finite radii*:

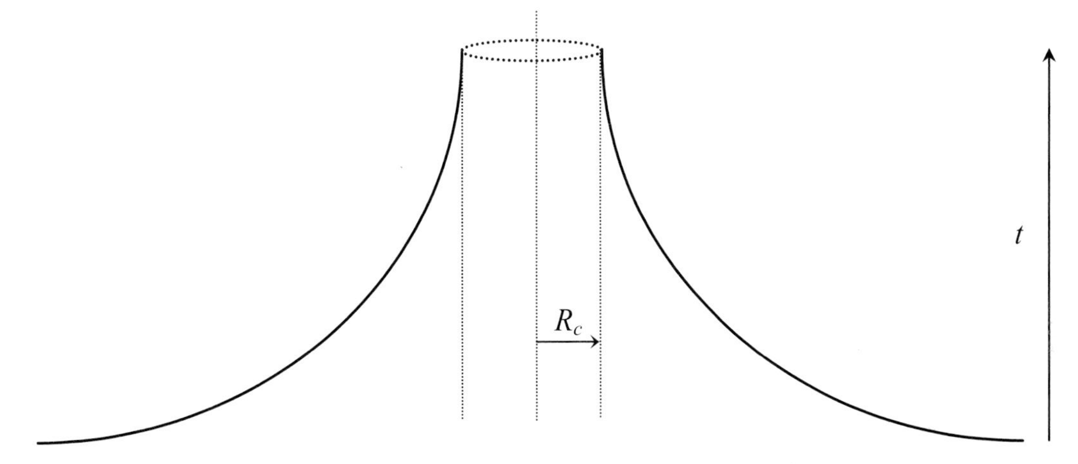

Figure (9.2) Finite rest energy defines a finite core radius

This will be referred to as a particle's *core*. It is the ultimate source of matter's long-range external deflection field.

Unit polarvolume is a *boundary condition*, and as such it represents a physical spatial boundary. The only way to symmetrically bound space along three dimensions is by physically removing it. *The volume within the cores of particles is not in our universe.* Space's four-dimensional size is by definition exterior to it. Within three-dimensional space, this four-dimensional exterior appears as the deflection of a spatial void - a resolute, four-dimensional vacancy. Particle cores contain no energy, deflection, *or space*.

They will be referred to as *hollow* to denote a greater degree of emptiness than vacuum:

Ψ THEOREM 9.2 - PARTICLE CORE {Ψ7.5, Ψ9.1}

ELEMENTARY PARTICLES HAVE A HOLLOW CORE REGION OF FINITE RADIUS THAT IS DEVOID OF SPACE AND REST ENERGY

A particle's core radius is given from its rest energy by a slight rearrangement and interpretation of Equation (9.15):

$$R_c = \frac{9\lozenge_q}{E_\wedge} = \frac{9\lozenge_q}{m_0 c^2} \tag{9.16}$$

In the case of a proton, whose rest energy is $1.5(10)^{-10}$ J, Equation (9.16) yields a core radius of 0.9464 Fermi, consistent with the scale of atomic nuclei. An electron, at ~1836 times lighter, has a correspondingly larger core radius of 1738 F, or about a thousandth of a nanometer. It is so small that it plays virtually no part in most atomic interactions.

9.5 PARTICLE CORE PROPERTIES

Particles' cores are not only devoid of their own field energy and space, they are completely and immutably empty of everything *from any source*. This is a natural consequence of their voided nature. No space means no energy from the fields of other particles or photons. This supremely vacant state will be referred to as *core purity*:

Ψ THEOREM 9.3 - PARTICLE CORE> PURITY {Ψ9.2}

PARTICLE CORES ARE DEVOID OF SPACE AND ENERGY DENSITY FROM ANY SOURCE

A particle's core is surrounded by an inconceivably large fourth-dimensional deflection, but contains not the slightest trace of energy or space. It is the only pristinely empty region there is. This means that the boundary between a particle's core and the rest of the universe is a singularity, like unit polarvolume itself:

Ψ THEOREM 9.4 - PARTICLE CORE> DISCONTINUITY {Ψ9.2}

A PARTICLE CORE BOUNDARY IS A DISCONTINUOUS STEP BETWEEN A HOLLOW VOID AND SPACE FILLED WITH A PARTICLE'S MAXIMUM ENERGY DENSITY

Maximum energy density is noted in (Ψ9.4) because the core boundary exists at a particle's minimum radius and therefore maximum external slope. *A core is a charged, submicroscopic hole in space and its surface is a physical spatial boundary.*

Reality only consists of a single universe, so it might seem odd to think of a particle's core (or any external deflection for that matter) as *outside* the universe. Since no other universes exist, there is literally no room for any region outside of space. What a particle core provides, like any other formulation of unit hypervolume, is direct access to our universe's *exterior* region. This exterior is closed in an existential singularity, and its surface is three-dimensional, but it is in fact exterior to space. The lack of additional universes doesn't imply that our universe has no exterior region - all it means is that there is, quite literally, nothing beyond this exterior.

CORE INTEGRITY

The only finite boundary in space's otherwise seamless continuum is the universe's four-dimensional size. Particles represent precisely half of this, so cores simply cannot intersect. If they could overlap, *boundary values other than unit polarvolume would exist*. Particle cores are the unblemished repositories of space-time's closure. Unit polarvolume's singularity demands an impenetrable particle core, a property that will be referred to as *integrity*:

Ψ THEOREM 9.5 - PARTICLE CORE> INTEGRITY {Ψ2.12, Ψ9.2}
PARTICLE CORES DO NOT OVERLAP

Core integrity is why, regardless of how much energy elementary particles acquire in our powerful accelerators or deep space, *they can't pass through each other*. The more energy colliding particles have, the more violent their collision becomes, creating progressively greater numbers of protons, antiprotons, mesons, and antimesons. No amount of energy can breach a particle's core boundary because it is part of reality's dimensional substructure. This is the validation for the adage *matter cannot occupy the same space at the same time*. There is one exception, however. Particle core boundaries collapse during matter-antimatter annihilation, producing two photons. *Balanced core cancellation is the only case that allows the release of trapped field energy and resolution of voided space.*

ISOEXTERNAL BOUNDARY

The external deflection at any point in space is the cumulative result of every field in the universe. This is also true of the deflection at a particle's core boundary, but this boundary has the unique distinction of being discontinuous. It is the connection between a core and the remainder of the universe, and it is also the source for the overwhelming majority of the boundary deflection. A particle core is a one-to-one relationship between external deflection and a volume of evacuated space. Variation of the magnitude of external deflection across its boundary would represent more than one relationship between space and time, so *external deflection is constant for all points on the surface of a particle's core boundary:*

Ψ THEOREM 9.6 - PARTICLE CORE> ISOEXTERNAL BOUNDARY {Ψ9.4}

PARTICLE CORE BOUNDARIES ARE SINGULAR IN SPACE AND TIME, WITH CONSTANT BOUNDARY DEFLECTION IRRESPECTIVE OF THE PRESENCE OF AMBIENT FIELDS

This is an extraordinarily stringent requirement. Superposition of exterior fields modulates a core boundary's peak deflection, so the core has to change shape in order for its deflection to remain uniform across its surface. Please see Appendix F for further discussion of the relationship between the shape of a particle's core and the field it generates.

The next step in our analysis is to move inside a particle's boundary for a closer look at its hollow core.

9.6 EVACUATED SPACE

A particle's core is a *spatial bubble* whose formation involves the internal displacement of the space it occupies. It is not a case of the core's innermost volume pushing the surrounding volume across a core boundary:

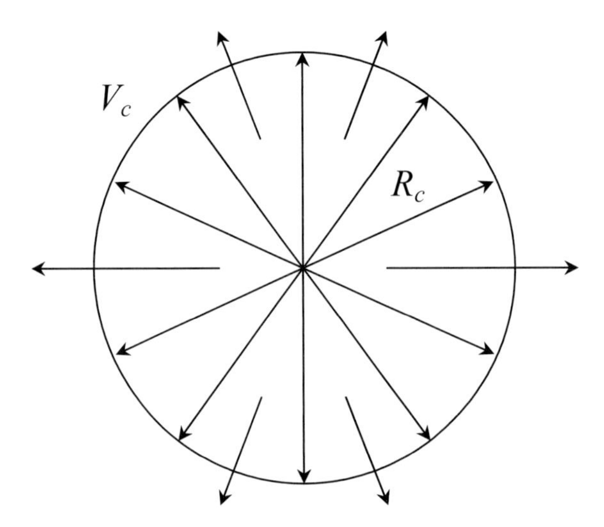

Figure (9.3) Inner volume *cannot* displace surrounding core volume across core boundary as shown

The displacement shown in Figure (9.3) is not viable, as *core volume pushed across a core boundary would violate its discontinuity with the presence of internal deflection.* There can be no connectivity between a core's interior and exterior. If there were, the core's boundary would not be spatially discontinuous, *and discontinuity is the sole requirement for its status as a spatial boundary.*

The only way to evacuate the space from a particle's core while preserving the integrity of its boundary is to transfer its interior volume directly *into* its boundary:

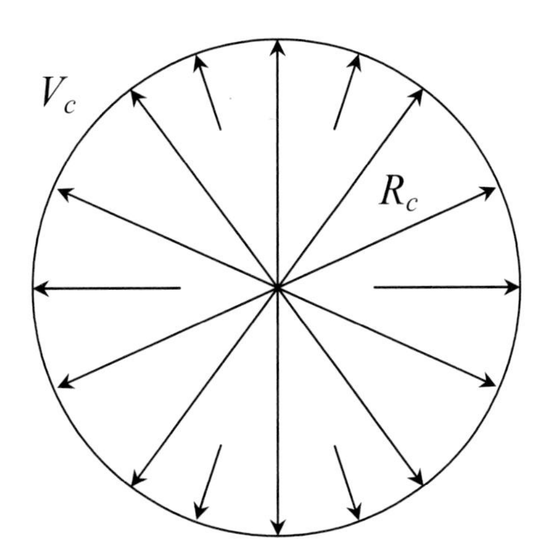

Figure (9.4) A core is evacuated when its interior volume is displaced into its boundary

Although a particle core's surface is only a spherical *area* in the third dimension, it is a three-dimensional *volume* in the fourth, as the product of spatial area and external deflection. This region has more than enough room to accommodate a particle's tiny core volume. When a core is evacuated, its interior space is infused directly into its four-dimensional boundary. Keep in mind, however, that externally deflected space must always adhere to the unit polarvolume criteria. The internal deflection that spatially depopulates a particle's core is perpendicular to and *independent* of external deflection. Since it references volumes within and therefore smaller than the core boundary, it necessarily defines external deflections greater than a particle's peak boundary deflection. What this means is that a particle's unit polarvolume distribution actually exists all the way to the center of its core, but *is transported into its core boundary by core evacuation.*

The space transferred into a core's boundary by evacuation has the unique distinction of existing parallel to the closure dimension, populating it with compositional points. It will be referred to as *degenerate space.* The peak external deflection of the normal space at a particle's core boundary is finite, but the peak deflection of its degenerate space is infinite, *because it corresponds to the unit polarvolume deflection of the particle's center* ($r = 0$). Finite particle energy requires a finite hollow core region. This doesn't occur by removing part of a particle distribution - it occurs by expanding the center of a particle distribution of infinite

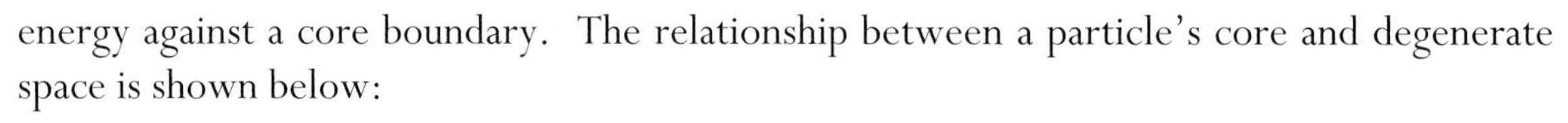

energy against a core boundary. The relationship between a particle's core and degenerate space is shown below:

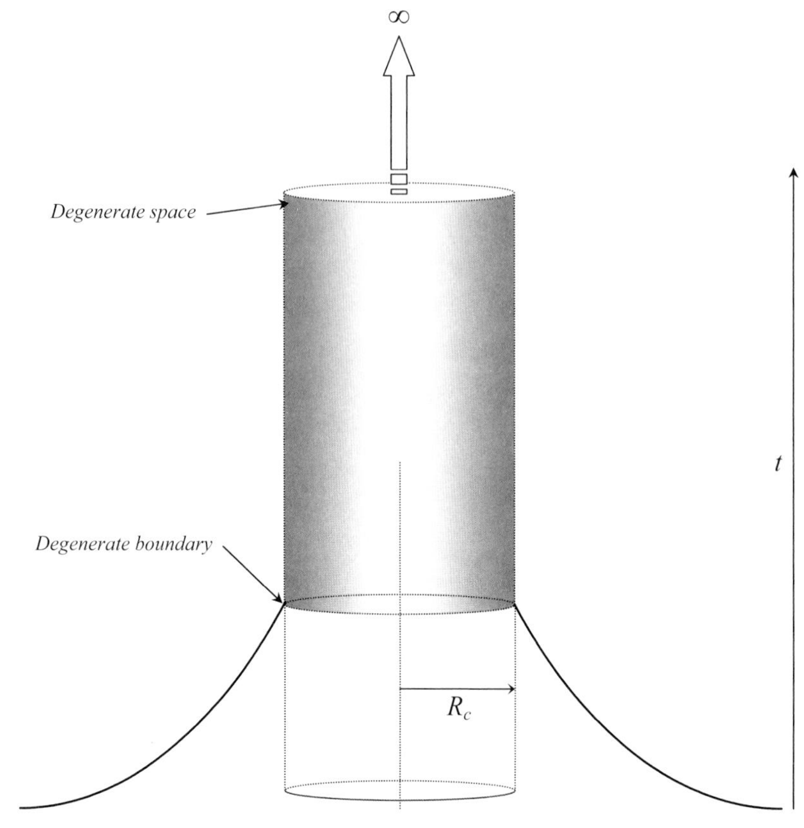

Figure (9.5) Degenerate space is shrink-wrapped against the inner surface of an extended particle core boundary

The boundary between degenerate and normal space will be called the *degenerate boundary*. It is located on a particle's core boundary at an elevation equal to the peak finite external deflection associated with its radius.

Spatial evacuation is responsible for achieving finite rest energy while maintaining a unit polarvolume distribution for the space that used to occupy a particle's core. A proton's energy field might begin at a fourth-dimensional elevation of a trillion light years, but its core boundary actually stretches to infinity. Ultimately, singularity *requires* infinity. A particle core can be portrayed as the ejection of *not-space* far into the fourth dimension, but this is an incomplete perspective of its singularity. The unit polarvolume distribution applies to space, not its absence. The formation of a spatial bubble only serves to reduce the energy of a *complete* (to $r = 0$) unit polarvolume distribution - its deflection profile is invariant.

Since a particle's degenerate surface extends to infinite height and is populated by a finite core volume, its spatial density becomes infinitely low at high fourth-dimensional elevation. This is consistent with the hollow core within it. The greater the deflection of degenerate space, the lower its associated spatial density, until it smoothly attains the spatial density of the core itself - zero. Think of a particle's core as a particle of zero radius and infinite energy hollowed out by its internal field, producing finite rest energy:

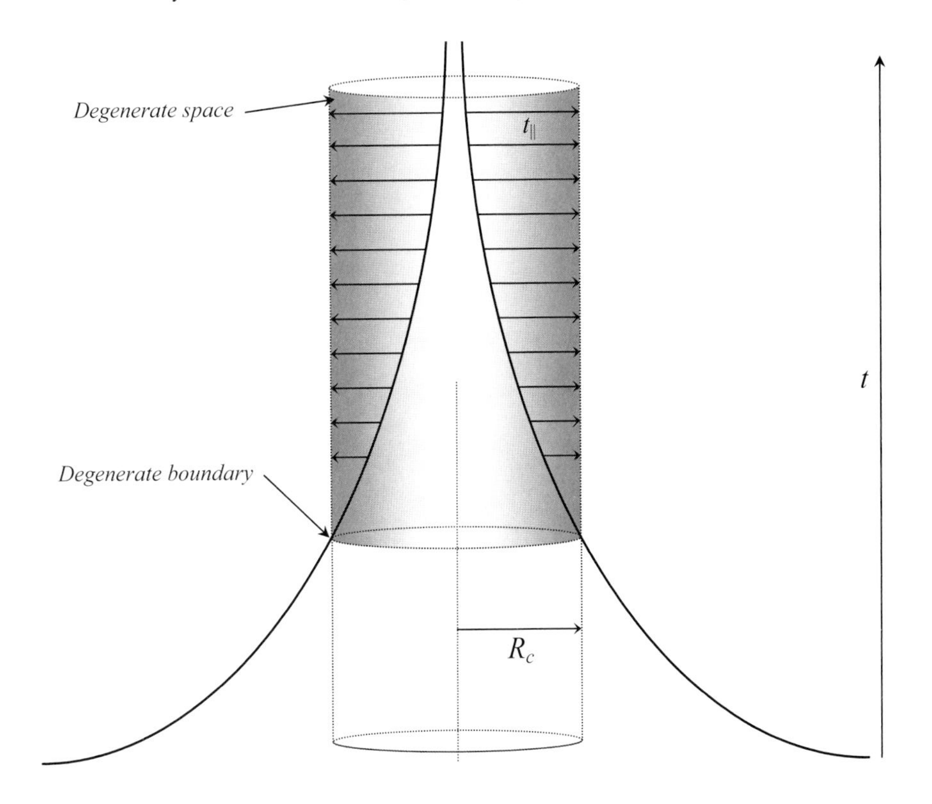

Figure (9.6) A particle core's internal field is the mechanism that truncates its field energy

This concept will be called the *unified particle core*:

Ψ THEOREM 9.7 - UNIFIED PARTICLE CORE {Ψ9.3, Ψ9.4}

A PARTICLE'S CORE IS A UNIT POLARVOLUME DISTRIBUTION OF INFINITE AMPLITUDE WHOSE CENTRAL CORE VOLUME IS COLLAPSED AGAINST ITS CORE BOUNDARY BY THE INTRINSIC INTERNAL DEFLECTION OF SPATIAL EVACUATION

An internal field is a particle's boundary condition for finite energy. Its core size provides the basis of this energy but the internal field is the agent by which it is facilitated. A core's evacuated volume *sheaths* its boundary in space - a dual singularity of internal and external deflection.

The unit polarvolume distribution is limited to the relationship between space and deflection normal to space. Since this is invariant under internal spatial displacement, spatial density anywhere on a particle's core boundary is defined by the absolute fourth-dimensional slope of its fourth-dimensional elevation:

$$\varphi_3 = \frac{1}{t'_a} = \frac{4\pi\rho_U R^4}{9\Diamond_q} = \left(\frac{4\pi\rho_U}{9\Diamond_q}\right)\left(\frac{3\Diamond_q}{4\pi\rho_U t}\right)^{\frac{4}{3}} \tag{9.17}$$

where φ_3 is space per unit volume and fourth-dimensional slope is subscripted with an a to denote absolute units of (distance/distance). This differs from the volumetric density of space introduced in Equation (7.8), φ_V, because volumetric density assumes a unity baseline spatial density. Note that the spatial density of Equation (9.17) goes to zero at infinite external deflection (t) on a particle's core boundary, where it merges seamlessly with its interior hollowness.

CORE BOUNDARY ENERGY

Degenerate space has the unique distinction of following a unit polarvolume distribution *along space*. It is stripped of nearly all of its composition and stretched to the absolute limit of geometric continuity. In normal space, fourth-dimensional slope is the ratio of *the change in fourth-dimensional deflection to the change in distance*. In degenerate space, both of these are parallel to space, so its fourth-dimensional slope is by definition unity and its deflection is by definition internal:

$$t'_{\parallel} = \frac{dt_{\parallel}}{dr} = 1 \tag{9.18}$$

Energy is composed of space, and in the case of normal space its magnitude is governed purely by fourth-dimensional slope because its spatial density is, with the exception of slight fluctuations due to gravitational fields, for all intents and purposes unity. Degenerate space is also composed of space, but its density is far below unity. Just as normal space has a density near unity and variable fourth-dimensional slope, degenerate space has a slope of unity and variable density. It is filled with *internal energy*.

The amount of internal energy in a core boundary's degenerate space will be referred to simply as *core boundary energy* and is driven entirely by its core volume.

Ψ THEOREM 9.8 - CORE BOUNDARY ENERGY {Ψ7.5, Ψ9.7}

DEGENERATE SPACE IS LOW DENSITY WITH A FOURTH-DIMENSIONAL SLOPE OF UNITY; ITS CORE ENERGY IS THE PRODUCT OF ITS UNITY SLOPE AND SPATIAL DENSITY, AS DEFINED BY ITS EVACUATED CORE VOLUME

A core boundary's energy density, which will also be called *degenerate energy*, is precisely equal to its spatial density:

$$\rho_{E_{\|}} = \left|\frac{dt_{\|}}{dr}\right|\varphi_3 = \varphi_3 \tag{9.19}$$

A core boundary's total (internal) energy is equal to the integral of spatial density over its surface volume. This is just the *energy equivalent* of the volume injected by core evacuation:

$$E_{\|} = \rho_U \int_{V_D} \varphi_3 dV = \frac{4\pi\rho_U R_c^3}{3} \tag{9.20}$$

where V_D is the infinite degenerate volume of a particle's core boundary. At a universal energy density (ρ_U) of $4(10)^{-10}$ J/m^3, a proton's core boundary energy is only $1.4(10)^{-54}$ J, far smaller than that of an electron, $8.8(10)^{-45}$ J.

CORE BOUNDARY DENSITY

A core boundary's energy is negligible by any standard, yet it maintains its integrity in environments as intense as atomic nuclei and cosmic rays. Cosmic ray collisions provide the most extreme challenge, as protons with energies in excess of 0.001 J have been observed. How does a core boundary preserve its integrity during impacts involving over *fifty* orders of magnitude more energy than its meager degenerate content?

Although a core boundary's internal energy is distributed across its degenerate volume at relatively low density, its spatial footprint in the third dimension is vanishingly small. A core boundary's effective energy density is given by the ratio of a core's volume and the (area)(thickness) this volume has been evacuated into. This is a core boundary's *density*, $\rho_|$:

$$\rho_| = \frac{E_{\|}}{\delta_| A_c} = \frac{\rho_U V_c}{\delta_| A_c} = \frac{\rho_U R_c}{3\delta_|} \tag{9.21}$$

where $\delta_|$ is a core boundary's *thickness* as measured along the third dimension. $\delta_|$ is infinitely thin, in keeping with a core boundary's discontinuous nature. This means that a core boundary's energy density is infinite even though its total internal energy is inconsequential. *Infinite boundary density is how a core preserves its integrity in any environment.*

If $\delta_|$ were finite but very small, a particle's core would still be discontinuous by virtue of its spatial void, but its boundary would be continuous, violating core integrity. Moreover, finite $\delta_|$ would make a particle's rest energy infinite, because its finite spatial footprint, as the product $(\delta_| A_c)$, would be distributed over the infinite external deflection of its degenerate space. The reason why particles have unbreachable core boundaries is because the energy density of their core boundary is infinite.

Ψ THEOREM 9.9 - CORE BOUNDARY DENSITY {Ψ9.7}

THE RATIO OF A CORE BOUNDARY'S ENERGY AND SPATIAL FOOTPRINT IS INFINITE

Infinite boundary density is core integrity's dynamic manifestation, and electrons have higher boundary density than protons even though their rest energy is much smaller.

9.7 ELECTRON CORES

Since a free electron's core is fairly large (3.3% of the size of a hydrogen atom), one is given to wonder why it hasn't made its presence known in physical interactions. It has. Evidence of electron cores has been well documented, and is referred to as *electron degeneracy pressure.* This is the effect cited as preventing the gravitational collapse of white dwarf stars, though in less severe environments it still has measurable consequences. Let's calculate a white dwarf's material density, assuming that its electrons retain the core radius defined by their rest energy [Equation (9.16)]. This will be referred to as a particle's *free-space* core radius, as we are to find (in the next chapter) that a particle's core radius varies when it is close to other particles.

Atoms have an average composition of 3 neutrons for every 2 protons, surrounded by 1 orbital electron for every proton. The average *atomic unit* is therefore 3 neutrons, 2 protons, and 2 orbital electrons. Nuclear core volume is negligible in comparison to that of free-space electrons, so an atomic unit's volume is approximately equal to the sum of the volumes of its two orbital electrons. At the tremendous pressure generated by a white dwarf's gravitational environment, electrons' spherical cores are compressed into their most spatially effective configuration, hexagonal-closest packing:

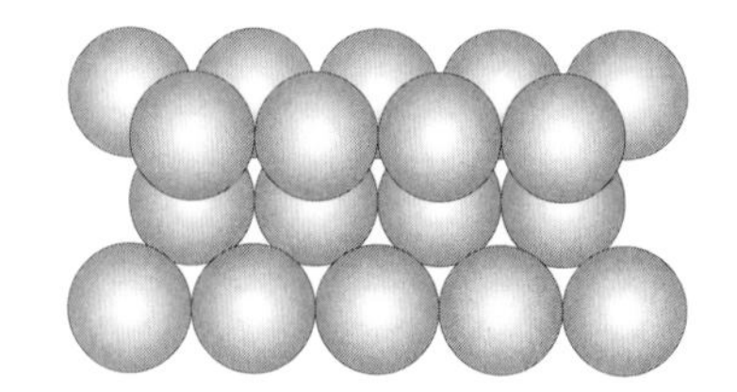

Figure (9.7) Hexagonally packed particle cores

In this arrangement, core volume fills a fraction of $(2\pi/9)$ of the available space, and the entire volume each particle subtends is related to the distance between them as $V = (3/4)d^3$.

Assuming hexagonal packing, an atomic unit's electronically degenerate density is:

$$\rho_{ed} = \left(\frac{2\pi}{9}\right)\frac{m_{au}}{2V_e} = \left(\frac{2\pi}{9}\right)\left(\frac{\left(3m_{n^0} + 2m_{p^+} + 2m_{e^-}\right)}{2\left(\frac{4\pi R_e^3}{3}\right)}\right) \cong \frac{5m_{n^0}}{12R_e^3} \qquad (9.22)$$

where m_{au} is its mass and V_e and R_e are an electron's free-space core volume and radius, respectively. Equation (9.22) evaluates to $1.3(10)^8$ kg/m^3 for the free-space electron radius of R_e = 1740 F given by Equation (9.16). This is consistent with the density of low-mass white dwarf stars.⟨6.15⟩ More massive dwarfs exert enough gravitational pressure to compress degenerate electrons to less than their free-space volume.

An electron's hollow core is not a quantum mystery, it is a physical boundary:

Ψ THEOREM 9.10 - ELECTRON DEGENERACY PRESSURE {Ψ9.2}

THE DEGENERACY PRESSURE OF ELECTRONS IS CAUSED BY THE UNBREACHABLE INTEGRITY OF THEIR HOLLOW CORES

Electrons' cores are small but unmistakable. They support a white dwarf's crushing density:

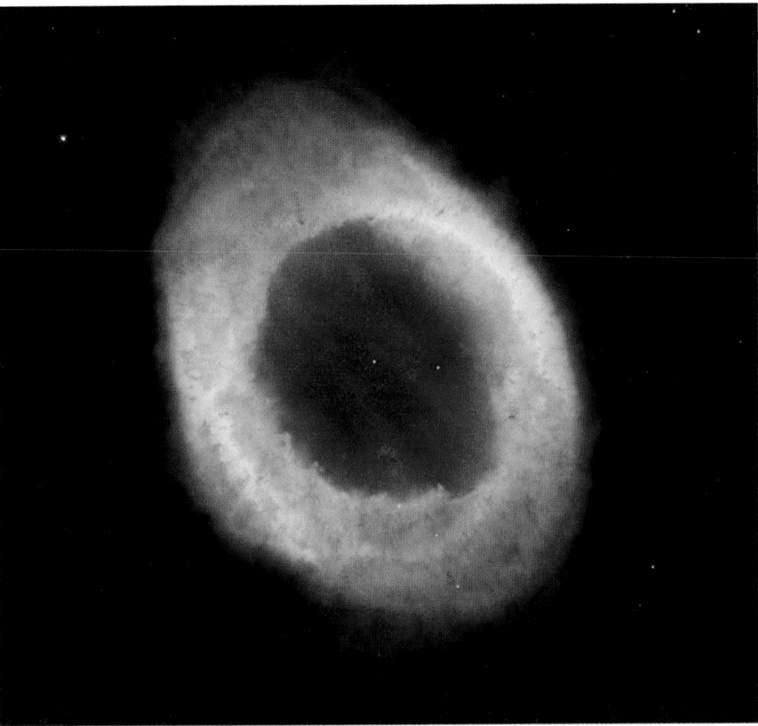

Figure (9.8) A white dwarf at the center of the Ring nebula, M57
(Courtesy NASA/Hubble Heritage Project)

9.8 NUCLEAR CONUNDRUM

Atomic nuclei are roughly spherical, with radii given by:

$$r = r_0 A^{\frac{1}{3}} \tag{9.23}$$

where A is nucleon count (usually referred to as mass number) and r_0 is observed to be ~1.25 Fermi.⟨18.1⟩ Lead, for example, at $A = 208$, has a nuclear radius of 7.4 F.

Average nucleon volume is given by nuclear volume divided by mass number:

$$V_n = \frac{\left(\frac{4\pi r_0^3 A}{3}\right)}{A} = \frac{4\pi}{3} r_0^3 \tag{9.24}$$

Thus r_0 represents the radius of the spherical volume a single nucleon subtends. It will be referred to as the *effective nucleon radius*.

The optimal spatial efficiency for spherical objects is hexagonal-closest packing, which is $(2\pi/9)$ filled. Proton cores packed to this density at their free-space core radius of 0.95 F have an effective volume radius of 1.066 F, small enough to fit into the observed effective nucleon radius of ~1.22 F. Moreover, the difference between these two values is only 17%, so *a 0.95 F proton core radius is consistent with nuclear density measurements*. The reason why atomic nuclei have such uniform density over such a wide range of mass number is because it is governed by a proton's free-space core size.

Although the correlation between proton size and nuclear density is reassuring, atomic nuclei are not composed purely of protons. There are neutrons to consider, and here the implications of core integrity become genuinely curious. Neutrons preserve the core integrity of their component protons and electrons, as evidenced by the tiny positive and negative space charge they exhibit in scattering experiments.⟨9.1⟩ These fields require a source, and the only viable source is a particle core. Hence an electron's negative core exists, in some capacity, within a neutron's substructure. One is given to wonder how this is possible when a free electron, at a radius of ~1740 F, *is two orders of magnitude larger than a typical nucleus*.

As it turns out, electrons in a nuclear environment are significantly smaller than in free space, as conclusively demonstrated in the next chapter.

9.9 GENERAL CONCLUSIONS

Applying unit polarvolume to matter produces intriguing and consistent results:

- Unit charge is the physical realization of a particle field's unit polarvolume quantization. The strength term of the Coulomb interaction, (q^2/ε_0), is four-dimensional, has the same units as unit polarvolume, and is about 18% of its value.

- Particles have a radially symmetric field of external deflection attenuating inversely as the *cube* of distance, symmetric about a positive or negative fourth-dimensional axis.

- The physical external deflection in a proton's field can be calculated from unit polarvolume and the universe's average energy density. At a distance of 1 Fermi from a proton's center, space is displaced a trillion light years into the fourth dimension.

- An elementary particle has a core region utterly devoid of energy and space whose diameter is inversely proportional to its rest energy. The core radius of a proton in free space is 0.9464 Fermi; an electron's core radius in free space is 1738 F, 3.3% of the size of a hydrogen atom in its ground state.

- A particle's core boundary is discontinuous; it is the embodiment of half of the four-dimensional size of the universe.

- A particle's core boundary is an immutable barrier. It cannot encroach on other core boundaries except during matter-antimatter annihilation, whereupon two boundaries mutually cancel to release energy and resolve their voided cores.

- A particle's core boundary maintains its integrity through infinite energy density.

- Electrons' hollow cores make their presence known as electron degeneracy pressure.

- Free-space electron core size corresponds to the material density of low-mass white dwarf stars, $1.3(10)^8$ kg/m^3.

Perhaps the best way to summarize this chapter is *matter is curved empty space, and it is curved immensely more than its gravitational field would lead us to believe.*

10. FIELD-CORE SUPERPOSITION

10.1 THE CONFLUENCE OF FIELDS

An isolated particle's isoexternal surfaces are perfect spherical shells, but become distorted by the presence of neighboring particles. The cross-sections of these shells are shown below for two charges of the same and opposite polarity:

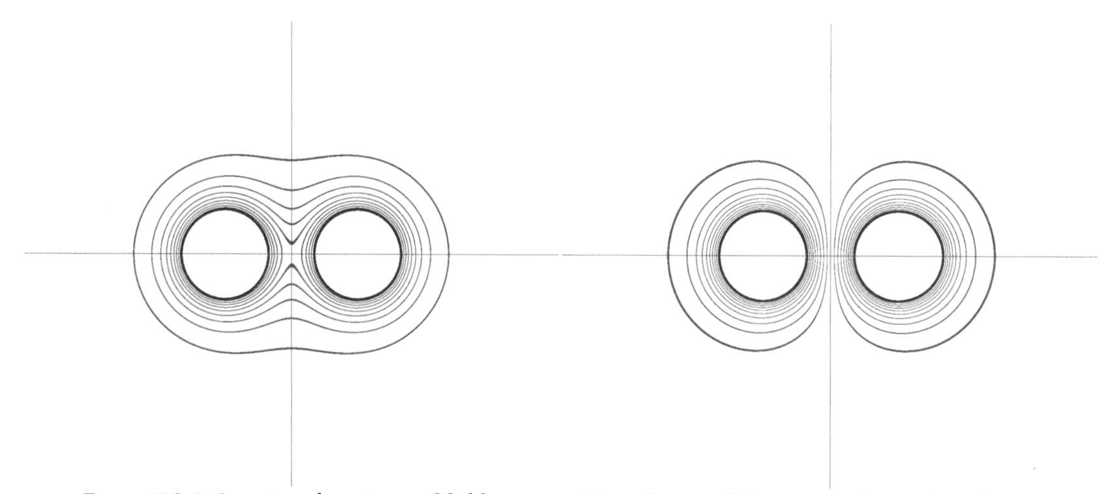

Figure (10.1) Isoexternal contours of field superposition, charges of the same and opposite polarity

Superposition of like charges is shown at left in Figure (10.1), and will be referred to as the (++) interaction. At large deflections its contours only encircle a single particle; at lower deflections they encircle both. On the right is the superposition of opposite charges. It has a markedly different appearance and will be called the (+–) interaction. No contours encircle both particles because their deflections lie in opposite directions of the fourth dimension.

CHARGE CONSERVATION

The external deflection at any point in space, to include the deflection at the core boundary of particles, is the cumulative product of all of the particle cores in the universe. Because of this, the (space)(time) product at any core boundary will never be precisely equal to unit polarvolume even though the core within it *retains constant unit polarvolume*. Charge is conserved because a particle's core, separated from the rest of the universe by its discontinuous boundary, is invariant under the superposition of exterior fields. This is why fifty protons in an atomic nucleus produce precisely fifty times the effect of a single proton.

Ψ THEOREM 10.1 - CHARGE CONSERVATION {Ψ9.6}
THE HYPERVOLUMETRIC CONTRIBUTION OF ANY PARTICLE'S CORE IS EQUAL TO UNIT POLARVOLUME

FIELD INTEGRITY

Field clarity (Ψ7.8) removes any direct association between field energy and the core that generates it, yet any given core has a quantized rest energy. This is because a particle's essence is not energy distribution; it is energy *difference* distribution. In an isolated particle this distinction is irrelevant because its energy difference is between empty space and the space deflected by its field. The same cannot be said of accumulations of particles. Each member of a multiple particle system represents a distributed energy density *difference*. This is the relationship between a core and its associated energy - its *field integrity*. Field integrity is the distributional equivalent of core integrity.

Ψ THEOREM 10.2 - FIELD INTEGRITY {Ψ7.8}
A PARTICLE IS A QUANTIZED DISTRIBUTION OF ENERGY DENSITY DIFFERENCE WHOSE INTEGRATED SUM IS EQUAL TO ITS REST ENERGY

10.2 CORE RECESSION

The superposition of external deflection from two different sources blends external slope and therefore energy density. As defined by geometry and vector addition, this is constrained by the following expression:

$$|t_1' + t_2'| \leq |t_1'| + |t_2'| \tag{10.1}$$

When two fields of external deflection superimpose, the magnitude of their combined slope at any point will always be less than or equal to the sum of their individual slopes.

This will be referred to as *superpositional slope loss*:

Ψ THEOREM 10.3 - SUPERPOSITIONAL SLOPE LOSS {Ψ7.5, Ψ7.8}

THE ENERGY CONTENT OF A SUPERIMPOSED COMBINATION OF EXTERNAL SLOPES IS LESS THAN OR EQUAL TO THE SUM OF ITS INDIVIDUAL COMPONENT SLOPES

The only place in the superimposition of two particle fields where no slope loss occurs is where the fourth-dimensional slopes of both sources are perfectly aligned:

$$|t_1' + t_2'| = |t_1'| + |t_2'| \qquad (10.2)$$

Since particle fields have intrinsic radial curvature, the amount of volume, and therefore energy, associated with this orientation is virtually nonexistent. *The sum of external deflection produces a net reduction of distributed energy density.* The *maximal* superpositional slope loss occurs when the fourth-dimensional slopes of two sources are perfectly misaligned:

$$|t_1' + t_2'| = \left||t_1'| - |t_2'|\right| \qquad (10.3)$$

As was the case with zero loss, the fraction of energy density from two sources that completely cancels is vanishingly small.

A particle's superpositional energy loss is the inevitable consequence of its position relative to other cores. This loss can be reconciled by radiative emission or a reduction in core size. Since the (++) interaction has positive potential and the (+−) interaction can occur adiabatically, there is only one possible destination for the energy deficit incurred in slope cancellation - the particle core. The shrinkage of particle cores in response to field superposition will be referred to as *core recession*. When particle fields superimpose, their cores must recede in order to offset the energy loss implicit in Equation (10.1).

Ψ THEOREM 10.4 - PARTICLE CORE> RECESSION {Ψ9.2, Ψ10.3}

LOSSES IN FIELD ENERGY INDUCED BY SUPERIMPOSITION CAUSE A REDUCTION OF PARTICLE CORE SIZE

Since particles of smaller core size have greater rest energy, the superposition of two particles with recessed cores can have the same total amount of energy as their nonrecessed, nonsuperimposed state.

CORE EXCISION

In addition to slope cancellation, a particle's field energy is also lowered by the proximity of nearby *cores*. Hollow cores punch holes in the fields of neighboring particles, reducing their distributed energy and inducing a compensatory core recession. At high particle density, this represents a far more substantial energy loss than slope cancellation. The energy deficit caused by the hollowness of proximate cores will be referred to as *core excision*:

Ψ THEOREM 10.5 - PARTICLE CORE> EXCISION {Ψ9.2, Ψ9.3}

CORE EXCISION IS A LOSS OF FIELD ENERGY INDUCED BY THE PRESENCE OF A HOLLOW CORE IN A PARTICLE'S FIELD

Slope cancellation and core excision reduce a particle's radius from its free-space value of R_c to its *recessed core radius*, R_{rc}.

10.3 NUCLEAR RECESSION CALCULATION

Consider n charged particles of equal core radius whose centers are located at coordinates of the form (x_n, y_n, z_n):

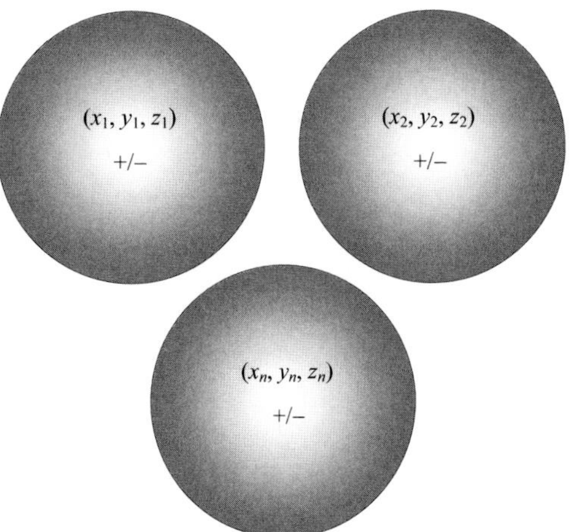

Figure (10.2) The juxtaposition of n particle cores is the sum of their superimposed fields

Exclusive of their hollow core regions, the superimposed external slope these particles generate at any spatial location (x, y, z) is given by:

$$\vec{t}' = -\left(\frac{9\Diamond_q}{4\pi}\right)\sum_{i=1}^{n}\left(\frac{B_i\left(x-x_i, y-y_i, z-z_i\right)}{r_i^5}\right) \tag{10.4}$$

where B_i is the charge of the ith particle, $B_i \in \{-1, 1\}$, and r_i is the radial distance from the point (x, y, z) to its center:

$$r_i = \sqrt{\left(x-x_i\right)^2 + \left(y-y_i\right)^2 + \left(z-z_i\right)^2} \tag{10.5}$$

Equation (10.4)'s components are:

$$t'_x = -\left(\frac{9\Diamond_q}{4\pi}\right)\sum_{i=1}^{n}\left(\frac{B_i(x-x_i)}{r_i^5}\right) \tag{10.6}$$

$$t'_y = -\left(\frac{9\Diamond_q}{4\pi}\right)\sum_{i=1}^{n}\left(\frac{B_i(y-y_i)}{r_i^5}\right) \tag{10.7}$$

$$t'_z = -\left(\frac{9\Diamond_q}{4\pi}\right)\sum_{i=1}^{n}\left(\frac{B_i(z-z_i)}{r_i^5}\right) \tag{10.8}$$

This system's total field energy is given by:

$$E(n) = \int_{V_U - V_{c_i}} \sqrt{(t'_x)^2 + (t'_y)^2 + (t'_z)^2}\, dV \tag{10.9}$$

where the notation on the integral indicates that the calculation is exclusive of the particles' core volumes.

For any group of proximate particles, there is a recessed core radius R_{rc} for which the energy of their superimposed fields is equal to the sum of their intrinsic rest energies and accumulated Coulomb potential:

$$\int_{V_U - V_{c_i}} \left(\sqrt{(t'_x)^2 + (t'_y)^2 + (t'_z)^2}\right)_{R_{rc}} dV = \sum_{i=1}^{n} E_{\wedge_i} + \Phi_q(n) \tag{10.10}$$

The R_{rc} subscript on the integrand denotes that the integration is a function of recessed core size, again exclusive of hollow core regions.

The Coulomb term is given by:

$$\Phi_q(n) = \sum_{i=2}^{n}\sum_{j=1}^{i-1}\Phi_{q_{ij}} \tag{10.11}$$

where $\Phi_{q_{ij}}$ is the Coulomb potential between the ith and j^{th} particle, and n is the total number of particles. *As it turns out, however, the Coulomb interaction's effect on core recession is negligible in comparison to slope cancellation and core excision.*

BOUND ELECTRON

Our recession analysis begins with the simplest case - two adjacent protons. While maintaining direct contact between their core boundaries, the protons' superposed energy is repeatedly integrated, and their core sizes reduced, until their total energy is equal to that of two free-space protons. The minimum recessed core size, R_{rc}, that meets this requirement is 0.8735 Fermi, 8.3% smaller than a proton's free-space radius:

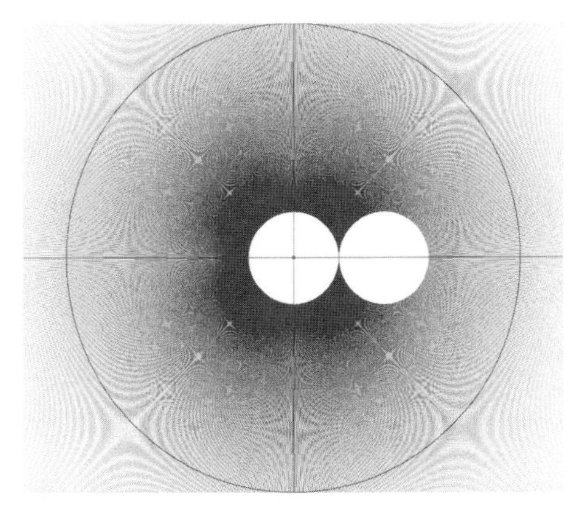

Figure (10.3) Core recession of two adjoining protons, recessed core size of 0.8735 F

A proton whose core recedes because of proximity to other protons will be called a *bound proton* or simply *proton* if its state is a given. This calculation assumes that bound protons retain their spherical symmetry. The source code used for this and other core recession calculations can be found at *www:nullphysics.com*. Note that the tiny radial patterns evident in Figure (10.3) (and similar images to follow) are graphical artifacts of the recession algorithm's fairly high spherical coordinate resolution.

Our next case is the combination of a single proton and electron - the enigmatic neutron. A neutron's formation from an electron and proton requires the addition of a small binding energy, but core integrity tells us the cores of both of these particles still exist within it. This is why high-energy electron scattering shows that neutrons contain both positive and negative fields.[9.1] Moreover, a neutron's binding energy is small in comparison to a proton's rest energy, so the lion's share of its energy originates from its proton component. In accordance with field integrity (Ψ10.2), the superpositional difference between the presence and absence of the proton's core is its rest energy. The same is true for the neutron's electron component. However, *a neutron in the absence of its electron is simply a proton*. This means a neutron's proton has virtually the same core radius as a proton in its free-space state. The only difference is neutron binding energy, which is about 50% greater than one electron mass. Even if all of this is attributed to the proton, its core radius would change by less than 0.001 F.

The negligible effect electrons have on the core recession of protons will be called *recessional exclusion*:

Ψ THEOREM 10.6 - RECESSIONAL EXCLUSION {Ψ10.2}

ELECTRONS INDUCE VIRTUALLY NO PROTON CORE RECESSION

Recession is a balanced interaction. Particles can only compress particles of similar or lower rest energy. As such, electrons have no tangible effect on protons.

The question our neutron recession calculation seeks to answer is:

> *What is the radius of the negative particle which, when superimposed at touching distance with a proton of fixed, free-space core radius R_p, produces a total system energy equal to the mass of a neutron?*

The answer is a particle of radius 1.70 F, about 70% larger than a proton:

Figure (10.4) Neutron as the core recession of a 1.70 F bound electron and proton of fixed free-space radius

An electron forced into the field of a single proton is compressed by a factor of a thousand, from 1740 F to 1.70 F. The accuracy of Figure (10.4) depends on the extent to which the electron's core remains spherical, but its relative size is a good estimate. Electrons compressed to nuclear density by a proton's fields will be referred to as *bound electrons*.

If a neutron were a static system, the fairly large (at least by nuclear standards) spacing between the bound electron and proton shown in Figure (10.4) would correspond to an electric dipole moment of $\sim 10^{-15}$ e m. However, a neutron is anything but static. Its bound electron orbits at relativistic speed, as evidenced by the peak kinetic energy (~0.78 MEV) they carry when emitted as a decay product. This is why its observed moment is a meager $<10^{-28}$ e m.[33]

The bound electrons that single protons produce are too large for a nuclear environment, but get progressively smaller in the presence of additional protons. Stable atomic nuclei have a ratio of about two protons to every three neutrons. This translates into *five bound protons for every three bound electrons*. Let's move an electron into the vicinity of two bound protons and measure its size.

Deuterium's bound electron, with a radius of 0.81 F, is only 93% the size of its two bound protons:

Figure (10.5) Deuteron as two 0.8735 F protons ($2R_{rc}$ spacing of 1.747 F) and a 0.81 F bound electron (bottom)

The deuteron, with a ratio of protons to bound electrons of (2:1), is fairly close to the ratio in heavy nuclei of (5:3), and its bound electron *is actually smaller than its protons*. This is well within the size requirement of the nuclear environment. Further, the inter-nucleon spacing of 1.747 F shown in Figure (10.5) is in excellent agreement with deuterium's triplet bound state, 1.748 ± 0.006.⟨18.3⟩

The last easily modeled nucleus is tritium. It can be represented as a triad of protons between two bound electrons. Numerical analysis puts the core radius of tritium's protons at 0.8194 F. Its bound electrons are smaller, at 0.61 F:

Figure (10.6) Tritium as three 0.8194 F protons with two 0.61 F bound electrons (centered on protons)

Tritium, with a proton/electron ratio of (3:2), is fairly close to the ratio of core species in heavy nuclei. Its bound electrons are about 75% of the size of its protons.

In general, the bound electrons of smaller nuclei have a size comparable to that of bound protons. There are reasons to suspect that this trend also holds for larger nuclei. The energy density generated by a core of a given size is inversely proportional to the fourth power of its radius. *The only way a nucleus can have a uniform energy density throughout its interior is if all of its internal particles are of a similar radius.* We know this is true for bound protons, because the total energy contribution of bound electrons is negligible. The reason why it is also true for bound electrons is because they are confined at nuclear density by bound protons. The only way to counter such intense positive fields is by generating comparable negative fields, and this requires similarly sized cores.

A neutron star is an extreme case of core recession. Here the number of bound electrons and bound protons are equal. Their radii have to be different in order for most of the energy contribution to be governed by protons, but not radically so. A good way to demonstrate this is with two protons and two bound electrons, a *neutron doublet* (^{2}n):

Figure (10.7) ^{2}n with two 0.871 F protons and two 0.879 F bound electrons (vertical pair)

This combination is not stable in nature. It is maintained by a neutron star's considerable gravitational potential.

The bound electrons in ^{2}n are *only 1% larger than the bound protons*. The slight difference in bound electron core size from 0.81 F in the deuteron to 0.88 F in ^{2}n is due to the interaction between the two bound electrons. Each bound electron represents an energy difference of one electron mass, but this difference is calculated with respect to the existence of the other particles, to include the other bound electron. It is interesting how parity between particle species translates into a slightly closer parity between their respective core sizes.

Particle field energy decreases fairly slowly as $(1/r)$, so the energy density throughout an atomic nucleus is uniform regardless of its core configuration:

Ψ THEOREM 10.7 - NUCLEAR UNIFORMITY {Ψ10.6}

AN ATOMIC NUCLEUS'S ENERGY DENSITY IS UNIFORM TO THE EXTENT ALLOWED BY ITS INTERNAL DISTRIBUTION OF HOLLOW PARTICLE CORES

This is consistent with observed nuclear density as well as the parity between the core radii of protons and bound electrons in our numerical recession calculations. The only way to maintain consistent energy density throughout a nucleus is with a lattice of similarly sized cores. *Differences in either core size or spacing produce significant variations in energy density.*

10.4 NUCLEAR DENSITY

Nuclear density is maintained by particles of similar sizes, but is ultimately driven by the native energy density of protons. Their *intrinsic density* is the ratio of their mass to core volume:

$$\rho_{\langle p\rangle} = \frac{\left(\dfrac{9\Diamond_q}{R_p c^2}\right)}{\left(\dfrac{4\pi R_p^3}{3}\right)} = \frac{27\Diamond_q}{4\pi c^2 R_p^4} \tag{10.12}$$

or $4.7(10)^{17}$ kg/m^3. Average nuclear density is lower than this due to the presence of bound electrons and the interstitial space between spherical cores. Let's calculate nuclear density taking both of these factors into consideration.

Nuclei have a large fraction of bound electrons, so their total core count significantly exceeds their mass number. As spherical objects at high density, these cores exist at or near the $(2\pi/9)$ spatial efficiency of hexagonal-closest packing. Applying these provisions to Equation (10.12) yields *nuclear density*, ρ_n:

$$\rho_n = \left(\frac{2\pi}{9}\right)\left(\frac{N_n}{N_c}\right)\rho_{\langle p\rangle} = \left(\frac{N_n}{N_c}\right)\frac{3\Diamond_q}{2c^2 R_p^4} = \left(\frac{N_n}{N_c}\right)\frac{m_p}{6R_p^3} \tag{10.13}$$

where N_n is the number of nucleons (mass number) and N_c is the number of cores. The $(2\pi/9)$ term accounts for the interstitial volume lost in hexagonal packing and the (N_n/N_c) term addresses the volume lost to bound electrons, which contribute negligible rest energy while using a considerable fraction of the available nuclear volume.

Stable nuclei have, on average, three bound electrons for every five protons. This amounts to five nucleons for every eight cores, so $(N_n/N_c) = 5/8$. Substituting this into Equation (10.13) yields:

$$\rho_n \cong \frac{5m_p}{48R_p^3} \qquad (10.14)$$

or $2(10)^{17}$ kg/m^3, consistent with the observed value.⟨18.2⟩ Even though a proton's field energy is distributed outside of its core, its free-space core volume ultimately limits the density at which it can be packed alongside other particle cores.

A nucleon's effective radius can be calculated directly in terms of a proton's core radius, packing arrangement, and bound electron density as follows:

$$r_0 = \left(\frac{9}{2\pi}\right)^{\frac{1}{3}} \left(\frac{N_c}{N_n}\right)^{\frac{1}{3}} R_p \qquad (10.15)$$

At $(N_c/N_n) = 8/5$ this amounts to 1.25 F, *within the range of the observed value for atomic nuclei* (1.2 – 1.25 F).⟨18.1⟩ *Protons compress electrons into the interstitial spaces between them at uniform energy density, optimizing the geometrically available space of a nuclear environment.*

If nuclei were composed entirely of neutrons, with one bound electron for every proton, they would be less dense. Substituting $(N_n/N_c = 1/2)$ into Equation (10.13) yields neutronium's native density:

$$\rho_{n_0} = \frac{m_n}{12R_p^3} \qquad (10.16)$$

or $1.6(10)^{17}$ kg/m^3, 80% of the average density of atomic nuclei.

In general, the distance between particle cores in a nuclear environment is governed by the energy density of hexagonally packed proton cores. Even though the addition of bound electrons increases inter-proton spacing, *inter-core* spacing remains constant regardless of the ratio of protons to bound electrons. Thus:

$$d_{cc} = 2R_p \qquad (10.17)$$

where d_{cc} is the distance between the centers of two adjacent cores. This is only true for nuclear material in the absence of external pressure. The inter-core spacing in a neutron star's interior is significantly smaller.

10.5 NUCLEAR SIZE LIMIT

Nuclear density is governed by protons' intrinsic density, so is defined by their free-space size. Although this accurately describes a nucleus's average inter-core spacing, it doesn't reveal much about *the distance between adjacent core boundaries.* This is set by slope cancellation and core excision. Just as core recession allows electrons to exist in a nuclear environment, it also has a profound and necessary influence on proximate protons.

Nucleons are packed to their maximum possible spatial density, but they also carry a significant amount of bound kinetic energy, moving at small fractions of the speed of light. *Hexagonal packing of free-space sized cores is simply too tight to allow nucleon motion.* Nuclear dynamics demands core recession. So even though nuclear density corresponds to a hexagonal packing at an inter-core spacing of twice a proton's free-space radius, the boundaries of these cores have receded enough to allow freedom of movement. When a nucleus gets too large, however, core recession begins to have a detrimental effect.

The difference between small and large nuclei isn't the effective volume of their nucleons; it is the size of their cores. Core recession increases with nucleon count, lowering core size while moving core boundaries further and further apart. The strength of the Strong force between nucleons, as quantified in Chapter 12, depends on core proximity as well as size. When cores get too small and their boundaries too far apart, this force attenuates, becoming too weak to allow a nucleus to maintain its mass number. The amount of inter-core spacing spanned by particle cores will be referred to as *nuclear retention*, η_r:

$$\eta_r = \frac{2R_{rc}}{d_{cc}} \tag{10.18}$$

η_r is unity when particle core boundaries touch. It is the fraction of inter-core distance that particle cores actually occupy, as shown below:

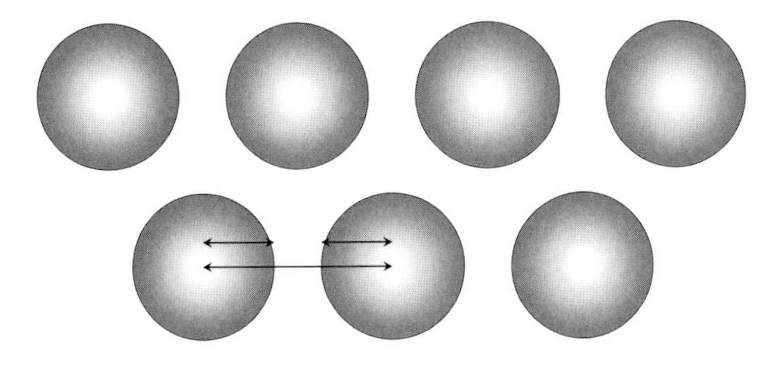

Figure (10.8) Nuclear retention as the ratio of voided distance to inter-core distance

The following shows nuclear retention as a function of core count for computer-generated clusters of hexagonally packed cores with an average of three bound electrons for every five protons. Cluster size varies from 12 to a maximum of 575 cores:

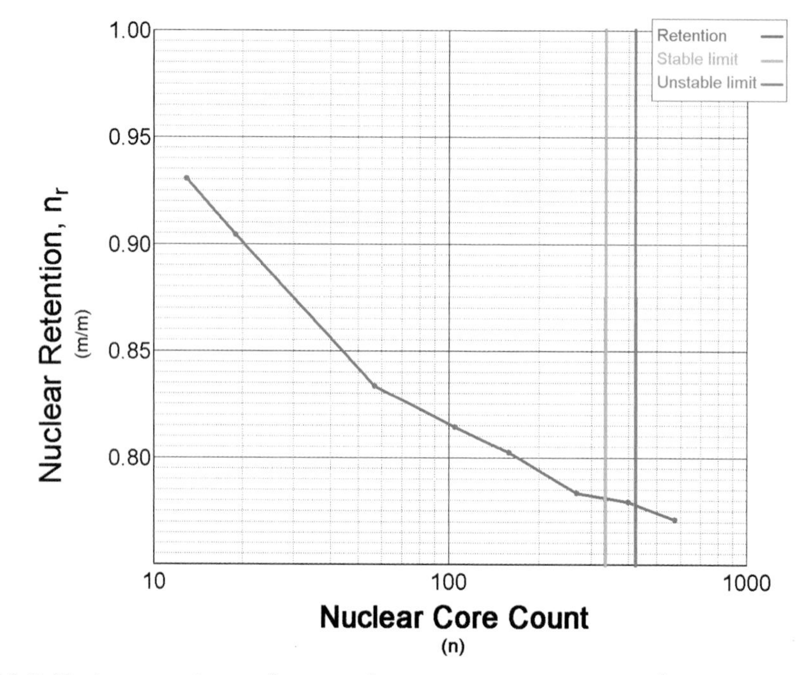

Figure (10.9) Nuclear retention as a function of core count, ~(5 protons/3 electrons), 12 to 575 cores

The approximate maximum core count for *stable* atomic nuclei is indicated by the orange vertical line. Lead, for instance, has one of the largest stable isotopes, ^{208}Pb, with 334 cores. One of the heaviest known unstable nuclei, ^{266}109, only lasts for a few milliseconds and has a core count of 423, shown as the vertical red line. No nuclei of any kind exist below a nuclear retention of 0.775, and as the green trace indicates, *retention is still decreasing at this core count*. This tells us that the concept of an island of nuclear stability, an ultra-heavy nucleus with a certain combination of protons and neutrons, is illusory. *Too many cores compress core size until the Strong force is too weak to control them, regardless of the ratio of bound electrons and protons.*

10.6 CORE HYPERDENSITY

As shown in the preceding calculations, core recession frees up a significant amount of inter-core space, so nuclear density is not limited by core integrity. However, if a significant external force is applied to nuclear material, as in the case of the crushing gravity of a neutron star or other massive compact object, particle cores are pushed closer to each other, increasing core recession. Smaller cores allow for even closer proximity, which in turn induces additional recession.

Two factors govern core recession, slope cancellation (Ψ10.3) and core excision (Ψ10.5). Their limitations are:

- Slope cancellation. Slope cancellation induces core recession, which in turn creates more intense fields with greater external slope. When recessed cores are forced together, the result is even greater slope cancellation and more recession. However, this effect attenuates rapidly in an extended electrically neutral matrix, such as a neutron star, because the net effect of a distribution of equidistant positive and negative particles is zero.

- Core excision. This is a direct consequence of the fixed geometric ratio of hexagonal packing. Packed cores fill a fraction of $(2\pi/9)$ of space, leaving the remaining $(1 - 2\pi/9)$ for field energy. A particle's transition from free space to a location in a packed array essentially removes all but $\sim(1 - 2\pi/9)$ of its field energy. The only way it can preserve its rest energy is with a core contraction to $(1 - 2\pi/9)$ of its former size. Hexagonally packed cores use the same fraction of space regardless of their size, so the loss of field energy by core excision has a hard upper limit.

Slope cancellation and core excision are inherently self-limiting, so the same is true of particles' energy loss due to close proximity. *Bound protons have a minimum size.* This means static material density, regardless of its gravitational environment, cannot exceed a certain immutable, maximum value. Although it is possible to achieve arbitrarily high energy density in the collision of relativistic particles, the same is not true if particles' rest energy is the source of their own energy density. The geometrically maximum density of elementary particle cores will be referred to as *hyperdensity*. This is the density where core boundaries are forced into direct contact. Such is the state of material in the central regions of massive black holes. *Hyperdense material has a nuclear retention of unity and represents the maximum energy loss associated with proximity effects.* Let's calculate its density.

In the absence of external pressure, an electrically neutral hexagonal distribution of protons and bound electrons has a density given by Equation (10.16):

$$\rho_{n_0} = \frac{m_n}{12R_p^3}$$

This takes into consideration both the ~30% free space of hexagonal packing as well as the 1:1 ratio between protons, which provide the lion's share of rest energy, and bound electrons, whose energy contribution is negligible. It also assumes a uniform size for all particles and therefore an inter-core spacing of $2R_p$, in accordance with Equation (10.17).

Core proximity reduces a system's total energy while core recession *increases a system's total energy inversely with core size.* This means that the cores of hyperdense material shrink in direct proportion to their proximity-induced energy loss. More specifically, the size of hyperdense cores is defined by the ratio between the energy density *remaining* after superposition at unity retention and the energy density *supplied* by a hexagonally packed lattice of particle cores. This is given by:

$$R_{|c|} = \left(\frac{\rho_{fc}}{\rho_{n_0}} \right) R_p \tag{10.19}$$

where $R_{|c|}$ is the hyperdense (minimum) core radius of protons and ρ_{fc} is the average central density of the superimposed fields from fixed cores packed into a large hexagonal lattice at a retention of unity. The greater the energy loss caused by maximum proximity, the smaller the lower limit of core recession. This limit is not maintained by core integrity; it arises from the particle distribution itself. *The only way to exceed hyperdensity is to increase the intrinsic rest mass of protons.* Our task now is to calculate ρ_{fc} from the unit polarvolume distribution.

SUPERIMPOSED DENSITY

Average fixed-core density ρ_{fc} can be determined with the following procedure:

a) Create 30 electrically neutral spherical *clusters* of hexagonally packed cores with total core counts that vary over a broad range. Core size is fixed and uniform, and nuclear retention is unity. The presence of bound electrons is simulated by giving all cores half of a proton's rest energy.

b) Numerically measure the average density of these clusters' central regions.

The size of the clusters will vary from the smallest, at 12 cores:

Figure (10.10) Schematic representation of smallest neutral core cluster used, $N_c = 12$ (layers shown by diameter variation)

to the largest, at 24,750 cores:

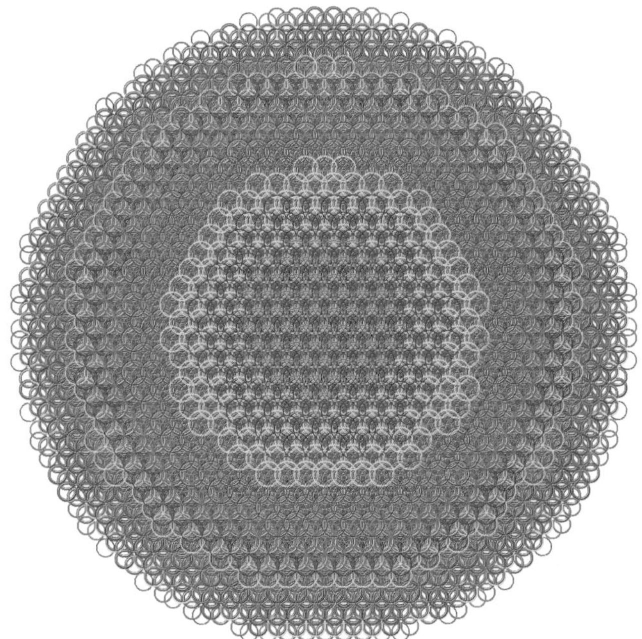

Figure (10.11) Largest neutral core cluster used, $N_c = 24,750$

Average density was measured for central regions of radii equal to $2R_c$ (green) and $4R_c$ (orange):

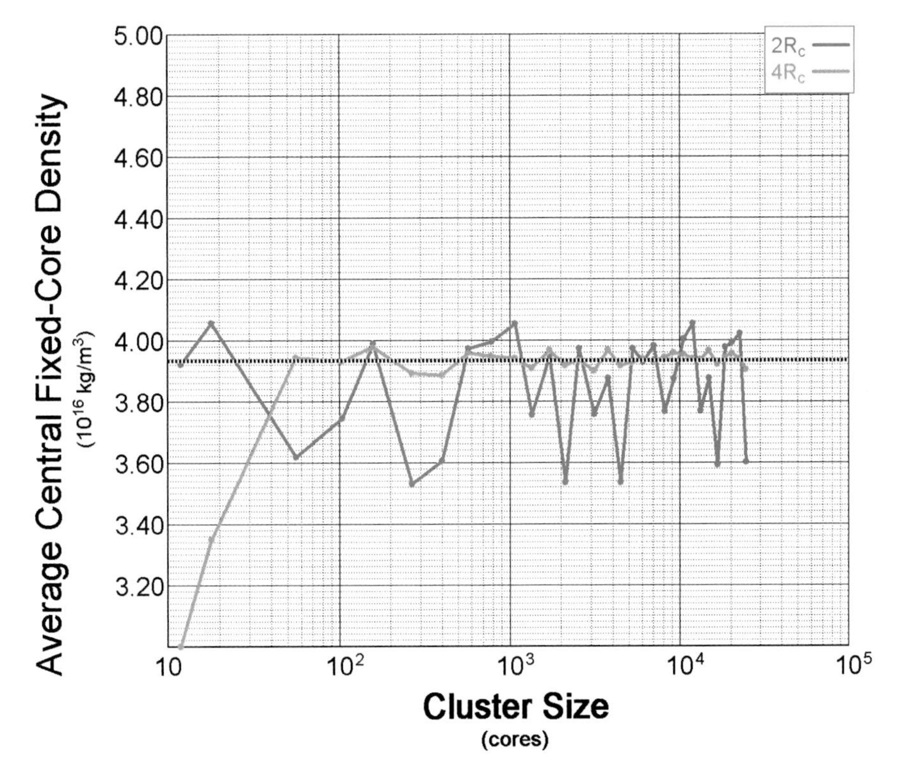

Figure (10.12) Average central fixed-core density ρ_{fc} as a function of cluster size, sample sizes $2R_c$, $4R_c$

As is evident from the average density traces of both sample regions, core excision and slope cancellation are short-range effects. Density fluctuation among the various clusters is a function of the (computer-generated) random distribution of positive and negative charges near their central cores, not their cluster size. This is why average central density is far more uniform in the larger ($4R_c$) sample size. The average density for all ($4R_c$) samples, as indicated by the horizontal line, is $3.94(10)^{16}$ kg/m^3.

Substitute Equation (10.16) into Equation (10.19):

$$R_{|c|} = \rho_{fc}\left(\frac{12R_p^4}{m_n}\right) \tag{10.20}$$

A ρ_{fc} of $3.94(10)^{16}$ kg/m^3 corresponds to a hyperdense proton size of only 0.23 Fermi, or a minimum inter-core distance of 0.46 Fermi.

The universe's maximum static material density follows from Equation (10.19):

$$\rho_{|c|} = \left(\frac{\rho_{n_0}}{\rho_{fc}}\right)^3 \rho_{n_0} = \frac{\rho_{n_0}^4}{\rho_{fc}^3} = \frac{m_n^4}{(12)^4 R_p^{12} \rho_{fc}^3} \cong \frac{6m_n}{R_p^3} \tag{10.21}$$

or $1.2(10)^{19}$ kg/m^3. Hyperdensity is about sixty times greater than the density of atomic nuclei. The inter-core spacing at this density is:

$$d_{|c|} = 2R_{|c|} = \frac{R_p}{\sqrt[3]{9}} \tag{10.22}$$

If the Earth were compressed to a spherical object of this density it would have a radius of about 49 meters.

SHRINKING CORES

In order to verify Equation (10.21), hyperdensity was tested by recursive recession analysis in electrically neutral core clusters over a wide range of size. As before, their 1:1 ratio between bound electrons and protons was simulated by giving all cores the same size and half of a proton's rest energy. In each case, the inter-core spacing of receding cores was repeatedly reduced until they were in direct contact and the system's total energy equal to the sum of the isolated cores. The inter-core spacing satisfying this requirement will be referred to as the *core minimum*, d_{cm}. Since the proximity effects on surface cores are far less than central cores, the core minimum for any given cluster will fall short of the separation associated with hyperdensity, but it should at least approach it with the decreasing (surface/volume) ratio of increasing cluster size. The processing of a small cluster is shown below:

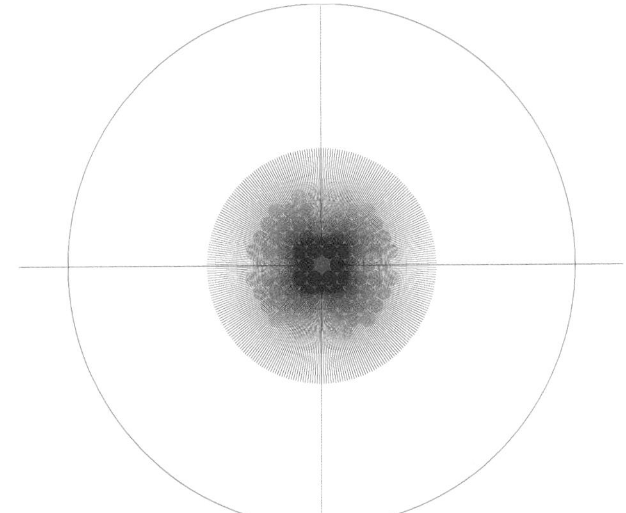

Figure (10.13) Iterative core recession applied to an electrically neutral, packed core cluster (N_c = 574)

The following depicts, as a function of core count, the core minimum for a number of electrically neutral clusters of hexagonally packed cores:

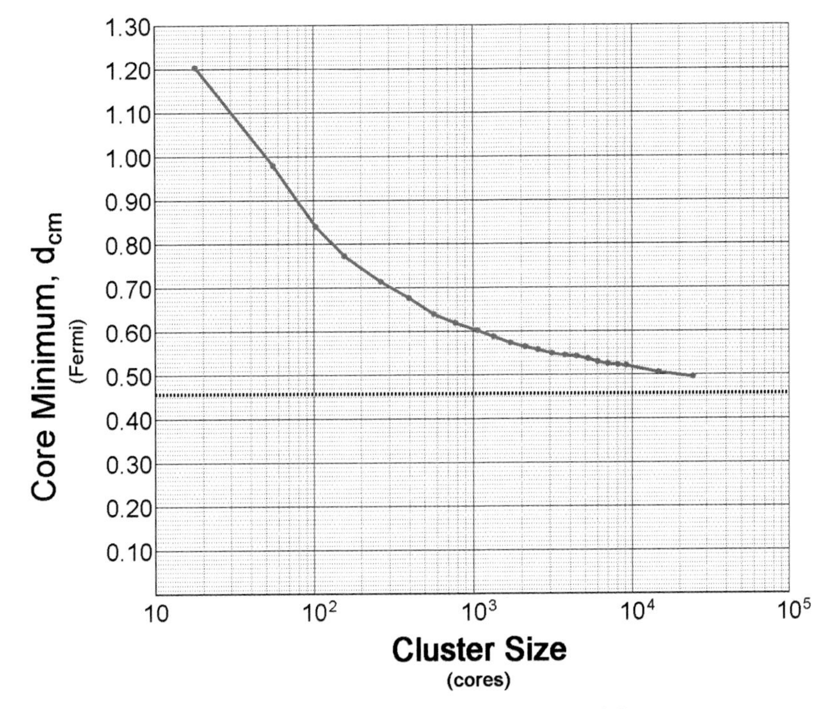

Figure (10.14) Core minimum as a function of cluster size

The particle cores in the clusters used in Figure (10.14) were artificially forced together with the same core recession algorithm used earlier to calculate nuclear retention. Core contact was achieved adiabatically at the spacing shown by the green trace. Note that the core minimum decreases monotonically with cluster size, as expected, asymptotically approaching the hyperdensity indicated by the horizontal line at 0.46 F. The source code used for all hyperdensity calculations can be found at *www:nullphysics.com*.

10.7 PARTICLE ZOO

Literally hundreds of different forms of elementary matter have been observed, understandably leading to the idea that more fundamental subunits exist within them. The quark concept was specifically created in response to this amazing subatomic variety. Particle physics' standard model uses quarks with properties such as *color* and *charm* to try to account for the teeming numbers of different elementary particles. Recently, physicists at one of the world's most powerful particle accelerators announced that they were able to identify a quark in its free state. This is not the case, because quarks do not exist. Anyone who claims to have found a quark is the victim of statistical analysis and wishful thinking.

There are only two stable particles, (electrons/positrons) and (protons/antiprotons). Particles such as mesons and kaons are composite systems like neutrons, but their subunits are not necessarily stable. They are combinations of stable particles and core-anticore resonances of various energies. The reason there are only two stable particle masses is because pair production only has two states able to generate *separate* particles. All other states collapse back into the release of two photons or other energy because the emergent particles are unable to escape their own mutual electrostatic attraction. Consider positronium. It consists of a positron and electron in a quasi-atomic orbit. Now suppose both of these particles were in a bound state, separated by nuclear distances. This is similar to the unit of composition found in most unstable particles, except that its cores have no free space analog. Unstable particles contain one or more core-anticore pairs whose components cannot exist singly in nature. This mirrored structure will be referred to as *core doublet*.

Take the case of the muon, $\mu^{\pm}$. It has a mass of ~206.77 electron masses and decays, on average, in a little over a millionth of a second. Muons are far more stable than other resonances and can even form transient atomic structures (muonium). A muon consists of a bound electron (positron) and a core doublet whose mass is ~205 electron masses. Although it is conceivable that a muon might contain multiple core doublets, this is unlikely since it is the lightest unstable particle.

Our core recession algorithm provides a closer look at a muon's substructure:

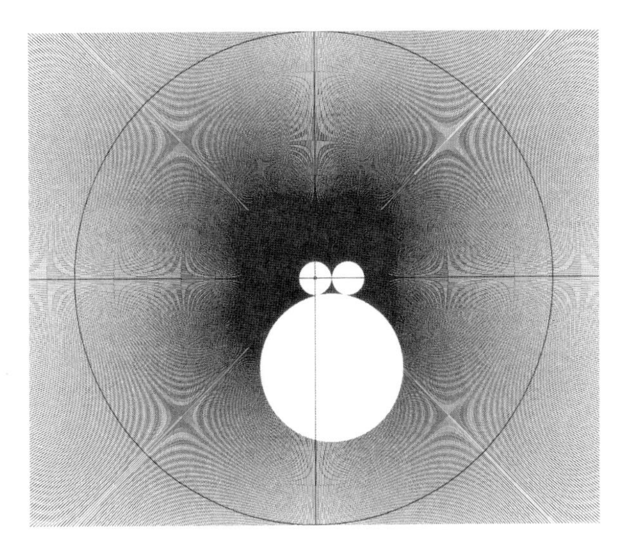

Figure (10.15) Muon as a combination of a 45.4 F bound electron and a 10.5 F core doublet

A muon's core doublet has components that are about ten times the size of a proton. The bound electron does not substantially contribute to the overall energy of the system, and has a size over four times as large as the core doublet.

10.8 GENERAL CONCLUSIONS

The interaction between particle cores and fields is introduced:

- A particle core's discontinuous boundary is responsible for charge conservation.
- The superposition of particle fields produces a distributed energy loss as a result of the cancellation of external slope and excision by exterior cores. This energy loss is reconciled by the recession of particle core boundaries.
- Neutrons are an elementary chimera of a proton and bound electron.
- Electrons are compressed to nuclear dimensions by exposure to protons' intense fields.
- The inter-nucleon spacing of deuterium's bound triplet state is 1.747 F.
- Electrons and protons have comparable sizes in a nuclear environment.

- The intrinsic energy density of protons at their 0.946 F free-space radius, in concert with comparably sized bound electrons, corresponds closely to the observed density of atomic nuclei, $\sim 2(10)^{17}$ kg/m^3.

- The effective nucleon radius in atomic nuclei is, on average, 1.25 F.

- Atomic nuclei become increasingly unstable as their total core count increases past a limit imposed by the core recession associated with nuclear proximity. There is no island of stability for ultra-heavy nuclei.

- The particle cores in the center of massive compact objects are significantly smaller than those of atomic nuclei.

- The universe's maximum static material density is $1.2(10)^{19}$ kg/m^3, about sixty times that of atomic nuclei.

- Unstable particles such as mesons and kaons contain paired cores and anticores that do not exist singly in nature.

11. PARTICLE FIELD FORCE

There are four universal forces: two long-range (Coulomb and gravitational) and two short-range (Weak and Strong). All involve the nonlinear superposition of two or more particle fields. However, before any of these interactions can be fully understood, it is first necessary to look more closely at the forces at work in an isolated particle's topology. This begins with their *dimensionality*:

Ψ THEOREM 11.1 - FORCE'S DIMENSIONALITY {Ψ7.1}
FORCE IS A TWO-DIMENSIONAL QUANTITY OF TIME-DISTANCE

Energy is three-dimensional and force is two-dimensional. They differ by only a single dimension of distance, but their operational distinction is far more extensive.

11.1 DEFLECTIONAL FORCE

A particle's temporal cross-section has units of time-distance and therefore represents force. It will be referred to as *deflectional force* and symbolized F_∇:

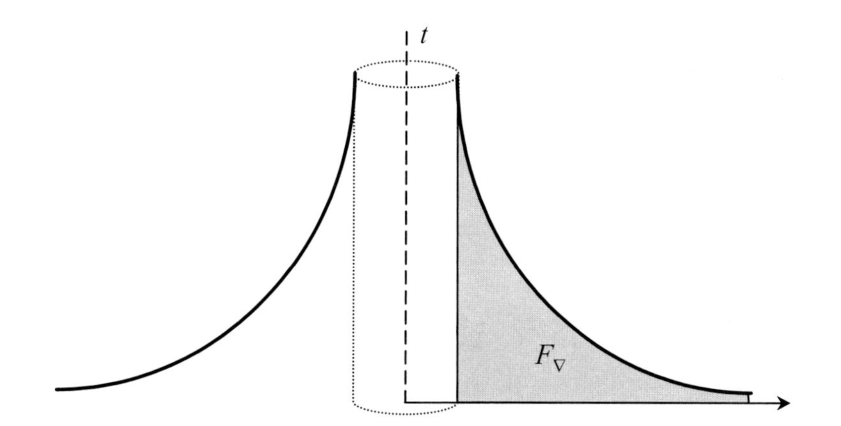

Figure (11.1) Deflectional force is the area of a spatial deflection's cross-section

Force is the universe's great mediator. It is a two-dimensional operator able to affect change while not subject to the demands of either composition or conservation. External deflection is one-dimensional and attenuates as the cube of distance; force is two-dimensional and falls off as the square of distance; energy, being three-dimensional, varies inversely with distance. They are all ramifications of pure dimensionality. Force's two-dimensional nature is essential to its operation. Unlike energy and space, two-dimensional entities are not bound to space's compositional surface. As such, force is the metric of the interface between elementary forms of energy.

Deflectional force blends seamlessly in accordance with field clarity (Ψ7.8), destroying any direct association between a cumulative deflection and the particles from which it originates. This means that the only location where force can physically act on a particle is against its core boundary, as it is the separation between the particle and the rest of the universe:

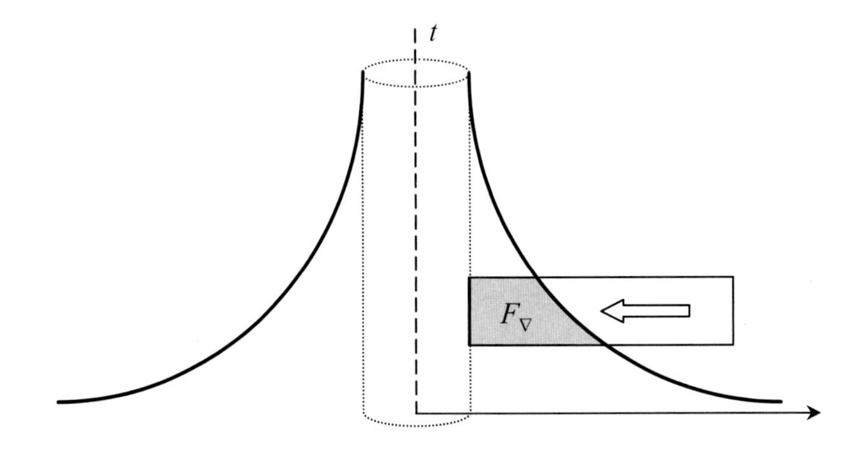

Figure (11.2) Deflectional force parallel to space

Deflectional force is the spatial *tension* within a particle's field.

Ψ THEOREM 11.2 - DEFLECTIONAL FORCE, F_∇ {Ψ11.1}

DEFLECTIONAL FORCE IS THE EXTRINSIC FORCE OF THE TEMPORAL CROSS-SECTION OF EXTERNAL DEFLECTION, ACTING ON THE PARTICLE CORE FROM WHICH IT ORIGINATES

The deflectional force of a particle's field from its core radius to infinity is its *total deflectional force*. This is the force acting against both sides of its core boundary:

$$F_{\nabla_T} = \left| 2 \int_{r=R_c}^{\infty} t dr \right| = \left| 2 \int_{r=R_c}^{\infty} \frac{3\Diamond_q}{4\pi r^3} dr \right| = \frac{3\Diamond_q}{4\pi R_c^2} \tag{11.1}$$

Total deflectional force increases as the square of a particle's core radius. A proton's total deflectional force in free space, for instance, is a staggering 4213 N (Newtons).

Deflectional force between any two radii ($R_2 > R_1$) follows from Equation (11.1):

$$F_\nabla[R_2 \to R_1] = \int_{r=R_2}^{R_1} t\,dr = \int_{r=R_2}^{R_1} \frac{3\Diamond_q}{4\pi r^3} dr = \frac{3\Diamond_q}{8\pi}\left(\frac{1}{R_1^2} - \frac{1}{R_2^2}\right) \tag{11.2}$$

The most visible evidence of a particle's deflectional force is the *Coulomb interaction*. Let's look more closely at external deflection's superposition to see how this interaction actually operates.

11.2 DEFLECTIONAL INTEGRITY

An isolated particle's deflectional force is balanced because it is radially symmetric with an equal magnitude in all directions. This changes when particle fields superimpose. Take the case of the (++) interaction. It has a temporal cross-section of the form:

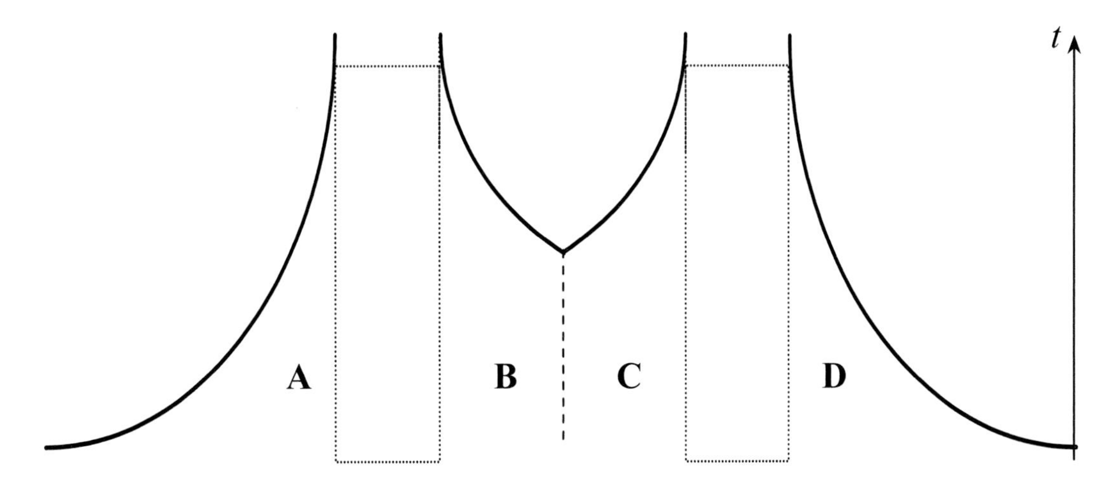

Figure (11.3) Temporal cross-section of the (++) superposition (two charges of same polarity)

The left core experiences a force defined by the difference of areas **A** and **B**. However, field clarity (Ψ7.8) destroys any association between deflectional force and the core(s) from which it originates. Since no discontinuity exists between **B** and **C**, it is not possible to calculate the *net* force on a given core from the combined cross-section against its boundary. Similarly, the force between the two cores in Figure (11.3) is not the difference between (**A**+**D**) and (**B**+**C**), because there is no longer a direct association between deflection and its source. *The only connectivity that remains in superimposed fields is that which exists between a core and the cross-sectional force it generates.*

When a core is injected into a region of high energy density, it recesses to preserve its field energy. The only way to evaluate this energy is to compare the system's total energy with and without the core. In precisely the same way, a core can be exposed to exterior forces, but the net force it experiences can only be evaluated in terms of the deflectional difference it generates *about its center* in its ambient environment. In field integrity, the energy associated with a particle is the energy difference that its core induces in space. An analogous relationship exists with its extrinsic force. It will be called *deflectional integrity*.

Ψ THEOREM 11.3 - DEFLECTIONAL INTEGRITY {Ψ7.8, Ψ10.2, Ψ11.2}

THE NET FORCE ON A PARTICLE CORE IN A GIVEN DIRECTION IS THE DIFFERENCE OF THE DEFLECTIONAL FORCE IT GENERATES ON EITHER SIDE OF ITS CORE BOUNDARY ALONG SAID DIRECTION

If the superposition of one particle field on another were linear, there would be no Coulomb force, because the distribution of deflectional force about the center of an isolated particle is perfectly balanced. There is only one possible source for the Coulomb force - *a departure from symmetry in the deflectional force a particle generates across its core boundary*. One particle's field cannot exert force on another *field* because field clarity makes them geometrically inseparable. It can, however, exert force on another *core*, but field clarity intercedes again and eliminates the causal link between field and core.

The only way to change a core's deflectional force distribution is to change its shape through the application of an exterior field. Induced core distortion is the Coulomb force's *precursor*:

Ψ THEOREM 11.4 - COULOMB PRECURSOR {Ψ11.3}

THE COULOMB FORCE IS CAUSED BY A SUPERPOSITION-INDUCED NONSPHERICAL DEFORMATION OF A PARTICLE'S CORE, WHICH IN TURN CAUSES VARIATION OF THE DEFLECTIONAL FORCE IT GENERATES ACROSS ITS CORE BOUNDARY

The exchange of photons is required to implement this force, but its underlying cause is purely geometrical. But how, exactly, does field superposition deform a particle's core? Let's derive the magnitude of this distortion.

11.3 COULOMB DISTORTION (LINEAR APPROXIMATION)

The Coulomb distortion has two components. The first is *applied force*, the action of an exterior field on a core's boundary. The second is the *resultant force*. This is the differential deflectional force generated by the core asymmetry caused by the applied force. It is possible to obtain a rough estimate of the magnitude of resultant force using simple analytical expressions, as will be presented below. A full quantification requires numerical analysis of a distorted core's nonlinear field, as described in Appendix F.

APPLIED FORCE

An exterior field engulfs a target particle from all directions, but the core asymmetry it induces is governed by the deflectional force acting on a target's core center:

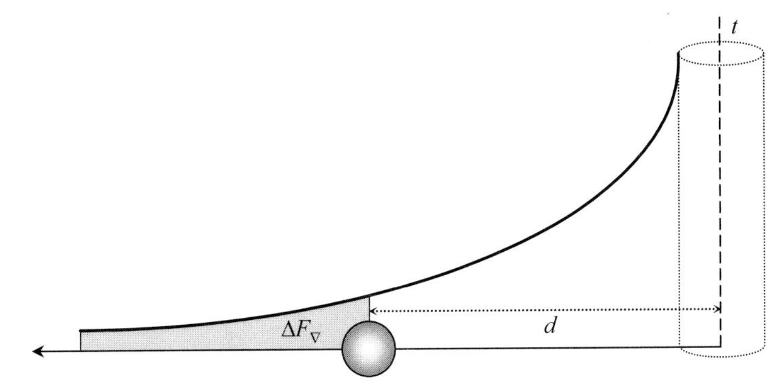

Figure (11.4) Deflectional force component applied by an exterior field at a distance d

The applied force's orientation is reversed between the (++) and (+−) interactions because the source's field either complements or attenuates the target's. Its magnitude is given by Equation (11.2) evaluated from d to infinity:

$$\Delta F_{\nabla} = F_{\nabla}[d,\infty] = \frac{3\Diamond_q}{8\pi d^2} \qquad \{d >> 2R_c\} \tag{11.3}$$

where the distance of separation d is large compared to the core size of the interacting particles. This simplifies the expression and eliminates core-related effects.

RESULTANT FORCE

The product of a particle's core radius and intrinsic energy density, from Equation (10.12), is its *radial force density*:

$$\rho_F = \rho_{\langle c\rangle} R_c = \frac{27\Diamond_q}{4\pi R_c^4} R_c = \frac{27\Diamond_q}{4\pi R_c^3} \tag{11.4}$$

This is the force per unit length long a particle's radius, and can be used to provide a reasonable estimate of the relationship between applied force and the core distortion it induces.

Assuming a linear relationship, the ratio of applied force to core asymmetry follows from Equation (11.4):

$$\rho_F = \frac{\Delta F_\nabla}{\Delta R_c} = \frac{\dfrac{3\Diamond_q}{8\pi d^2}}{\Delta R_c} = \frac{27\Diamond_q}{4\pi R_c^3} \tag{11.5}$$

where ΔR_c is the difference in a particle's core radii along the axis of the applied force.

Solve for ΔR_c:

$$\Delta R_c = \frac{3\Diamond_q}{8\pi d^2}\frac{4\pi R_c^3}{27\Diamond_q} = \frac{3R_c^3}{54d^2} \tag{11.6}$$

The differential deflectional force that a particle core generates due to unequal core radii along a given direction is, in accordance with the definition of force, equal to the product of the difference between the core radii and the core's peak external deflection, as given by Equation (9.6):

$$\Delta F_\nabla = \Delta R_c t = \Delta R_c \frac{3\Diamond_q}{4\pi R_c^3} \tag{11.7}$$

Substitute the radial difference of Equation (11.6) into Equation (11.7) and simplify:

$$\Delta F_\nabla = \left(\frac{3R_c^3}{54d^2}\right)\left(\frac{3\Diamond_q}{4\pi R_c^3}\right) = \frac{\Diamond_q}{24\pi d^2} \tag{11.8}$$

Equation (11.8) is a linear approximation of the Coulomb force:

$$F_q \cong \frac{\Diamond_q}{24\pi d^2} \tag{11.9}$$

This is about 9% smaller than the observed force, which is a reasonable correlation for a linear approximation of a highly nonlinear interaction.

For the most accurate assessment of a core's Coulomb distortion, simply use the strength of the observed force. From Equation (11.7):

$$F_q = \frac{q^2}{4\pi\varepsilon_0 d^2} = \Delta R_c \frac{3\Diamond_q}{4\pi R_c^3} \tag{11.10}$$

Solve for the distortion:

$$\Delta R_c = \frac{R_c^3 q^2}{3\Diamond_q \varepsilon_0 d^2} \tag{11.11}$$

Compare this to core recession. The algorithm used for the nuclear density calculations of Chapter 10 indicates that the core recession of two superimposed protons falls off inversely with the distance between them as:

$$\delta R_{rc} \cong (0.172) R_p^2 \left(\frac{1}{d}\right) - (0.027) R_p^3 \left(\frac{1}{d^2}\right) - (0.070) R_p^5 \left(\frac{1}{d^4}\right) \tag{11.12}$$

For an atomic distance of (d = 0.1 nm), the recession between two protons is $1.5(10)^{-21}$ m whereas their Coulomb distortion is nearly a million times smaller, at $5.2(10)^{-27}$ m.

The Coulomb force is a second-order superpositional effect, much smaller than recession:

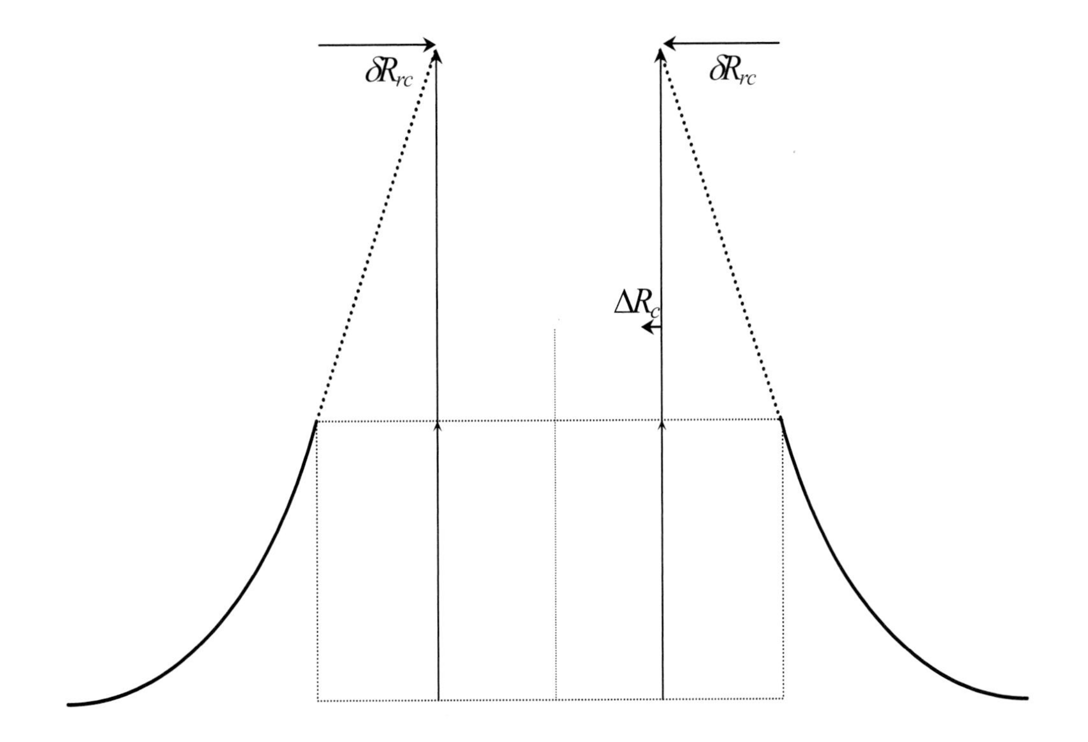

Figure (11.5) Schematic comparison of core recession and Coulomb distortion

11.4 CORE INTEGRITY

The core distortion produced by the Coulomb interaction follows the inverse square relationship of Equation (11.11) as long as interacting core boundaries are not in direct contact. As soon as $d < \sim 2R_{rc}$, core boundaries begin to flatten against each other along their axis of separation, and the displacement of the particles' radii comes against a hard geometric limitation:

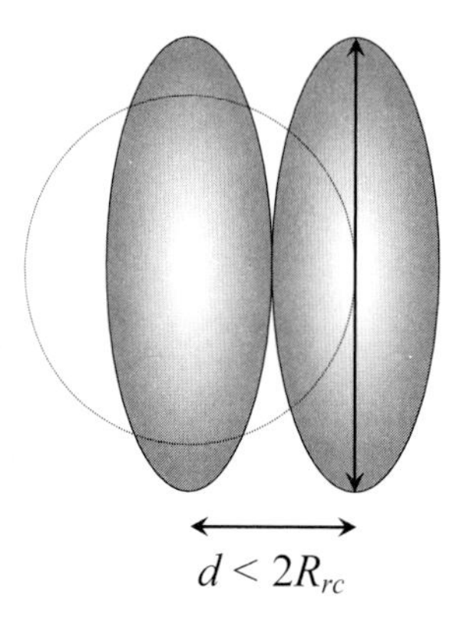

Figure (11.6) Deformation force begins when core integrity governs physical proximity

The force generated by this compression is a first-order effect that is far more powerful than the Coulomb interaction. It will be called the *core deformation force*. The magnitude of this markedly nonlinear force can be approximated as the difference of total deflectional force between a core's compressed and uncompressed states. This is evaluated along the distance separating the interacting cores, *d*:

$$F_{cd}\left(d \leq 2R_{rc}\right) \approx \left(\frac{3\lozenge_q}{4\pi}\right)\left(\frac{1}{d^2} - \left(\frac{1}{2R_{rc}}\right)^2\right) \qquad (11.13)$$

where the $2R_{rc}$ term represents the maximally recessed, uncompressed state. Deformation force is zero when cores first touch, at $d = 2R_{rc}$. Equation (11.13) is considered an approximation because the field that a deformed core generates is different from the normal spherical field of its free-space state.

11.5 COULOMB POTENTIAL

Field superimposition does more than cause core recession and nonspherical deformation; its effect also accumulates as *Coulomb digression*. Coulomb potential energy is stored in or released from a particle during interactions with other particles. The only mechanism available for this process is a change in the interacting particles' core radii, δR_{cq}. This change is necessarily smaller than superimposition-induced core recession because the Coulomb force is a second order effect. When $\delta R_{cq} << R_c$ the per-particle change in core radius related to the Coulomb potential is given by:

$$\frac{9\Diamond_q}{R_c^2}\delta R_{cq} = \frac{q^2}{4\pi\varepsilon_0 d} \tag{11.14}$$

Solving for the core's differential change of radius yields:

$$\delta R_{cq} = \frac{q^2 R_c^2}{36\Diamond_q \pi\varepsilon_0 d} \tag{11.15}$$

The Coulomb digression of protons is about 100 times smaller than their core recession.

MAXFIELD POTENTIAL

The superposition of a proton and bound electron of radius 1.70 F has the energy of a neutron, as shown earlier in Figure (10.4):

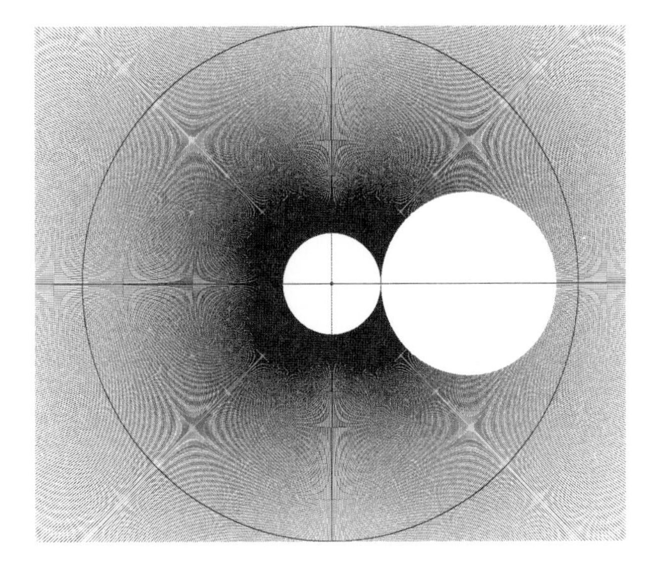

If the radius of a neutron's bound electron is adjusted to 1.701 F, a difference of only a thousandth of a Fermi, *its combination with the neutron's bound proton has precisely the same total energy as an isolated proton.* This is the bound electron's Maxfield potential. It is the mathematical extrapolation of field integrity.

ELECTROMAGNETIC EFFIGY

The electron-proton (or positron-antiproton) combination of Figure (10.4) is the only superposition able to exemplify Maxfield because other combinations of opposites, such as protons and antiprotons, result in annihilation. A particle at a Maxfield Coulomb potential will be referred to as an *electromagnetic effigy*. Effigies do not exist in nature because particle cores need to have a nonzero effect on their environment. If the difference between the presence and absence of a particle were in fact zero energy, its existence would not be quantifiable and it would not exist. Particles generally tend to maintain their free-space field energy in environments that induce core recession, seldom relinquishing even a small fraction, with the sole exception of annihilation. The mass fraction lost by protons in atomic nuclei, for instance, is only ~0.7% even when their cores recess by nearly 50%. Bound electrons lose none of their field energy; instead they tend to augment it with binding energy exceeding the mass of their free state.

What is particularly interesting about our numerical simulation of a neutron is its 2.65 Fermi inter-particle separation, as shown in Figure (10.4). This has a Coulomb potential of 546 KEV, comparable to an electron's 511 KEV rest energy, and therefore its Maxfield potential.

11.6 GENERAL CONCLUSIONS

The forces between particles present a formidable analytical challenge, but a few aspects of their curious nature emerge:

- External deflection's temporal cross-section is equal to the force it exerts on space.

- The Coulomb interaction is phenomenologically linear yet fundamentally nonlinear.

- The Coulomb force is caused by a nonspherical core distortion that modulates the external deflection that a particle generates across its own core boundary.

- When particle cores are exceptionally close ($d<2R_{rc}$), they preserve their core integrity by exhibiting nonspherical deformations more pronounced than those of the Coulomb interaction.

Let's return now to the atomic nucleus, to investigate how positively charged protons, which ought to repel each other, are confined at such extreme density.

12. CORE INTERACTION

Unit polarvolume is the governing dynamic of matter's quantization, so is ultimately linked, in some way, to each and every one of the four universal forces. Since an elementary particle consists of only two regions, an unbounded field and a finite hollow core, the inevitable conclusion is the long-range Coulomb and gravitational forces originate in a particle's field while the short-range Strong and Weak forces are directly associated with its core. Henceforth, Coulomb and gravitational interactions will be referred to as *field interactions* whereas Strong and Weak forces will be called *core interactions*. Our focus returns once more to the cores of elementary particles.

12.1 SHORT RANGE FORCES

The universe contains only two stable particles, the electron and the proton (and their antimatter twins). This provides the potential for three different types of core interaction:

- Proton-proton.
- Electron-proton.
- Electron-electron.

The first provides the Strong force, the second is responsible for the Weak force, and the final permutation, the electron-electron, has no physical expression for reasons demonstrated later in this chapter.

The Strong force arises when proton cores are in close proximity to each other:

Ψ THEOREM 12.1 - STRONG PRECURSOR {Ψ9.2}

THE INTERACTION BETWEEN HOLLOW PROTON CORES GOVERNS THE STRONG FORCE

The Weak force is more complex than the Strong because it occurs between cores of opposite sign and different size.

Ψ THEOREM 12.2 - WEAK PRECURSOR {Ψ9.2}

THE INTERACTION BETWEEN HOLLOW PROTON CORES AND HEAVILY COMPRESSED ELECTRON CORES GOVERNS THE WEAK FORCE

In light of our recently acquired knowledge of the particle field, the short-range nature of the Strong and Weak forces leaves little doubt that core interaction is responsible for their existence. But how it works is not as clear. The remainder of this chapter will be devoted to the simpler of the two core interactions - the Strong force. It occurs between cores of equal size and polarity. When one proton's hollow core is near another, *a strong attraction pulls them together*. This, as it turns out, is a direct consequence of core purity (Ψ9.3).

STRONG ATTRACTION

A particle's core boundary is discontinuous and pristinely devoid of space and energy. Since it disallows superposition, its intrusion near another core nullifies a segment of the target core's deflectional force. As derived in the previous chapter, the net force a particle experiences is the sum of the deflectional force it generates in space. The presence of a spatial void prevents the existence of deflectional force, thus creating an imbalance about a particle's core. This in turn causes a force differential across the target core's boundary, producing an attractive effect:

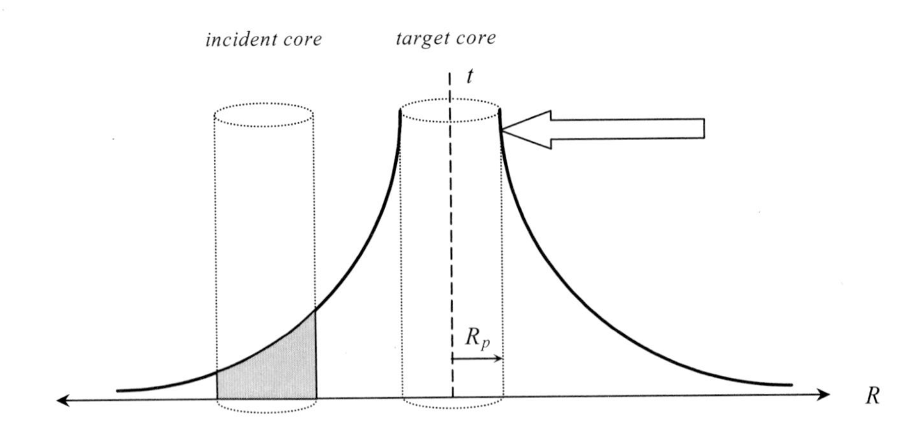

Figure (12.1) Effect of one hollow core on another

The core imposing on a target core's field will be called the *incident core*. Since their combination is symmetrical, a target core also induces an attractive force on an incident core. This is the underlying cause of the Strong force. Let's calculate its magnitude.

12.2 STRONG INTERACTION (LINEAR CORE APPROXIMATION)

As shown in Chapter 10, a particle's core changes size when it is near other cores, producing distortions too complex to be fully defined by simple analytic expressions. This is a pressing concern for our Strong force calculation since this force only exists when cores are close to each other. As it turns out, however, the recessed core size of two protons separated by a distance d can be accurately approximated using Equation (11.12):

$$R_{rp}(d) \cong R_p - (0.172)R_p^2\left(\frac{1}{d}\right) + (0.027)R_p^3\left(\frac{1}{d^2}\right) + (0.070)R_p^5\left(\frac{1}{d^4}\right) \qquad (12.1)$$

This, and the simplifying provision that core recession abruptly ceases as soon as cores are so close that their boundaries are in direct contact, constitutes the *linear core approximation*.

When a proton's core boundary is overlaid on another proton's field, it defines an amount of deflectional force that varies with distance of separation in a relatively straightforward way:

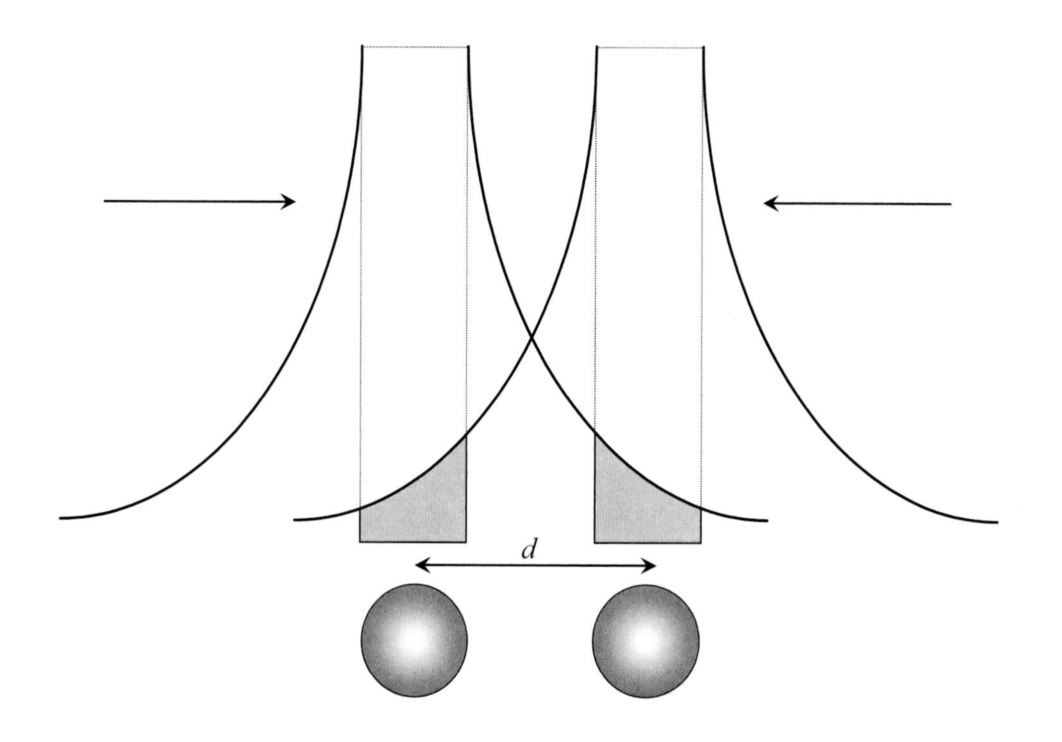

Figure (12.2) The Strong force is generated when hollow cores excise deflectional force from neighboring fields

The shaded areas in Figure (12.2) depict the deflectional force lost when two protons superimpose. Its magnitude is given directly by their core/field topology and will be called the *Strong force*, F_s:

$$F_s\left(d \geq 2R_{rc}\right)=2\left(\int_{r=d-R_{rp}(d)}^{d+R_{rp}(d)} t\,dr\right)=-\frac{3\Diamond_q}{4\pi}\left(\frac{1}{\left(d-R_{rp}(d)\right)^2}-\frac{1}{\left(d+R_{rp}(d)\right)^2}\right) \tag{12.2}$$

where d is the distance between the centers of the two protons, R_{rp} is defined by Equation (12.1), and R_{rc} is the minimum recessed core radius of two adjacent protons in free space.

When protons are separated by a distance significantly larger than their core size, the Strong force of Equation (12.2) falls off rapidly, becoming:

$$F_s\left(d >> R_p\right) \cong -\left(\frac{3\Diamond_q R_p}{\pi\, d^3}\right) \tag{12.3}$$

Unlike the Coulomb force, it decreases as the *cube* of distance. This is one of the reasons it is not a long-range effect. Conversely, when proton cores touch:

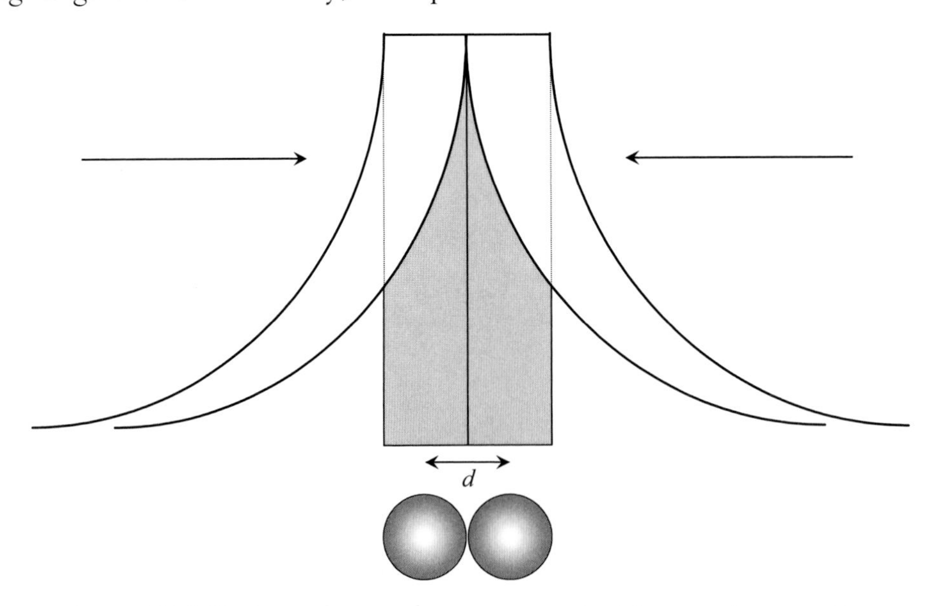

Figure (12.3) The Strong force reaches a maximum at minimum proton separation

Equation (12.2) gives the attraction between them as:

$$F_s(d=2R_{rc})=-\frac{2\Diamond_q}{3\pi R_{rc}^2} \tag{12.4}$$

or –4400 N at the deuteron's recessed core radius of 0.8735 F (Figure 10.5).

Equation (12.4) is the Strong force's peak value for a system of two protons because it is the range where their cores excise the maximum amount of deflection from each other's external fields. The ratio between this and the Coulomb force at the same range is given by:

$$f_{s/q} = \frac{\left| -\dfrac{2\Diamond_q}{3\pi R_{rc}^2} \right|}{\left| \dfrac{q^2}{4\pi\varepsilon_0 (2R_{rc})^2} \right|} = \frac{32\varepsilon_0 \Diamond_q}{3q^2} \cong 58.2 \qquad (12.5)$$

This is consistent with the Strong force's observed strength, approximately two orders of magnitude greater than the Coulomb. Although the strength ratio of Equation (12.5) appears to be directly proportional to unit polarvolume, if $\Diamond_q$ had a different size so would the Coulomb strength term, (q^2/ε_0). In short, the ratio of Strong to Coulomb force is independent of unit polarvolume's value, as both are derived from different components of the same topology.

At spacing less than $2R_{rc}$, core integrity causes the protons' cores to flatten against each other on their shared side, invoking a powerfully repulsive core deformation force that offsets the attraction of core excision:

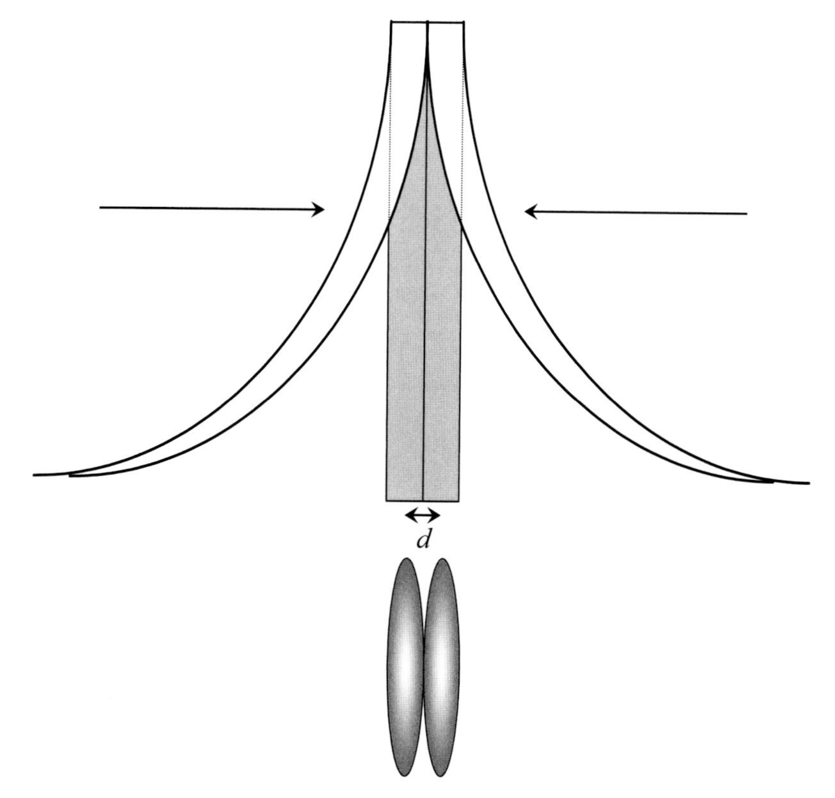

Figure (12.4) The Strong force of flattened cores

The attraction between flattened cores can be estimated from the spherical case because charge conservation keeps the protons' deflection fields comparable to their free-space profiles. This means that their core volume remains roughly constant, and so does their peak isoexternal boundary deflection:

$$t_c \cong \frac{3\Diamond_q}{4\pi R_{rc}^3} \tag{12.6}$$

Thus when cores flatten during a separation of ($d<2R_{rc}$), the variable radial terms of Equation (12.2) (underlined):

$$F_s\left(d \geq 2R_{rc}\right) = 2\left(\int_{r=d-R_{rp}(d)}^{d+R_{rp}(d)} t\, dr\right) = -\frac{3\Diamond_q}{4\pi}\left(\frac{1}{\left(d-\underline{\underline{R_{rp}(d)}}\right)^2} - \frac{1}{\left(d+\underline{\underline{R_{rp}(d)}}\right)^2}\right)$$

become asymmetrically fixed at the maximally-recessed proton core radius, R_{rc}:

$$F_{sa}\left(d \leq 2R_{rc}\right) \cong -\frac{3\Diamond_q}{4\pi}\left(\frac{1}{R_{rc}^2} - \frac{1}{\left(d+R_{rc}\right)^2}\right) \tag{12.7}$$

When ($d<2R_{rc}$), core excision is defined entirely by distance of separation. It decreases with increased proximity, eventually falling to zero at ($d = 0$) since completely flattened cores lose their capacity for field excision. Equation (12.7) is identified as F_{sa} as it is the attractive *component* of the Strong force of flattened cores.

The Strong attraction of core excision attenuates as cores collapse against each other while the repulsion of core distortion rapidly increases. The confluence of these two effects can be described by combining the deformation of Equation (11.13) with the attraction of Equation (12.7):

$$F_s\left(d \leq 2R_{rc}\right) \approx -\frac{3\Diamond_q}{4\pi}\left(\frac{5}{4R_{rc}^2} - \frac{1}{\left(d+R_{rc}\right)^2} - \frac{1}{d^2}\right) \tag{12.8}$$

where d is again the distance between the protons' centers and R_{rc} is the size of two recessed spherical proton cores at first contact. Although additional recession may occur when cores flatten against each other, our linear core approximation defines R_{rc} as minimum core size at any distance of separation. Core deformation is nonlinear, so Equation (12.8) is only an approximation of the actual Strong force of flattened cores, but it provides an accurate description of its general character. Note that F_s diverges to infinity at ($d = 0$).

STRONG FORCE

From Equations (12.2) and (12.8), the Strong force between two protons is shown below for a range of separation from 0 to 5 Fermi. The dotted trace shows what the Strong force would look like in the absence of its core deformation component:

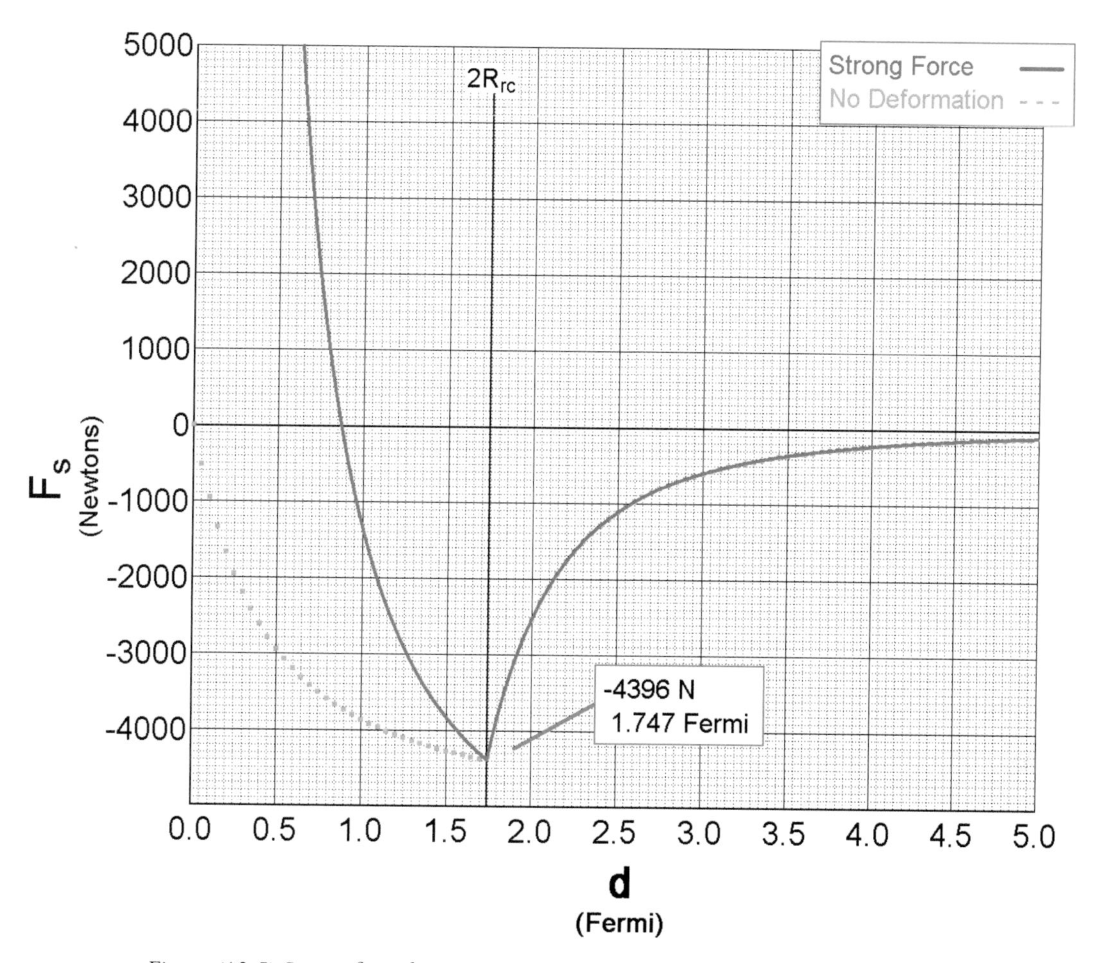

Figure (12.5) Strong force between two protons, with and without core deformation

The Strong force in a system of two protons peaks at –4400 N near ~1.75 F, defined by the relatively mild core recession in deuterium. The enhanced core recession found in heavier nuclei corresponds to a much more powerful force that reaches its maximum at an inter-nucleon spacing closer to 1.2 F. One of the reasons the nuclear force is so difficult to characterize is because it varies with the core size of interacting particles, and their core size varies with their proximity and density. F_s becomes repulsive in a system of two protons near a separation distance of 0.8 F, diverging rapidly to infinity at a separation of (d = 0). *Particle cores cannot occupy the same space at the same time.*

STRONG POTENTIAL

The Strong *potential* at some distance of separation *d* between two protons is the integral of their Strong force from *d* to infinity:

$$\Phi_s(d) = \int_{r=d}^{\infty} F_s(r)dr \tag{12.9}$$

This must be evaluated numerically because of the complexity introduced by core recession. It is shown below in units of millions of electron volts (MEV), for two protons at a range of separation from 0 to 5 Fermi. As before, the dotted trace shows what the Strong potential would look like in the absence of its core deformation component:

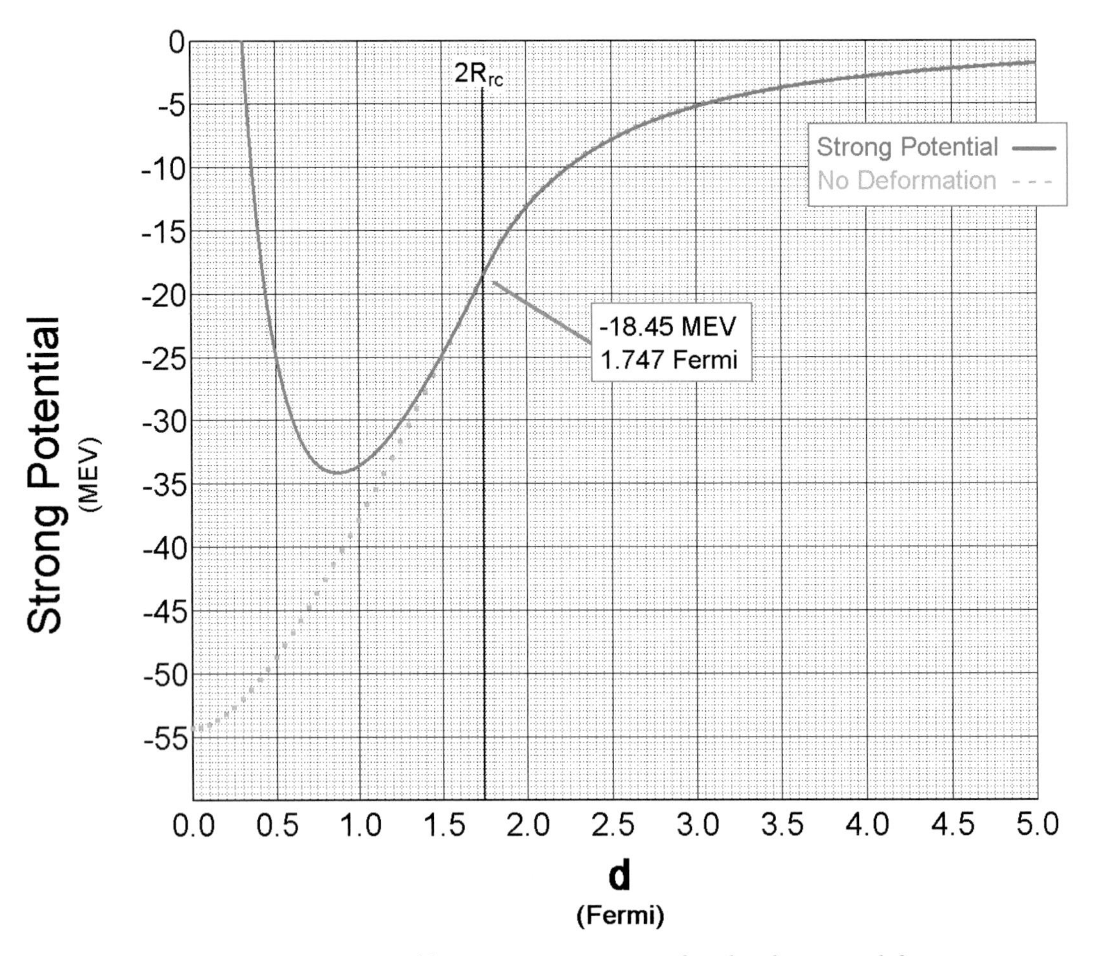

Figure (12.6) Strong potential between two protons, with and without core deformation

The potential when two recessed proton cores first touch is −18.45 MEV, in excellent agreement with the estimated energy of a deuteron's loosely bound triplet state ($l = 0$), which is ~ −18 MEV[9.5] (this is calculated as ~ −20 MEV in its unbound ($l = 1$) energy level less its −2.22 MEV binding energy). *Unit polarvolume provides both the inter-nucleon spacing and energy of a deuteron's bound triplet state.*

The peak attractive potential that a system of two protons can source is −34.5 MEV, occurring at $d \cong 0.8$ F. Given our simple approximation of the nonlinear region ($d<2R_{rc}$), this is exceptionally close to the observed maximum depth of a deuteron's nuclear potential, ~ −35 MEV.[18.5] The distance where the Strong potential becomes intensely repulsive, 0.3 F, is consistent with the observed size of the Strong force's repulsive center ~0.5 F.[9.3] As shown by the dotted trace, in the absence of core deformation the peak attractive potential between two protons would be ~ −55 MEV, occurring at an inter-particle separation of zero.

Figure (12.6) indicates a potential of ~ −2 MEV at 5 Fermi, which is certainly a measurable value. Since no effect is evident at any distance of separation greater than about 2 Fermi, *the Strong interaction commences upon the immediate proximity of particle core boundaries.* It is *literally* a core interaction, as opposed to the Coulomb potential's field interaction, and it requires the direct association of particle cores in order to transfer energy and maintain potential. This is because the particle carrying this force is another core: the electron. In atomic nuclei, the Strong force is carried exclusively by bound electrons. This is why a certain ratio of neutrons to protons is required for stability, and why neutrons are thought to be stable in the nuclear environment. In free space, neutrons decay in about 10.3 minutes; in a nucleus they are continuously transmuting between protons and neutrons at a rate sufficient to carry the Strong potential.

ELECTRON "STRONG" FORCE

Like protons, electrons have cores of voided space, yet they generate no attractive potential. It is fairly easy to show why this is the case. Using an electron analogy of Equation (12.9):

$$\Phi_{s_e}(d) = \int_{r=d}^{\infty} F_{s_e}(r)\,dr \qquad (12.10)$$

the maximum depth of an electron-electron "Strong" potential amounts to ~ −29.7 KEV, precisely (m_p/m_e) smaller than the −54.5 MEV shown in Figure (12.6). This is far less than the ~ −1 MEV required to produce electron-positron cores. As such there is no core exchange and no observable force. The potential, but not the actuality, exists. A core interaction is only possible through the exchange of core energy. The Strong force is a negative potential *by-product* of rest energy. An electron's potential is by definition too small to support its entire mass, and electrons are the lightest available particles. The only universal forces that electrons exert on each other are Coulomb and gravitational.

ELECTRON COMPRESSION ENERGY

Atomic nuclei are more than just the interaction of bound protons, and energy is required to compress their bound electrons to nuclear dimensions. A neutron's binding energy of 0.78 MEV is required to compress its electron to 2.7 F, and *energy scales inversely with a particle's core radius*. The relationship between a bound electron's radius and its *core compression* energy follows as:

$$E_{ec} = \frac{n_b R_{en}}{R_{ce}} \tag{12.11}$$

where n_b is a neutron's binding energy, R_{en} is the radius of its bound electron (2.7 F), and R_{ce} is the radius of a nuclear electron. Bound electrons in atomic nuclei have radii comparable to bound protons, ~0.75 F, as derived by applying Equation (10.17) to the nuclear retention profile of Figure (10.9). This corresponds to electron compression energies in the neighborhood of 3 MEV, consistent with the energy level that electrons often carry away from the beta decay of radioactive isotopes.[9,4] In stable nuclei, electron compression energy is sourced from the Strong potential of protons. Deuterium's bound electron, for example, has a radius of 0.81 F (Figure (10.5)) and therefore a compression energy of 2.6 MEV. Since this is necessarily sourced from the nuclear potential of deuterium's protons, a fraction of their bound kinetic energy is present in their bound electron companion.

12.3 SATURATION

The Strong interaction's *saturation* aspect becomes evident in heavy nuclei, where binding energy per nucleon begins to fall off for isotopes with a mass number greater than ~56 (iron). Saturation exists because the Strong interaction has such a short range - nucleons can only exchange bound electrons with a limited number of adjacent nucleons. The proton to neutron ratio of stable nuclei is evidence of this, as too many bound electrons can spoil the mix by physically interfering with inter-proton proximity.

A nucleus's protons exchange bound electrons at a rate sufficient to maintain nuclear potential. Protons are continuously converting into neutrons while neutrons flux back into protons. The reason why there is no such thing as ^{2}He (two protons) is because it has no bound electron to transfer energy, and therefore no Strong interaction to overcome the protons' Coulomb repulsion. The reason why there is no such thing as ^{2}n (two neutrons) is because neither of its protons are available to accept a bound electron. Nuclei are a blend of geometry and dynamics:

Figure (12.7) Bound protons exchange bound electrons to maintain stability

On first inspection, the frenetic bound electron exchange required to support the nuclear potential might seem at odds with beta decay. When neutrons decay in free space, they emit neutrinos. So if protons and neutrons continuously exchange roles in a nucleus, why don't they lose neutrinos in the nuclear equivalent of the Rutherford catastrophe? *For the same reason an atom's electrons don't emit photons and crash into its nucleus.* The Rutherford catastrophe is avoided because an orbital electron's net acceleration is zero within its corresponding photon emission time. The nuclear neutrino catastrophe is avoided because *for every decaying neutron another is formed, and the time between these two events is shorter than the time necessary to emit a neutrino.* Moreover, since the bound electrons exchanged within a nucleus remain in their bound state, never expanding to their free state, beta decay is avoided entirely. The only case where neutrinos are released in a nuclear interaction is when a permanent change of nuclear charge occurs, such as the net loss or gain (by positron emission) of an electron.

In summary, the absence of energy and space at the heart of a proton's unimaginable energy density creates an attractive potential deep enough to hold atomic nuclei together. Protons' hollow cores are directly responsible for the Strong force. It has nothing to do with illusory phantasms called quarks or the invisible dimensions envisioned by string theorists.

12.4 GENERAL CONCLUSIONS

The unit polarvolume distribution is successfully applied to nuclear phenomenon:

- Hollow proton cores provide an interaction that corresponds closely to the strength and range of the Strong force, as well as the spacing and bound energy of deuterium.

- Proximate electron cores do not experience an interaction analogous to the Strong force because their core potential is too small to produce an exchange particle.

- Neutrons continuously form and decay within atomic nuclei as protons exchange bound electrons to sustain their nuclear potential. No neutrinos are released in this process because the time needed to transfer a bound electron in a nuclear environment is shorter than the time required for neutrino emission.

Although it is easy to be misled by calculations, unit polarvolume provides far too many compelling correlations to be easily dismissed. The results presented in this chapter are derived through the extension of first principles with a logical basis. They are grounded in the dimensional structure of energy, and there is only one step to get from this to the size of proton cores. It all begins with unit polarvolume, *and this provides everything from the densities of atomic nuclei and white dwarfs to the magnitude and range of the Strong potential between nucleons.*

There is a reason why positively charged nuclei stay together; the same reason why particles have a finite rest energy. *Hollow elementary cores.* Understanding is the most important thing; it is the foundation preceding any real growth in science. *Null Physics* not only gives us the strength and range of the Strong force but also explains the Coulomb force, the Weak force, and the astonishing complexity of elementary particles. The universe doesn't have seven or nine or eleven submicroscopic dimensions as string theorists would have us believe. The forces and complexity of nature arise from the simple fact that the energy of an individual particle's field is distributed into space in a precise way.

What is less easy to explain, and what has always haunted theoretical physicists in their quest for grand unification, is why matter's structure seems so overwhelmingly and inconceivably *appropriate.* Protons and electrons have a mass ratio that provides atomic structure and stable nuclei. Matter generates forces with strengths necessary for the formation of molecules, planets, and stars. Stars burn hydrogen at a rate that enables life to evolve and at some stage, after billions of years, attain enough consciousness to marvel at this long string of suspiciously fortuitous relationships. Nature has no coincidences. Every aspect of the universe follows directly from the inevitability of nothingness. This isn't because it was designed by some all-powerful and omniscient being; it is because totality's relentless boundlessness requires spectacularly complex mechanisms for its own self-nullification. Our universe is the only equation that *works.*

13. GRAVITATION

13.1 GRAVITATION IN REAL SPACE

The last few chapters have focused on the external deflection a particle core generates and its relationship to the Coulomb, Weak, and Strong forces. Our analysis now turns to the fourth universal force: gravitation. It is wholly unlike the others because it is the result of an entirely different form of spatial deflection - internal. As it turns out, however, internal and external deflections share quite a few attributes. One of their most basic similarities is that neither can *arbitrarily* change its value.

A change of deflection must always be accompanied by a change in the substrate upon which it acts. Spatial deflection has to *vary as space* regardless of its orientation. Space is continuous and conserved. Its dimensionality is the property governing how deflections are distributed within it. Spatial deflections attenuate for a reason, *and the reason is their response to increasing volume.* Internal deflection, like its external analog, attenuates into space as volume:

Ψ THEOREM 13.1 - INTERNAL DEFLECTION ATTENUATION {Ψ3.14}

INTERNAL DEFLECTION'S MAGNITUDE ATTENUATES AS SPATIAL VOLUME

Symbolically:

$$\left|t_{\parallel}(r)\right| \propto \frac{1}{r^3} \tag{13.1}$$

This is one of many things that all spatial deflections have in common.

Unless otherwise specified, the term *internal deflection* will refer to the long-range fields outside a particle's core, not the deflection associated with its spatial evacuation. Also, although internal deflection is a vector quantity, the majority of our calculations are only concerned with its magnitude, so its directional nature will be referenced only when necessary. In all other cases its value will be symbolized as $|t_{\parallel}|$.

13.2 INTERNAL DEFLECTION

There is no geometric restriction to the magnitude of external deflection because *it exists perpendicular to space*. Indeed, a deflection of a trillion light years near a proton's core can attest to this. The same is not true of internal deflection. Space's three-dimensional continuity imposes restrictions on its variation across distance.

Internal slope is the rate of change of internal deflection along space. Take the case of two closely spaced points **A** and **B**. **A** is deflected to **A′** and **B** to **B′** as follows:

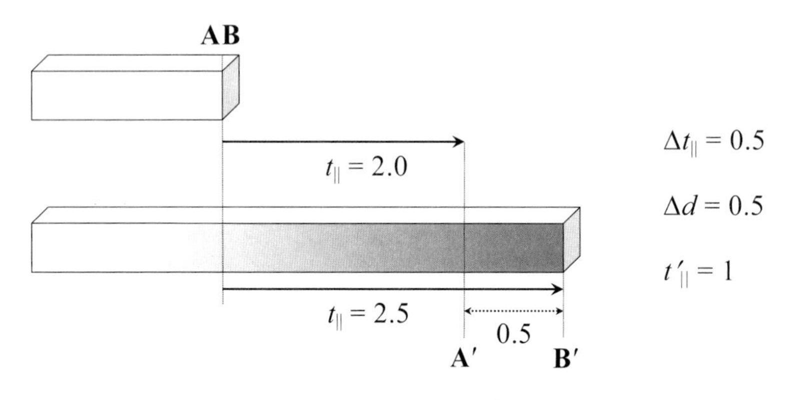

Figure (13.1) Continuity limit

In space with no *external* slope, which will be referred to as *flat space*, the rate of change of internal deflection per unit distance is limited to unity. If the deflection at **A′** is 2.0 and **A′** and **B′** are separated by 0.5, a deflection at **B′** any greater than 2.5 would change the *order* of **AB** in the source metric. This would create a spatial fold, violating point conservation. The maximum internal slope of flat, *continuous* space is arbitrarily close to, but less than unity:

$$\left|\frac{\Delta t_{||}}{\Delta d}\right| = \left|t'_{||}\right| < 1 \qquad \{t' = 0\} \qquad (13.2)$$

The slope of *external* deflection has no limitation in the particle field, while the slope of *internal* deflection in flat, continuous space is confined to less than unity:

Ψ THEOREM 13.2 - INTERNAL SLOPE LIMIT {Ψ2.6}

THE MAGNITUDE OF INTERNAL SLOPE IN FLAT, CONTINUOUS SPACE IS LESS THAN UNITY

When unity slope is reached, space becomes discontinuous. The only environment where this occurs is in the boundary of a particle's evacuated core. For space with nonzero external slope, the internal slope limit applies only to its flat component, even though the actual internal displacement *along* space is amplified by the external slope.

INCOMPRESSIBLE SPACE

Just as there is a limit to spatial stretching, there is a limit to its compression, and it is far more severe. Space and energy are both conserved quantities. If internal deflection were to force a region of space into a smaller footprint, its volume would necessarily be preserved throughout the process. However, the only way to compress space is by the differential infusion of internal energy, and internal energy has volume. *Energy adds volume to space.* This is all well and good in external deflection because energy's additional volume is distributed perpendicular to space. External slope always *stretches* space along the fourth dimension, adding varying amounts of its temporal component. The same is not true of internal slope, because the temporal component it adds must lie within space. So while it is possible to stretch flat space, where its internal energy offsets the loss of spatial density, it is not possible to compress it. Flat space is *incompressible*:

Ψ THEOREM 13.3 - SPATIAL INCOMPRESSIBILITY {Ψ7.5}
FLAT SPACE IS INCOMPRESSIBLE

Flat space is never compressed, even when a particle's core is evacuated. A core's evacuation represents extraordinary spatial stretching, not compression, as its volume is transferred into the fourth-dimensional extent of its boundary. Space is not a material substance, but it still has a remarkable number of properties.

ORIENTATION

Although much can be said of the similarity between external and internal spatial deflection, where they differ, they differ markedly. Wholly unlike an external field, an internal field has an *orientation* with respect to its source. But do elementary particles push space outward or pull it inward? The incompressibility of flat space quickly resolves this issue. If a particle's internal field were oriented outward, it would increase spatial density away from its core, and the density of the empty space ($t' = 0$) near massive objects would be above unity. This is not viable. Flat space is inherently incompressible, so internal fields have an inward orientation:

Ψ THEOREM 13.4 - INTERNAL ORIENTATION {Ψ13.3}
INTERNAL DEFLECTION IS ORIENTED TOWARD ITS SOURCE

Gravity's inward orientation is consistent with its observed effect, but somewhat counterintuitive in terms of matter's elementary structure. A particle's core can only be evacuated by pushing space outward from its center, so it is reasonable to think that perhaps its gravitational field is a remnant of the creation of its hollow core. This is not the case, because an outwardly oriented field would violate core integrity just as surely as an evacuation that crosses its core boundary, as shown previously in Figure (9.3):

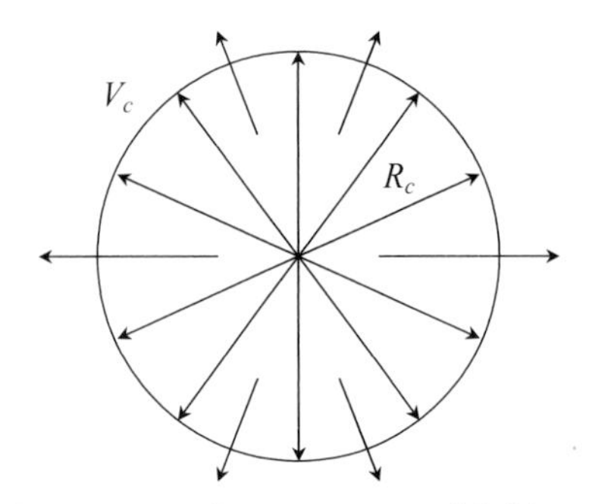

Figure (13.2) An outward orientation for a gravitational field would violate core integrity

The only way to have outward orientation for a core's gravitational field is for its core boundary to be the *destination* of its peak internal deflection. If this were true, the space immediately exterior to the core boundary would have been transported there by internal deflection from within the boundary, violating core integrity. There can be no connectivity between the interior and exterior of a core boundary, as it is discontinuous. The only other option is that all interior deflection terminates at a core's boundary, which is the default situation in the unified particle core. But this results in an internal deflection of *zero* at the degenerate spatial boundary, as shown earlier in Figure (9.6). Gravity pulls space *inward*.

13.3 INTERNAL DISTRIBUTION

The magnitude of any field of spatial deflection, external or internal, is expressed in terms of a (deflection)(volume) relationship. Since volume is the only agent that causes the attenuation of spatial deflection, the product (deflection)(volume) generated by a source field is necessarily constant. In the case of internal deflection, it will be called *internal hypervolume*:

Ψ THEOREM 13.5 - INTERNAL HYPERVOLUME, $\Diamond_{||}$ {Ψ13.1}

THE PRODUCT OF INTERNAL DEFLECTION AND THE SPATIAL VOLUME IT SUBTENDS IS CONSTANT FOR A GIVEN SOURCE

Unlike internal deflection, internal hypervolume is not a vector quantity because the directional components of all of the deflections of which it is composed sum to zero.

Internal hypervolume has the same units as unit polarvolume, time-distance^3, though it will sometimes be expressed as quadric distance since its associated deflection is along space. Its deflection distribution has the form:

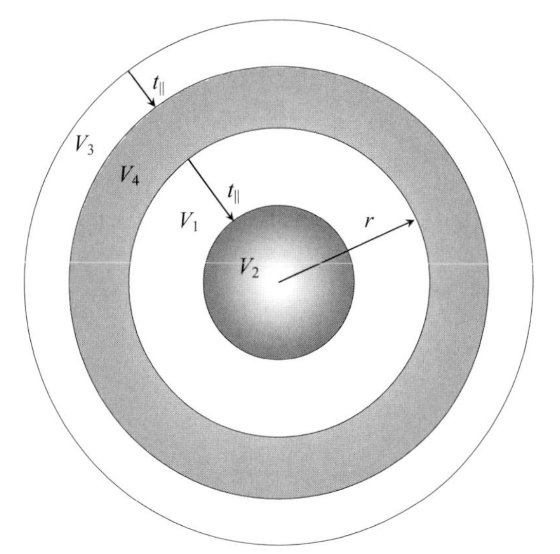

Figure (13.3) Internal deflection relative to a common center

where internal deflection's magnitude attenuates with increasing volume. Space is shown before (white) and after (gray) its internal deflection. This form of deflection distorts space's basic metric, but its volume is always conserved. Space is never created or destroyed under the action of either internal or external deflection. As such, ($V_1 = V_2$) and ($V_3 = V_4$) in Figure (13.3), at least in terms of actual magnitude. The only thing internal deflection alters is spatial density. Deflection, regardless of its orientation, *does not alter the quantity of space it deflects.*

A field of internal deflection has a form similar to that of external deflection, except for the provision that the magnitude of its internal slope can never meet or exceed unity:

$$|t_{\|}| = \frac{3\Diamond_{\|}}{4\pi r^3} \qquad \left\{|t'_{\|}| < 1 \; (t' = 0)\right\} \qquad (13.3)$$

$|t_{\|}|$ is the magnitude of the internal deflection at some *undeflected* radius r. In other words, r represents the spatial *source* of the internal deflection. Internal slope is the magnitude of Equation (13.3)'s first derivative:

$$|t'_{\|}| = \frac{9\Diamond_{\|}}{4\pi r^4} = \frac{3\Diamond_{\|}}{Vr} \qquad \left\{|t'_{\|}| < 1 \; (t' = 0)\right\} \qquad (13.4)$$

Since internal hypervolume is the product of volume and internal deflection, and internal slope can never exceed unity, Equation (13.4) puts an immutable limit on a spherical field's internal deflection at any radius:

$$\left| t_{\parallel}(r)_{\max} \right| = \frac{r}{3} \tag{13.5}$$

Note that Equations (13.3), (13.4) and (13.5) are based purely on geometry and are true regardless of gravity's underlying nature.

ISOINTERNAL CONTOURS

An object's internal deflection field is given by Equation (13.3):

$$\left| t_{\parallel} \right| = \frac{3\Diamond_{\parallel}}{4\pi r^{3}} \qquad \left\{ \left| t'_{\parallel} \right| < 1 \; (t' = 0) \right\}$$

The superposition at any location (x, y, z) of two objects of equal internal hypervolume located at (x_1, y_1, z_1) and (x_2, y_2, z_2) is given by:

$$\vec{t}_{\parallel} = \left(\frac{3\Diamond_{\parallel}}{4\pi} \right) \left(\frac{\vec{r_1}}{r_1^3} + \frac{\vec{r_2}}{r_2^3} \right) \tag{13.6}$$

where:

$$\begin{aligned} \vec{r_1} &= (x - x_1, y - y_1, z - z_1) \quad & r_1 &= \sqrt{(x - x_1)^2 + (y - y_1)^2 + (z - z_1)^2} \\ \vec{r_2} &= (x - x_2, y - y_2, z - z_2) \quad & r_2 &= \sqrt{(x - x_2)^2 + (y - y_2)^2 + (z - z_2)^2} \end{aligned} \tag{13.7}$$

Like external deflection, the superposition of internal deflection results in equideflectional spatial contours, which will be called *isointernal contours*. However, unlike external deflection, which only occurs in positive and negative forms, internal deflection has an infinite number of different orientations with respect to space. This is why vector notation is necessary in Equation (13.6).

The total internal deflection due to the superimposed fields of Equation (13.6) is given by:

$$\left| t_{\parallel} \right| = \frac{3\Diamond_{\parallel} \sqrt{\left(r_{1_x} r_2^3 + r_{2_x} r_1^3 \right)^2 + \left(r_{1_y} r_2^3 + r_{2_y} r_1^3 \right)^2 + \left(r_{1_z} r_2^3 + r_{2_z} r_1^3 \right)^2}}{4\pi r_1^3 r_2^3} \tag{13.8}$$

where:

$$r_{1_x} = x - x_1,\ r_{1_y} = y - y_1,\ r_{1_z} = z - z_1$$
$$r_{2_x} = x - x_2,\ r_{2_y} = y - y_2,\ r_{2_z} = z - z_2 \qquad (13.9)$$

Equation (13.8)'s *isointernal* contours appear as:

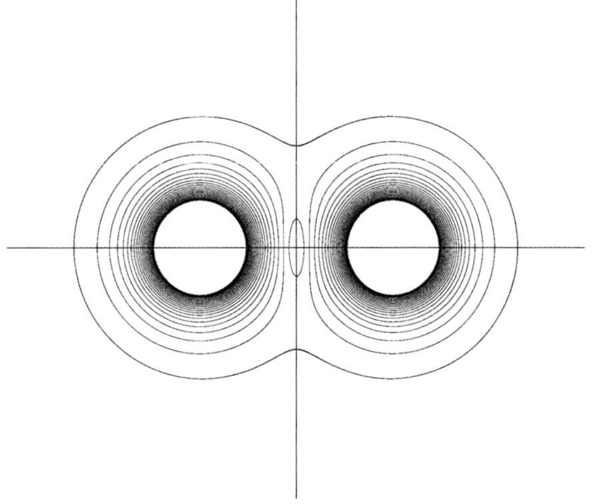

Figure (13.4) Isointernal contours of a gravitational superposition

These contours have the same shape for any value of $\Diamond_{\|}$, *even one that is infinitely small*. They share characteristics of both the (++) and (+−) external interactions, yet are uniquely different. In the (+−) interaction, for example, any location equidistant from both charges has an external deflection of zero. In a gravitational superposition, there is only a single location where internal deflection is zero - the system's center of gravity. The centers of the isointernal contours closest to both sources are displaced away from their shared side, indicating an attractive force consistent with the observed phenomenon. Compare Figure (13.4) to the contours of the attractive (+−) interaction shown earlier in Chapter 10:

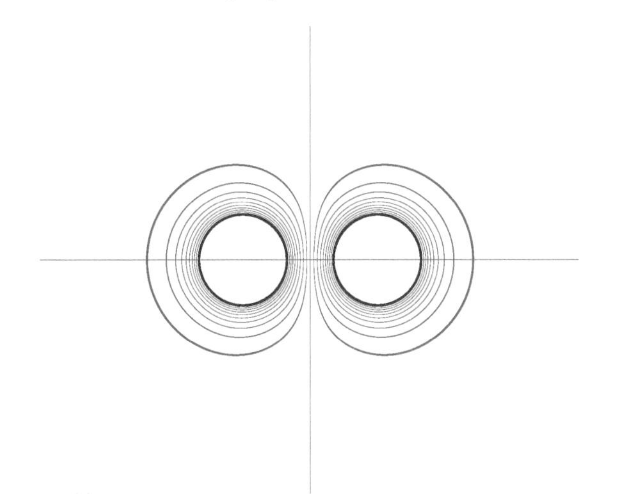

Figure (13.5) Isoexternal contours of the field superposition of unlike charges [From Figure (10.1)]

13.4 GRAVITATIONAL SOURCES

Although we've been able to characterize internal deflection based on its associated geometry, the most important question remains - what causes it? Our investigation begins with the phenomenon that launched the General Theory of Relativity into the spotlight - the gravitational deflection of light.

DEFLECTING PHOTONS IN A VACUUM

Light is deflected by the gravitational field an object generates, and when this field is strong, as in the case of our sun, this deflection is readily observable. It is an elastic interaction, so the conservation of momentum demands that for each change in photon momentum there be a compensatory change in the momentum of the object that deflects it. The only way deflected light can change the momentum of a nearby mass is if it has a gravitational field of its own. It could be argued that massive objects *induce* gravitational fields in photons, which in turn enable them to transfer momentum, but this is unlikely because the applied field is independent of a photon's native fields. Moreover, if gravitational fields induced compensatory fields in the objects on which they acted, the field generated by an object would be greater than the sum of its component's fields, not proportional to its mass.

The crux of the argument is field integrity. Particles induce changes in other particles only to the extent that their fields interact with the *source* of other particles' fields. Matter's fields are distributed derivatives of the presence of cores in space, and particles react to the asymmetry of their own fields, and nothing else. A planet, composed of particle cores, can only respond to a photon's presence if the photon's fields *induce* a force. *Photons have long-range gravitational fields*. Further, photons of any energy follow the same curved trajectory near a massive object. Since their change in momentum is proportional to their energy, and since they must induce a corresponding change in the massive object they pass, their gravitational fields are proportional to their energy. So although photons' electromagnetic fields are bundled by unit hypervolume, their gravitational fields are far less constrained, and are the bridge that unifies the universe's fundamental energy forms:

Ψ THEOREM 13.6 - GRAVITATIONAL LIGHT {Ψ7.10, Ψ13.5}

PHOTONS HAVE INTRINSIC GRAVITATIONAL FIELDS WHOSE STRENGTH IS PROPORTIONAL TO THE PHOTON'S ENERGY CONTENT

This is a pivotal consideration. Photons and particles share many attributes, and both are distributed regions of external deflection with variable energy density. Gravitation being common to all forms of energy reveals a great deal about its underlying nature.

It follows that neutrinos, like photons, also have intrinsic gravitational fields:

Ψ THEOREM 13.7 - GRAVITATIONAL NEUTRINOS {Ψ13.6}

NEUTRINOS HAVE INTRINSIC GRAVITATIONAL FIELDS WHOSE STRENGTH IS PROPORTIONAL TO THE NEUTRINO'S ENERGY CONTENT

A neutrino's gravitational field scales with its energy for the same reason a photon's does - the gravitational deflection of its path through space. Neutrinos are fantastically ethereal, able to penetrate light years of lead with no interaction, so there are no direct measurements to support Theorem (13.7). But it is well-grounded nonetheless. Neutrinos move through space at the speed of light, and space's geometry is single-valued. As such they must take the same paths that light takes, and incur the same change to their momentum. *Anything that has energy has gravity*. This is consistent with the stress-energy tensor of the General Theory of Relativity.

ELEMENTARY PARTICLE FIELDS

Gravitation is so weak that it only makes its presence known in macroscopic objects, and the strength of their fields is proportional to the amount of material of which they are composed. As was the case in photons, the conservation of momentum can again be used to show that this proportionality extends all the way down to individual elementary particles.

In a perfect vacuum, protons and electrons fall at the same rate in Earth's gravitational field. This means that if Earth were to capture either particle at rest from deep space, they would both have the same speed, as they fall inward, at any given radius above Earth's surface. Thus, in accordance with the Maxfield constraint, electrons and protons exposed to the same gravitational potential convert the same fraction of their rest energy into kinetic energy. Their momentum, however, is a different issue. At any given speed, a proton has 1836 times as much momentum as an electron. This means that the gravitational interaction between Earth and these particles can only satisfy momentum conservation if a proton induces a change in Earth's momentum 1836 times as large as the effect an electron invokes. This in turn means that a proton's gravitational field is 1836 times more powerful than an electron's.

Gravitational field strength is proportional to internal hypervolume much as Coulomb field strength is proportional to unit polarvolume, and since all fundamental forms of energy (photons, neutrinos and elementary particles) exhibit proportionality between their energy content and intrinsic gravitational field, it follows that internal hypervolume is proportional to energy at the compositional level.

The constant of proportionality between energy and its innate internal hypervolume will be called the *gravitational index*, $\kappa_{||}$:

Ψ THEOREM 13.8 - GRAVITATIONAL INDEX, $\kappa_{||}$ {Ψ13.6}

AN OBJECT'S INTERNAL HYPERVOLUME IS PROPORTIONAL TO ITS ENERGY, AND THE CONSTANT OF THEIR PROPORTIONALITY IS THE GRAVITATIONAL INDEX

Symbolically:

$$\lozenge_{||} = \kappa_{||} E \tag{13.10}$$

The gravitational index differs markedly from the gravitational constant G normally used in physics because $\kappa_{||}$ is based on gravity's substructural foundation, whereas G originates from empirical measurements made in the absence of any knowledge of energy's dimensional composition. That said, Equation (13.10) is still consistent with the latest gravitational data.⟨21-31⟩

BACKGROUND GRAVITATION

Each bit of energy carries its own gravitational field. Since energy is conserved and eternal, it follows that gravity is as well:

Ψ THEOREM 13.9 - ETERNAL GRAVITATION {Ψ13.8}

GRAVITATIONAL FIELDS HAVE INFINITE AGE, RANGE, AND CONNECTIVITY

Gravity is driven by energy - it is not limited to matter. Unlike electrostatic fields, gravitational fields have infinite history and therefore infinite range and connectivity. But this does not mean that their accrued *effect* is unbounded. Gravitational force decreases as the square of distance and the universe's unlimited energy distribution increases as the cube, so at first glance it seems like there should be an infinite amount of gravitational force acting on any particle in space:

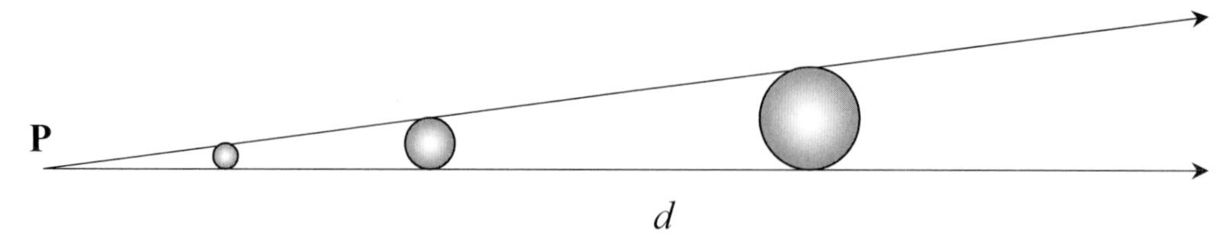

Figure (13.6) Unchecked accumulation of gravitational force on particle **P**?

The reason this is not the case is because gravity's physical spatial deformation (internal deflection) decreases as the cube of distance, and accumulated *differences* in internal deflection decrease as the fourth power of distance. Take any huge spherical sample of the universe of radius *R* with central location *C*. The sum of the internal deflection at *C* due to the energy within *R* asymptotically approaches the value due to the local energy distribution near *C* as *R* goes to infinity. This is in keeping with gravity's customary operation. Most of the universe's galaxies move no faster than a few hundred km/s in response to variations in the cosmic distribution of matter - ponderously slow in relation to their size.

13.5 GRAVITATIONAL INDEX

The natural inclination, when investigating gravity, is to compare it to the other universal forces. As it turns out, this isn't very helpful. Nongravitational interactions are all based on unit polarvolume, which provides clear, crisp interactions like the Strong force, emanating directly from a well-defined boundary condition. Unit polarvolume's forces are not subtle; they are powerful, two-dimensional slices of a four-dimensional whole. And then there is gravity. Its effect is governed by energy, not unit hypervolume, and is nearly forty orders of magnitude weaker than the other universal forces. Gravitation is a deeply enigmatic phenomenon. The size of its field can be accurately described using internal hypervolume, but the *cause* of this field is the more compelling question.

Rearrange Equation (13.10):

$$\kappa_{\parallel} = \frac{\Diamond_{\parallel}}{E} \tag{13.11}$$

Energy has a *fixed internal hypervolumetric density*. This means that space, the substrate in which energy resides, is similarly disposed. But we already knew this. Gravitational fields pervade the entire universe, filling it with an imperceptible density of internal energy. Internal energy requires internal slope, which in turn requires internal deflection. Internal deflection is space's *internal hypervolumetric density*, and by Equation (13.11) is directly proportional to energy density:

$$|t_{\parallel}| = \frac{\Diamond_{\parallel}}{V} = \frac{\Diamond_{\parallel}}{E}\left(\frac{E}{V}\right) = \frac{\Diamond_{\parallel}}{E}\rho_E = \kappa_{\parallel}\rho_E \tag{13.12}$$

The universe's average energy density is unity, so the gravitational index is the *average internal deflection of space*:

$$\overline{|t_{\parallel}|} = \kappa_{\parallel}\frac{\overline{\rho_E}}{\rho_U} = \kappa_{\parallel} \tag{13.13}$$

where $|t_{\parallel}|$ is expressed in units of distance instead of seconds or Joules/m^2.

The relationship between internal deflection and the universe's energy will be called the *gravitational foundation*:

Ψ THEOREM 13.10 - GRAVITATIONAL FOUNDATION {Ψ13.8}

INTERNAL DEFLECTION IS PROPORTIONAL TO ENERGY DENSITY AND HAS A UNIVERSALLY AVERAGE VALUE EQUAL TO THE GRAVITATIONAL INDEX

The question now is - *How large is the universe's average internal deflection?* As noted earlier, the gravitational force is far weaker and markedly different from the other universal forces, and as it turns out, deriving the value of its index is a grueling contest.

GRAVITY'S ELEMENTARY CAUSATION

As our investigation digs more deeply toward gravity's ultimate source, its underlying nature starts to get truly fascinating. Particles and photons *generate* gravitational fields, and momentum conservation requires exact proportionality between their energy content and the gravitational fields they produce. Energy's inherent hypervolumetric density is *caused by some governing agent*. Three candidates for this agent are reviewed below, along with their justification and contentious issues:

1. Evacuated particle cores

   Premise: The presence of a universal density of hollow particle cores scattered across space imbues both the unit hypervolume of photons and the unit polarvolume of elementary particles with trace amounts of internal hypervolume.

   Rationale: Distributed internal fields, like external fields, have to start at some boundary, and the only boundaries the universe has to offer are unit hypervolume and its charged half, unit polarvolume. If a particle or photon exhibits finite internal hypervolume, it represents a modification of its bounding condition. Unit hypervolume is infused with internal hypervolume *because space is*. Energy, composed of space and governed by unit polarvolume, has the same tiny inherent fraction of internal hypervolume, as shown by its gravitational index.

   Problem(s): The first problem with this premise is its asymmetry, where photons' gravitational fields are caused by the existence of hollow particle cores instead of the existence of the photons themselves. Photons' intrinsic gravitational fields are an active component of the universal gravitational field. The photons that help to generate this all-encompassing field need to be part and parcel of its basis, regardless of the nature of said basis. The second problem with this idea is it does not explain *why* the presence of particle cores in space alters the (photon/unit hypervolume) and (particle/unit polarvolume) relationships, both of which are based solely on the universe's four-dimensional size. Space and energy are preserved in all interactions. Altering energy's form does not alter its underlying amount, so unit hypervolume should be as immutable, from any perspective, as the existential substrate of which it is composed. In a more general sense, the idea that finite internal deflections could compress a particle's core boundary is inimical to its singular nature. Nor does it make much sense. There is, quite literally, nothing inside of a particle core, and its boundary has infinite energy density. How could it be compressed? It might be possible to force the field immediately exterior to a particle's core into its core boundary with internal deflection, similar to the mechanism by which its interior volume is evacuated, but this would blur the core boundary's discontinuous nature.

Determination: Gravitation caused by evacuated cores is artificial and unviable.

2. Temporal closure
Premise: Gravitational field strength is a temporally closed parameter (Ψ4.9).

Rationale: Just as elementary particles cannot exist in the absence of photons, and photons can only be emitted in the presence of elementary particles, gravitation is a temporally closed parameter that provides a deep connection between all fundamental forms of energy. The strength of the universal gravitational field affects unit hypervolume, which in turn defines the strength of elementary fields, which in turn define the strength of the universal gravitational field. *Unit hypervolume and nothingness give the universe very few degrees of freedom in which to achieve self-nullification.* We already know of the temporally closed link between the genesis and destruction of photons and particles, but their codependence does not stop there. Photons and elementary particles are *complementary* entities. Light is the way it is because it has to be fully compatible with matter, and vice versa. Perhaps gravity provides the operational link that, along with the relative concentrations of all forms of energy, allows the universe to sum to zero.

Problem(s): The reasoning behind this premise is disturbingly close to the style used in the anthropic argument, and might serve as a perfect example of the misapplication of temporal closure. Poor universe! It needs gravity to make everything work! Temporal closure should only be used once a phenomenon's underlying dynamics are fully understood, and as it stands, very little is currently known about the interaction between two sources of internal deflection. Temporal closure cannot tell us how, for instance, stars produce heavy elements - it just tells us the fact that stars can't form without stellar by-products *is not a paradox*. Also, as was the case with the prior scenario of evacuated cores as gravity's progenitor, it is unclear (a) why the universal gravitational field would alter the relationship between energy and its quantization agents, and (b) how the value of a particle's unit polarvolume could be modified by a finite internal deflection at its core boundary. One of the challenges facing the continued development of *Null Physics* is to know where geometry ends and temporal closure begins. In the absence of evidence to the contrary, geometry should be assumed to be the governing mechanism.

Determination: Gravitation from temporal closure is unviable and untestable.

3. Intrinsic component

Premise: Energy's substructure contains a trace amount of internal hypervolume.

Rationale: This is in fact the case, although whether or not it has anything to do with gravity remains to be seen. Energy is sloped space, with additional volume per unit volume. The only way that extraspatial volume can be stored within space's compositional substrate is by increasing its four-dimensional *thickness*. Empty space has a four-dimensional thickness of netherspace, $\delta_3 = (1^4/\infty^3)$, and energy density increases this in direct proportion:

$$\rho_\Diamond = \left(\frac{\rho_E}{\rho_U}\right)\delta_3 = \left|t_{||}\right| \qquad (13.14)$$

Since this extra volume exists within space's geometric fabric, it is by definition *internal*. But even though energy's substructure *does* contain a small density of internal hypervolume - it is an *infinitely* small density. That said, this idea is still (a) consistent with the definition of the gravitational index, where ($\kappa_{||} = \delta_3$) in Equation (13.12); (b) decidedly different from the basis of the other universal forces; and (c) symmetric, since photons and particles both produce gravitational fields directly from their innate energy density. Also, infinitely small fields are consistent with the incompressibility of flat space, (Ψ13.3), in that energy is only the slightest bit more compressible than the spatial substrate of which it is composed.

Problem(s): The central issue with this concept is operational. How can an infinitely small internal field possibly generate a finite amount of force? *Perhaps by symmetry*. If it takes a finite object to produce an infinitely small field, and this field interacts with the infinitely small field of another finite object, it should produce a finite result. The ratio of any two gravitational fields is finite, regardless of their underlying magnitudes. If gravitational interactions involve applied and resultant internal forces, analogous to the Coulomb interaction, the only way to modulate the strength of an internal field is with a finite deformation of the energy density from which it is generated.

Determination: Interesting possibility.

The one thing particles and photons have in common, other than gravity, is energy, and since energy has internal hypervolumetric density by default, it would appear that the strength of gravity is defined purely by energy's spatial distribution:

Ω HYPOTHESIS 13.1 - GRAVITATIONAL PRECURSOR {Ψ13.10}

ELEMENTARY GRAVITATIONAL FIELDS ARE CAUSED BY THE INFINITELY SMALL INTERNAL HYPERVOLUMETRIC DENSITY OF ENERGY'S EXTRASPATIAL VOLUME

If this hypothesis is true, the superposition of arbitrarily small internal fields will produce a finite interaction, and gravity's strength is *independent* of the size of the gravitational index. This can be tested as soon as elementary gravitation can be simulated. Let's turn now to gravity's practical application - its day to day role in the preservation of our eternal universe.

13.6 INTERNAL FORCE

As was the case with the Coulomb force, gravity cannot exist without the generation of differential deflectional force across a particle's asymmetrical core boundary. Gravitational fields induce nonspherical core distortions, altering the internal and external deflection that a core generates. A particle's internal field is far smaller (perhaps infinitely smaller) than its external field, so it represents an insignificant contribution to induced core asymmetry, even though it is its ultimate causation. Let's use the known strength of gravity to calculate the magnitude of gravitational core distortion.

For weak gravitational fields, the force between a particle of rest mass m_0 and source M is given by:

$$F = \frac{Gm_0 M}{d^2} \tag{13.15}$$

Equation (11.10) from our Coulomb analysis can be modified, using Equation (13.15), to show the core asymmetry caused by weak gravitational fields:

$$\frac{Gm_0 M}{d^2} = \Delta R_c \frac{3\Diamond_q}{4\pi R_c^3} \tag{13.16}$$

Solving for gravitational core distortion and substituting Equation (9.16) for a particle's rest energy (mass) yields:

$$\Delta R_c = \frac{4\pi Gm_0 MR_c^3}{3\Diamond_q d^2} = \frac{12\pi GMR_c^2}{c^2 d^2} \tag{13.17}$$

Equation (13.17) represents a remarkably small core distortion. Two protons separated by a meter alter their respective radii by only $4.2(10)^{-83}$ m. A proton located on the sun's surface ($d = 7(10)^8$ m) incurs a $1.0(10)^{-43}$ m distortion, unimaginably small even by nuclear standards. As it turns out, however, the core asymmetry described by Equation (13.17) is not the most interesting aspect of matter's elementary gravitational behavior. Neutron stars and black holes have a far more profound effect on a particle's *size* than its shape.

13.7 CORE EXPANSION

The potential energy of a particle of rest mass m_0 in the weak gravitational field of a mass M is given by:

$$\Phi_g = \frac{Gm_0 M}{d} \tag{13.18}$$

This is the amount of rest energy that a particle loses to its gravitational interaction with another object. It is either converted to kinetic energy or lost as radiation. Since internal and external deflections do not superimpose directly, the only way a particle can lose rest mass within a gravitational field is by an increase in its core size. This will be referred to as *gravitational core expansion*. Its magnitude is given by applying Equation (9.16) to Equation (13.18):

$$9\Diamond_q \left(\frac{1}{R_c} - \frac{1}{R_{xc}} \right) = \frac{Gm_0 M}{d} \tag{13.19}$$

where a particle's initial rest energy is defined by its core radius in free space R_c, and R_{xc} represents the expanded radius induced by the applied gravitational field. In a weak field, $(R_{xc} - R_c) << R_c$, so Equation (13.19) reduces to:

$$\frac{9\Diamond_q}{R_c^2} \delta R_c = \frac{Gm_0 M}{d} \tag{13.20}$$

Solving for the differential change in core radius and simplifying using Equation (9.16) yields:

$$\delta R_c = \left(\frac{GM}{c^2 d} \right) R_c \tag{13.21}$$

where the term (GM/c^2) has units of distance. A proton on our sun's surface experiences a core expansion of about $2(10)^{-21}$ m. A proton on a neutron star's surface expands ~20% or more. Yet this is only a hint of the size that protons can attain in the intense gravitational field of a black hole.

CORE EXPANSION IN A STRONG FIELD

Equation (5.19) gives gravitational potential as a function of enclosed mass in a strong, radially symmetric, nonrotating gravitational field:

$$\Phi_g = m_0 c^2 \left(e^{\Phi} - 1\right) = m_0 c^2 \left(\sqrt{1 - \frac{2GM}{c^2 r}} - 1\right)$$

Gravitational potential is the fraction of rest energy that a particle loses in exchange for its position in a gravitational field. The amount it retains is its *residual rest energy*:

$$E_{\triangle} = m_0 c^2 \left(\sqrt{1 - \frac{2GM}{c^2 r}}\right) \tag{13.22}$$

Light's energy is not affected by gravitational potential. Gravitational potential reduces the rest mass of elementary particles, which in turn emit light of lower energy.

A particle's residual rest energy is related to its expanded core size as:

$$\frac{9\lozenge_q}{R_{xc}} = m_0 c^2 \left(\sqrt{1 - \frac{2GM}{c^2 r}}\right) \tag{13.23}$$

Solve for expanded core size:

$$R_{xc} = \frac{9\lozenge_q}{m_0 c^2 \left(\sqrt{1 - \frac{2GM}{c^2 r}}\right)} = \frac{R_c}{\sqrt{1 - \frac{2GM}{c^2 r}}} \tag{13.24}$$

Core expansion is another reason why the gravitational force is so phenomenologically distinct from all other forces. Whereas the superposition of external fields causes particle cores to shrink, the superposition of internal fields causes them to *expand*.

The sharp contrast between these two effects will be called *core response dichotomy*:

Ψ THEOREM 13.11 - PARTICLE CORE> RESPONSE DICHOTOMY {Ψ5.4, Ψ10.4}

THE SUPERPOSITION OF EXTERNAL FIELDS REDUCES A PARTICLE'S CORE RADIUS, WHEREAS THE SUPERPOSITION OF INTERNAL FIELDS EXPANDS IT

13.8 GALACTIC BLACK HOLES (ADIABATIC APPROXIMATION)

The core hyperdensity calculations of Chapter 10 provide compelling evidence that the material in black holes is in a degenerate state similar to that of a neutron star, albeit at a higher gravitational potential. The other possibility is plasma, and it is fairly easy to show that a black hole's material is not in this state. Gravitational potential near a black hole is given by Equation (5.19):

$$\Phi_g = m_0 c^2 \left(\sqrt{1 - \frac{2GM}{c^2 r}} - 1 \right)$$

Its magnitude at a distance of 5 Schwarzschild radii ($R_S = 2GM/c^2$) is ~ –0.1. Freely moving particles with kinetic energy of only 90% of their rest mass achieve a distance of 4 R_S beyond their veneer. Thus if the material inside a black hole were in a state of hot plasma, its kinetic energy would transport it outside of its veneer, where it would freely radiate. This is not viable. *Black holes are composed of degenerate matter confined by gravitation and nuclear potential.*

Ψ THEOREM 13.12 - BLACK HOLE COMPOSITION {Ψ5.6}

BLACK HOLES ARE COMPOSED OF DEGENERATE NUCLEAR MATTER, RESTRAINED BY GRAVITATION AND NUCLEAR POTENTIAL

Plasma would thermalize a black hole's veneer and in so doing radiate profuse amounts of energy. This might be a transient condition for a few of the particles on a black hole's surface, but is certainly not a stable global configuration.

Before any attempts are made to describe a black hole's unfathomably harsh interior, we must decide:

a) how much energy its material lost as a result of gravitational capture.

b) whether or not this loss is relevant to a black hole's properties and cosmic functionality.

There are two reasons to suspect that this energy loss is significant. First, the Virial Theorem⟨6.26⟩ tells us that any gravitationally-bound system loses half of its gravitational potential to radiation. A black hole with a near-unity gravitational potential represents a huge energy deficit. Second, the intense radiation given off by accretion disks, though as yet not unambiguously observed, has been well documented in terms of basic physics.

As it turns out, however, radiative loss during accretion has no bearing on a black hole's primary function: cosmic equilibrium. The key and only consideration is the Maxfield criterion. A black hole can exchange elementary particles with the rest of the universe if and only if they regain their full complement of rest energy. Regardless of how much energy a proton does or does not lose on its journey down a black hole's gravitational well, it can only reemerge with its free-space rest energy fully intact. For this reason, the contents of a black hole will be viewed as an adiabatic collection of degenerate material, where the total energy present is defined by the number and species of its elementary particles.

Applying a gravitational field to a nuclear matrix induces core expansion, thereby increasing inter-core separation and lowering material density. If this occurs adiabatically, 100% of the field energy lost by expansion is recaptured as kinetic energy. This means that *the effective material density of a gravitationally expanded substrate is purely a function of its increased core size.* The adiabatic relationship between gravitational potential and material density is given by Equation (13.24), the relationship between free and expanded core size:

$$\rho = \left(\frac{R_p}{R_{xc}}\right)^3 \rho_{n_0} = \left(\sqrt{1-\frac{2GM}{c^2 r}}\right)^3 \rho_{n_0} = \left(\frac{\Phi_g}{m_0 c^2}+1\right)^3 \rho_{n_0} \tag{13.25}$$

where ρ_{n_0} is electrically neutral nuclear density at zero gravitational potential and zero pressure, given by Equation (10.16):

$$\rho_{n_0} = \frac{m_n}{12R_p^3}$$

The density of atomic nuclei on a neutron star's surface ($\Phi_g/m_0c^2 \sim -0.2$), for instance, is about half that of the same nuclei on Earth.

Pressure compresses particle cores while gravitational potential expands them. Indeed, the only agents that can counter the incredible pressure inside a black hole are *core integrity and gravitational potential*. Cores exposed to this potential expand accordingly, so the interior density profile of a massive compact object has the form:

$$\rho = \left(\sqrt{1-\frac{2GM}{c^2 r}}\right)^3 \rho_{bn}(r) \tag{13.26}$$

where $\rho_{bn}(r)$ is the baseline nuclear density at the pressure that exists at radius r. This, like ρ_{n_0}, represents an electrically neutral degenerate matrix.

Gravitational potential is a function of the material within r; pressure is a function of the material inside and outside of r. The material density at any internal radius is a product of both, where pressure sets the baseline density. For the sake of simplicity, we will assume that baseline density varies linearly with r, peaking to $\rho_{\max}$ at $(r = 0)$:

$$\rho_{bn}(r) = \rho_{\max} - \left(\frac{\rho_{\max} - \rho_{n_0}}{R_{sr}} \right) r \tag{13.27}$$

where R_{sr} is the radius of a compact object's surface. Since pressure is a function of exterior material, an object's nuclear baseline density returns to ρ_{n_0} at its surface:

$$\rho_{bn}(R_{sr}) = \rho_{n_0} \tag{13.28}$$

The mass at any radius R is a function of the density profile for $(r < R)$, so rewrite Equation (13.26) to define the density at a radius R in terms of the total mass of the adiabatic density profile below it:

$$\rho(R) = \left(\sqrt{1 - \frac{2G\left(\int_{r=0}^{R} \rho(r) dV \right)}{c^2 R}} \right)^3 \rho_{bn}(R) \tag{13.29}$$

This nonlinear equation has boundary conditions of:

$$\int_{r=0}^{R_{sr}} \rho(r) dV = M \tag{13.30}$$

$$\rho(r > R_{sr}) = 0 \tag{13.31}$$

$$\rho(r = 0) = \rho_{\max} \tag{13.32}$$

where M is the mass of the compact object. For any given object, there is a density profile $\rho(r)$ that is consistent with the density at each radius R as defined by Equation (13.29), and which integrates to a total mass M over the volume defined by the radius of the object's surface, R_{sr}.

A galaxy's central black hole will be referred to as its *core*. The mass of the Milky Way's core is about three million times that of our sun.⟨1.9⟩ Its interior density profile is shown below:

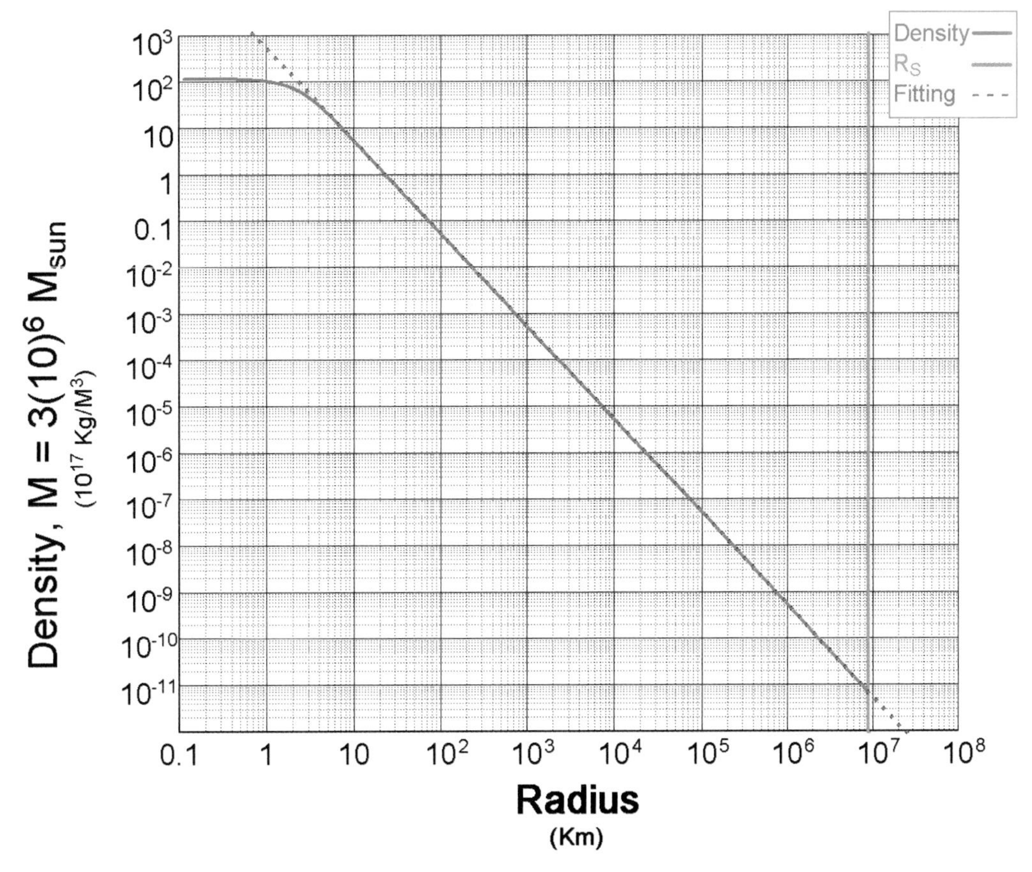

Figure (13.7) Density distribution of the black hole at the center of the Milky Way

It is assumed that a black hole this heavy can compress its innermost matter to hyperdensity ($1.2(10)^{19}$ kg/m^3). However, although the material at the center of the Milky Way's core is in all likelihood hyperdense, its peak interior density has no effect on the density profile in its outer layers or at its surface. The Milky Way core's Schwarzschild radius is shown as a orange vertical line near 10^7 km. Its material surface extends slightly beyond this, but the deviation is too small to appear in Figure (13.7). Most significantly, *the ratio between the total mass and volume of Equation (13.29)'s distribution is within 6 parts per billion (ppb) of a Schwarzschild mass/volume ratio for the galactic core.* Even though matter is not infinitely compressible, when black holes are large enough for their surfaces to decouple from the hyperdense material deep within their interior, they approach a Schwarzschild relationship. The source code for this calculation can be found at *www:nullphysics.com*.

Our calculations were simplified by reducing baseline density linearly with radius, but Figure (13.7)'s density profile emerges *for virtually any baseline density function*. So although

Equation (13.29) is a fairly complex nonlinear expression, a black hole's interior density profile can be accurately described with a simple fitting expression of the form:

$$\rho \cong \frac{5.37(10)^{25}}{R^2} \tag{13.33}$$

in units of kg/m^3. This is shown as the dotted trace in Figure (13.7).

Compare our fitting function to the *average* density of a black hole of Schwarzschild radius:

$$\rho_S = \frac{M}{V} = \frac{M}{\left(\frac{4\pi}{3}R_S^3\right)} = \frac{\left(\frac{c^2 R_S}{2G}\right)}{\left(\frac{4\pi}{3}R_S^3\right)} = \frac{3c^2}{8\pi G R_S^2} = \frac{1.61(10)^{26}}{R_S^2} \tag{13.34}$$

also in units of kg/m^3. The ratio between Equation (13.34) and Equation (13.33) is equal to 3, as is typical for the ratio of instantaneous and average values in a spherical distribution. The material density of a large black hole's surface becomes progressively closer to one third of its Schwarzschild density as its mass goes to infinity. Black holes are not *singularities*, gravitational or otherwise, regardless of their size. They are merely compact objects with deep gravitational potentials. The only singularities the universe has to offer are geometric points and the cores of elementary particles.

SURFACE DENSITY

The surface density of a massive compact object will be called its *veneer density*:

$$\rho_\vee \to \frac{\rho_S}{3} = \frac{c^2}{8\pi G R_\vee^2} \qquad \{M \to \infty\} \tag{13.35}$$

where $R_\vee$ is the object's veneer radius - the size of its physical surface. Varying peak central density has virtually no effect on this profile; all it does is alter the radius, deep within a black hole, where this profile becomes an accurate representation of the true material density. In the Milky Way's core, this profile begins about 5 km from its center. Indeed, if matter were infinitely compressible, the radius of any black hole would equal its Schwarzschild value, but its surface density would still be a third of its average density.

Even removing our adiabatic constraint has no tangible effect on a black hole's internal density profile. The *non-adiabatic* relationship between core expansion and material density includes the volume adjustment of Equation (13.26) but has an additional expansion term to account for the overall energy loss due to the core expansion of individual particles:

$$\rho_{na} = \left(\sqrt{1 - \frac{2GM}{c^2 r}} \right)^4 \rho_{bn}(r) \tag{13.36}$$

where ρ_{na} is subscripted to denote non-adiabatic density. *This produces a density profile indistinguishable from the one shown in Figure (13.7).*

The near-unity gravitational potential of a black hole's veneer, its *veneer potential*, is given by a slight rearrangement of Equation (13.25), adapted for the veneer:

$$\frac{\Phi_\vee}{m_0 c^2} = \left(\frac{\rho_\vee}{\rho_{n_0}} \right)^{\frac{1}{3}} - 1 \tag{13.37}$$

Substitute Equation (10.16) for ρ_{n_0} and Equation (13.35) for veneer density:

$$\frac{\Phi_\vee}{m_0 c^2} = \left(\frac{12 R_p^3 \rho_\vee}{m_p} \right)^{\frac{1}{3}} - 1 = \left(\frac{12 R_p^3 c^2}{8\pi G m_p R_\vee^2} \right)^{\frac{1}{3}} - 1 \quad \{R_\vee \geq 100\,\text{km}\} \tag{13.38}$$

The radius of a galactic black hole is very close to its Schwarzschild value, ($R_\vee \cong R_S$), so substitute R_S (as $2GM/c^2$) to yield its veneer potential as a function of mass:

$$\frac{\Phi_\vee}{m_0 c^2} = \left(\frac{3 R_p^3 c^6}{8\pi G^3 m_p M^2} \right)^{\frac{1}{3}} - 1 \cong \left(\frac{5.3(10)^{20}}{M^{\frac{2}{3}}} \right) - 1 \quad \{M >> M_{sun}\} \tag{13.39}$$

The veneer potential of the Milky Way's black hole, at $M = 6(10)^{36}$ kg, is –0.99984.

Veneer redshift affects light emitted from the material exposed to this potential. Its magnitude is given by:

$$z_\vee = \left(\frac{1}{\frac{\Phi_\vee}{m_0 c^2} + 1} \right) = \left(\frac{8\pi G^3 m_p M^2}{3 R_p^3 c^6} \right)^{\frac{1}{3}} \cong 1.9(10)^{-21} M^{\frac{2}{3}} \quad \{M >> M_{sun}\} \tag{13.40}$$

Light radiating from the surface of our galaxy's black hole is redshifted by a factor of ~6300. It reduces X-rays to infrared and infrared to radio waves.

RADIATIVE ENERGY LOSS

A black hole is not perfectly black, so it radiates energy based on its temperature and radius. We will assume that its veneer is in thermal equilibrium with electromagnetic radiation because the bound electrons and protons of which it is composed interact with each other at radically different energy levels. A black hole's radiant power loss will be referred to as its *veneer luminosity*. In a galactic black hole, where ($R_\vee \cong R_S$), it is given by:

$$L_\vee = \frac{\sigma T^4 A_\vee}{z_\vee} = \left(\frac{8\pi G\sigma}{c^2}\right)\left(\frac{3R_p^3}{\pi m_p}\right)^{\frac{1}{3}} T^4 M^{\frac{4}{3}} \qquad \{M >> M_{sun}\} \quad (13.41)$$

This is in units of power, where $A_\vee$ is the area of a galactic core's veneer, approximated to its Schwarzschild area. Substituting the various constants yields:

$$L_\vee \cong 8(10)^{-40}\left(T^4 M^{\frac{4}{3}}\right) \qquad \{M >> M_{sun}\} \quad (13.42)$$

If the core of our galaxy were 280,000 °K, (as derived in Part IV), it would have a luminosity of $6(10)^{31}$ W ($M = 6(10)^{36}$ kg). This is nearly 200,000 times the power output of our sun but only ~4 ppm of our galaxy's total output.

The peak energy density per wavelength of the heavily redshifted spectrum that a veneer emits is given by:

$$\lambda_\vee = \left(\frac{z_\vee}{4.9651}\right)\left(\frac{hc}{kT}\right) \quad (13.43)$$

Again using the Milky Way's black hole as an example, at 280,000 °K *its peak radiative wavelength would be 0.06 mm, in the far infrared band.* With a weak infrared glow and radius only slightly larger than its Schwarzschild value, the Milky Way's core lays hidden in the center of our galaxy:

Figure (13.8) X-ray imaging of the center of the Milky Way
(Courtesy NASA/CXC/MIT/F.K. Baganoff et al.)

VENEER CAPACITY

The total flux of material moving at the surface of a large black hole's veneer will be called its *veneer capacity*. It is given by its density, average particle velocity, and area as:

$$C_{\vee} = \frac{\rho_{\vee}}{4}\overline{\mathrm{v}}A = \frac{\rho_{\vee}}{4}\overline{\mathrm{v}}4\pi R_S^2 = \rho_{\vee}\overline{\mathrm{v}}\pi R_S^2 \qquad \{M >> M_{sun}\} \quad (13.44)$$

where radius is again approximated by the Schwarzschild value. Substituting Equation (13.35) for veneer density yields:

$$C_{\vee} = \frac{\overline{\mathrm{v}}c^2}{8G} \qquad \{M >> M_{sun}\} \quad (13.45)$$

The total material flow of a galactic core's surface is dependent only on the average speed of its surface material. As a black hole increases in mass, its veneer density drops in inverse proportion to its increased area, keeping its capacity constant for any given surface material speed.

Average material speed can be no greater than c, so black holes have an upper limit of mass loss, regardless of their temperature. Maximum capacity is given from Equation (13.45) with an average velocity of c:

$$C_{\|\vee\|} = \frac{c^3}{8G} \qquad \{M >> M_{sun}\} \quad (13.46)$$

or $5(10)^{34}$ kg/s. This is a prodigious rate, on the order of a small star cluster per second.

It will be called the *veneer capacity limit*:

Ψ THEOREM 13.13 - VENEER CAPACITY LIMIT {Ψ5.6}

ALL LARGE BLACK HOLES HAVE THE SAME MAXIMUM VENEER CAPACITY

A galactic core's veneer capacity limit is fixed, but enormous. A hot black hole has the ability to spew torrents of mass into space.

A more detailed analysis of the relationship between a black hole's mass loss and temperature is presented in Part IV. Black holes are not a celestial oddity. Their operational parameters (as derived by the Maxfield constraint) are absolutely essential to the maintenance of an eternal universe.

13.9 GENERAL CONCLUSIONS

Gravity and its relationship to energy are investigated:

- Internal spatial deflections attenuate as the cube of distance - the same profile as external deflections.
- Space with no fourth-dimensional slope is incompressible, but can be stretched to a maximum of one unit per unit distance.
- Like elementary particles, photons and neutrinos also have intrinsic gravitational fields.
- The root cause of gravitational fields is the hypervolumetric density associated with the extraspatial volume in energy density.
- Particle cores expand in the presence of gravitational potential. This expansion is consistent with the low material density of large black holes.
- A black hole is slightly larger than its Schwarzschild radius would suggest because the General Theory of Relativity does not take the maximum compressibility of degenerate nuclear matter into consideration.
- Black holes are not utterly black, but approach blackness with increasing mass. The Milky Way's black hole, at an estimated mass of $\sim 3(10)^6$ times that of our sun, has a surface potential of ~ -0.99984, redshifting the light it emits by a factor of ~ 6300.
- The available material flux (capacity) of any large black hole's gravitational veneer has a maximum value of $5(10)^{34}$ kg/s.

And so ends Part III, the substructure of matter. The next stop is physical cosmology, to drift once more among the stars. This is a far more leisurely experience, as investigating things that are large enough to see is immeasurably easier than exploring the four-dimensional landscape of the elementary particle field. Part IV portrays the seamless merger of ultrastasis and the universe at large - a celestial ecology of matter and light.

PART IV: COSMOLOGY
BACK TO THE STARS

The universe's large-scale structure and function

- *Cosmic power cycles*
- *Intergalactic redshift and the Cosmic Microwave Background*
- *Olbers' Paradox*
- *Universal equilibrium*
- *Galactic dynamics*
- *The galactic core*
- *Dark matter*
- *Matter and antimatter*

At the last dim horizon, we search among ghostly errors of
observations for landmarks that are scarcely more substantial.
The search will continue. The urge is older than history.
It is not satisfied and it will not be oppressed.

Edwin Hubble, 1889-1953

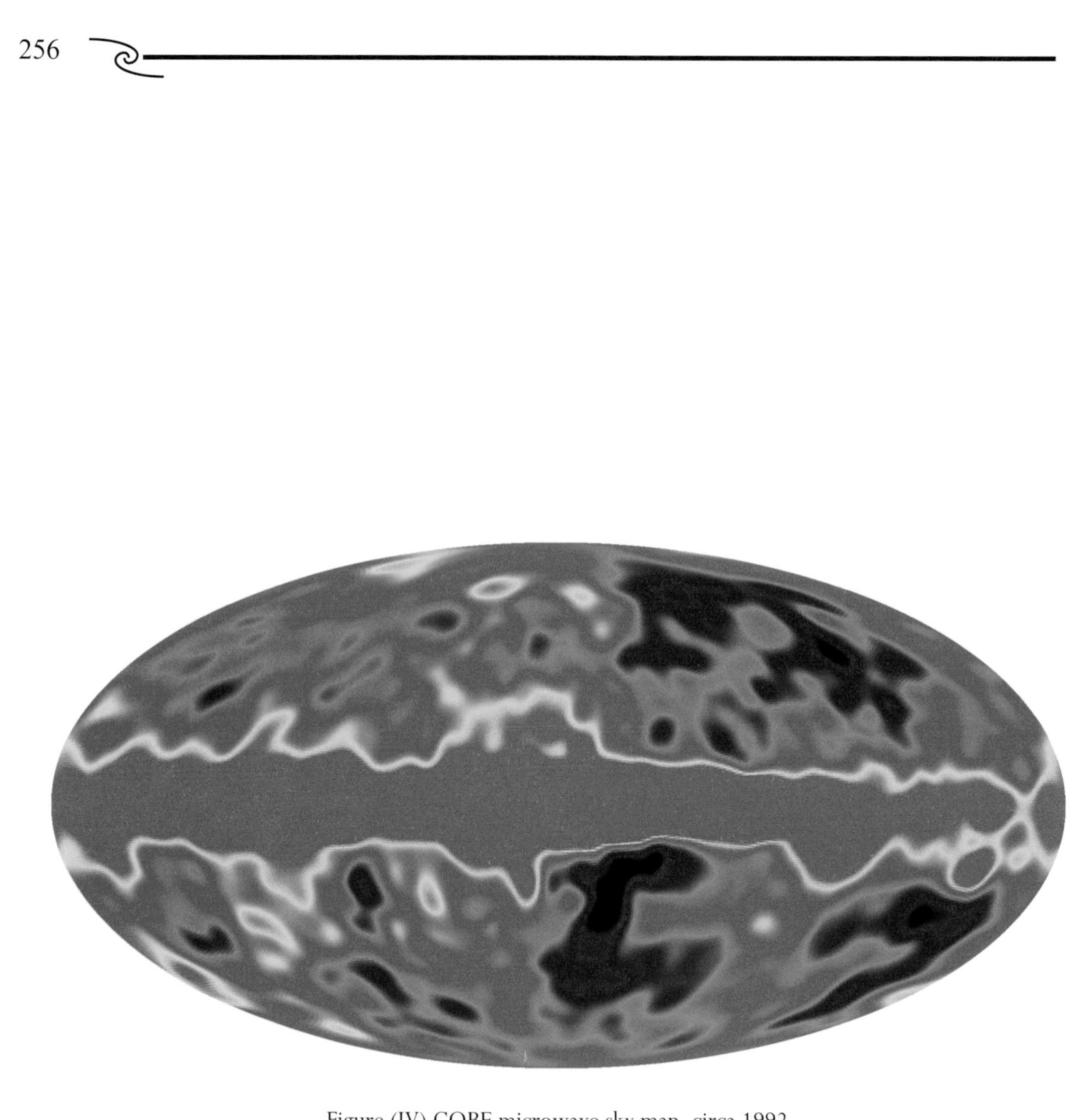

Figure (IV) COBE microwave sky map, circa 1992
(Courtesy NASA/WMAP Science Team)

Prerequisite Concepts for Part IV

The following pivotal concepts are carried forward from Parts I, II, and III:

- The universe sums to zero and is therefore a distributed form of zero - the Null Axiom.
- The universe is limitless in order to express the full extent of nothingness.
- The universe has no net change because nothingness cannot change; it is frozen in ultrastasis on the largest scale.
- In accordance with the Maxfield concept, if a particle's total energy exceeds its rest mass it can escape from a black hole.
- Black holes are not completely black; their luminosity is defined by their temperature and surface potential. Light's energy is not affected by gravitational potential. Gravitational potential reduces the rest mass of elementary particles, which in turn emit light of lower energy.
- Photons, elementary particles and neutrinos are real objects with discrete spatial volume and finite energy density distributed over said volume.
- Like elementary particles, photons and neutrinos have gravitational fields.
- A photon's polarization is the result of the asymmetry of its spatial footprint.

14. PHYSICAL NULL COSMOLOGY

Our investigation has now come full circle. The Null Axiom forms the basis for an infinite universe full of space and energy. Its associated geometry provides the quantization of light and matter and the four universal forces of nature. Let's now return to the stars to learn how the universe's structure operates on the intergalactic scale. Like many of the other discoveries presented in *Null Physics*, it is truly remarkable - an eternal dance of energy and matter.

14.1 THE OBSERVED UNIVERSE

Our analysis begins with a quick review of the known characteristics of the observable universe. The *Cosmological Principle* will be assumed in this assessment. Its premise is that our local cosmic neighborhood out to a few billion light years is an accurate portrayal of the nature of the universe everywhere. Anything more distant than this may appear different to our instruments, but this is only because of the signal distortion caused by such phenomenal range. The Big Bang requires an evolving universe, so cosmologists have used laborious statistical analysis on the blurry images of distant galaxies to try to demonstrate universal evolution in terms of star formation rate, galactic merger rate, and variation in galactic type abundance. This is a misguided venture, as there is currently no way of knowing how to properly correct these images for distance-related losses. Crossing deep space has a devastating effect on an optical signal's content, as derived in Appendix M. A striking example of this loss is provided in Figure (14.1) below. It contains Hubble images of portions of three galactic clusters, shown left to right at increasing range from Earth, in units of Mly (million light years). A Hubble constant of 60 Hz-km/Mpc is assumed.

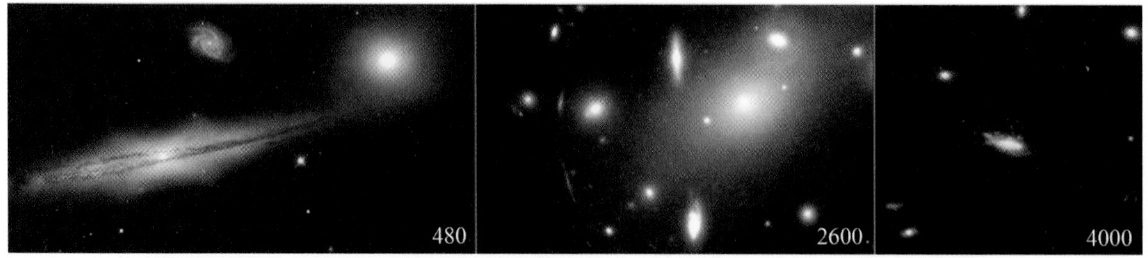

Figure (14.1) HCG87 (~480 Mly), Abell 2218 (~2600 Mly), Abell 2125 (~4000 Mly)
(Courtesy NASA/Hubble Heritage Project)

Even with the Hubble telescope, whose resolution at ~0.1 arc second is typically 10 times that of the best ground-based instruments, galaxies lose most of their detail beyond a range of about 2 billion light years. This is the result of signal degradation and observational limitations - it is not caused by a maturing universe. Deep-space galactic redshift surveys, as shown below in Figure (14.2), reveal the universe's uniform large-scale material distribution. *The universe does not evolve - it is ultrastatic.*

MATTER

GENERAL

Stars are the building blocks of galaxies, which are the building blocks of groups, which form clusters, which ultimately form superclusters of tens of thousands of galaxies. So-called *rich* superclusters also contain vast amounts of hydrogen, and galaxies represent only ~20% of their total mass. Cosmic structure does not end with superclusters, however. They form the skeleton of a gigantic foam-like construct with interconnected voids notably empty of galaxies or other material. These cells are about 100 Mly in diameter and have diaphanous walls composed of superclusters, filaments of galaxies, and galactic clusters:

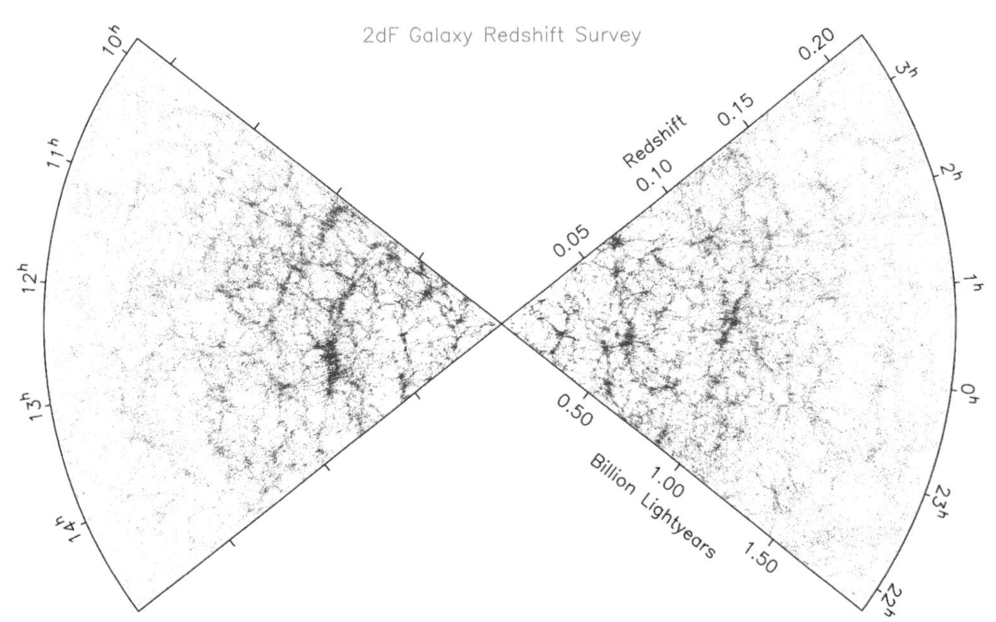

Figure (14.2) A tiny fraction of the universe's innumerable galaxies - the 2dF galactic redshift survey (Courtesy 2dF GRS Team, shown previously as Figure (1.2))

There appears to be no level of structure beyond this, as the universe's material distribution becomes progressively more uniform with scale.⟨8.1⟩ If space were sectioned into a population of equally sized cubes, the standard deviation of their masses would be inversely proportional to their volume.⟨8.2⟩

The average density of luminous material in space is $\sim 10^{-28}$ kg/m^3, and is composed of about ~70% hydrogen, ~25% helium, and ~5% various other elements.⟨1.1⟩ Gravitational interactions between galaxies suggest the existence of a large fraction of nonluminous material, so the total density of luminous and nonluminous matter might be as high as $\sim 5(10)^{-27}$ kg/m^3. This represents a contribution to universal energy density, ρ_U, of as much as $\sim 4.5(10)^{-10}$ J/m^3. Plasma is the most prevalent state of the universe's visible matter, which is not surprising since nuclear fusion is the primary source of celestial luminosity.

STARS

Stars vary from faint red dwarfs to super-massive blue-white giants. Our sun is a yellow dwarf near the midpoint of this range. Its output in the visible band is $3.8(10)^{26}$ watts.⟨1.20⟩ This is often used as a unit for the luminosity of other celestial objects, to include galaxies, and will be symbolized L_{sun}. Similarly, our sun's mass, at $2(10)^{30}$ kg, is denoted as M_{sun} and is the astrophysical standard for this parameter. Stellar luminosity varies over ten orders of magnitude $\{5(10)^{-4}\,L_{sun} - 6(10)^{6}\,L_{sun}\}$, while the range of a star's mass is more constrained at four orders of magnitude, $\{0.08\,M_{sun} - 150\,M_{sun}\}$.⟨6.4⟩

A star's lifespan depends on its luminosity. Supergiants last for only a few million years, while a star like our sun will burn for over ten billion years. Extreme longevity belongs to red dwarfs, whose dim glow continues for tens if not hundreds of billions of years. Stars whose mass is $\sim 4\,M_{sun}$ or greater end their lives in violent explosions, leaving behind hot white dwarfs, neutron stars, or, in the case of the most massive stars, black holes.⟨6.25⟩ These remnants are typically surrounded by a nebula of expanding debris.

GALAXIES

There are four basic types of galaxies: spiral, like our own Milky Way; elliptical; lenticular; and irregular. Galaxies are identified not only by their morphology but also by the stellar populations they contain. Lenticular galaxies have a structure intermediate between spirals and ellipticals, and irregulars have generally distorted shapes. Elliptical galaxies tend to contain a uniform distribution of older, cooler stars, whereas irregular and spiral galaxies have a small fraction of young hot stars responsible for most of their luminosity. The size of spiral galaxies varies over three orders of magnitude $\{10^{9}\,M_{sun} - 10^{12}\,M_{sun}\}$. Elliptical galaxies have a wider range of size at six orders of magnitude, $\{10^{7}\,M_{sun} - 10^{13}\,M_{sun}\}$. The galactic population of rich clusters is dominated by ellipticals and lenticulars, but as a universal average, spirals and lenticulars are the most common isolated galaxies, called *field galaxies*, and are scattered at random within the foam-like material distribution of deep space.

INTERGALACTIC MATERIAL

Intergalactic material, referred to as IGM, is the matter dispersed in the space between galaxies. It is generally dark and thin, but will periodically contain isolated stars presumably ejected from their parent galaxy by collisions with other galaxies or some other violent process. The IGM's average material density amounts to at most a few particles per cubic meter, existing at a purity of vacuum not currently possible to duplicate on Earth.

The IGM's energy content is defined by the CMB's 2.7 °K temperature. It is so cold and thin it has virtually no energy, far less than even the radio wave background of space. The average kinetic energy of a particle in deep space is derived in Appendix P as:

$$\overline{KE}_{IGM} = \frac{3kT_{CMB}}{2} \tag{14.1}$$

At an average density of ~3 particles per cubic meter, this amounts to an energy density of $1.7(10)^{-22}$ J/m^3, about 250 million times less than the CMB.

ELECTROMAGNETIC ENERGY

GENERAL

Signals from distant objects in all bandwidths in all directions have redshifts roughly proportional to their distance from Earth. This is known as *intergalactic redshift*, and the ratio between redshift and distance is the *Hubble constant*, H_0:⟨7.1⟩

$$H_0 \cong \frac{cz}{d} \qquad \{z < 0.2\} \tag{14.2}$$

where redshift is quantified by the factor z:

$$(z+1) \equiv \frac{\lambda}{\lambda_0} \equiv \frac{E_0}{E} \tag{14.3}$$

A z of unity corresponds to a photon with twice its original wavelength and half its original energy. This represents a distance on the order of 10 Gly (ten billion light years).

The universe's electromagnetic energy can be globally characterized in terms of two densities: luminous and luminosity. Luminous density is the amount of light energy per unit volume. It has units of energy/distance^3 (J/m^3). Luminosity density is light energy *output* per unit volume, with units of power/distance^3 (W/m^3). The three most significant spectral backgrounds of deep space, in order of increasing energy density and decreasing photon energy, are:

- Integrated starlight
- Infrared background
- CMB (Cosmic Microwave Background)

These are by far the most energetic backgrounds. By comparison, space contains only trace amounts of radio waves and gamma radiation. The radio band, for example, has about a millionth of the CMB's energy density. The peak of *luminous energy density* lies in the central portion of the electromagnetic spectrum, the CMB. This decreases in the infrared, attenuating still further in the visible band.

INTEGRATED STARLIGHT

The majority of the light in space originates from a small fraction of hot young stars. This is balanced by the diffuse light produced by the rest of the universe's stars, which are generally cooler and far more numerous. The combined luminous output of all stars is referred to as *integrated starlight*. It has a spectrum similar to a 10,000 °K blackbody because its source is the averaged output of the universe's hot luminous objects. Our sun, by comparison, has a surface temperature of ~6,000 °K.

INFRARED BACKGROUND

The infrared spectrum, to include near and far infrared, encompasses all electromagnetic radiation in the range [8–1000 μm]. This represents a significantly broader wavelength scale than either the visible band [290–800 nm] or the CMB [0.3–10 mm], and overlaps the CMB's high-energy region. The universe's infrared density is about 30% of the CMB, an order of magnitude greater than integrated starlight.⟨2.1⟩

CMB

As viewed by the COBE and WMAP satellites, the CMB conforms to an ideal 2.724 °K blackbody spectrum to the accuracy with which it can be measured. Its intensity, however, is not perfectly uniform. It is marked by slight temperature fluctuations on the order of 18

± 1.6 μK. It also has a dipole anisotropy (opposing redshift-blueshift) due to Earth's motion through it at 370 km/s:⟨36⟩

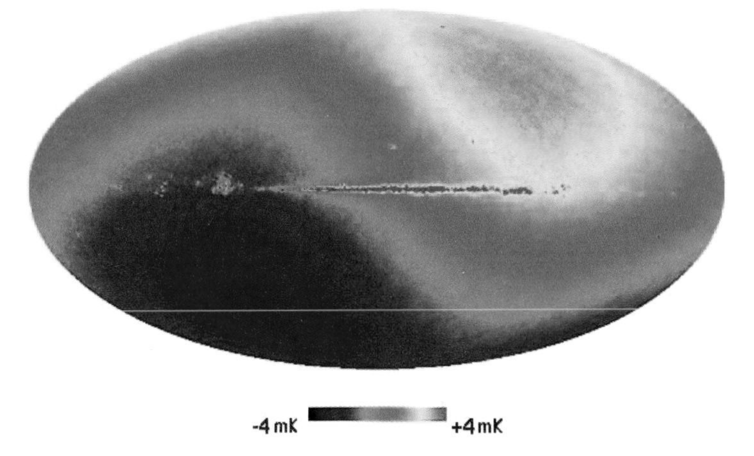

Figure (14.3) WMAP with Earth's motion
(Courtesy NASA/WMAP Science Team, shown previously as Figure (1.1))

The CMB's spectral appearance has the following form:

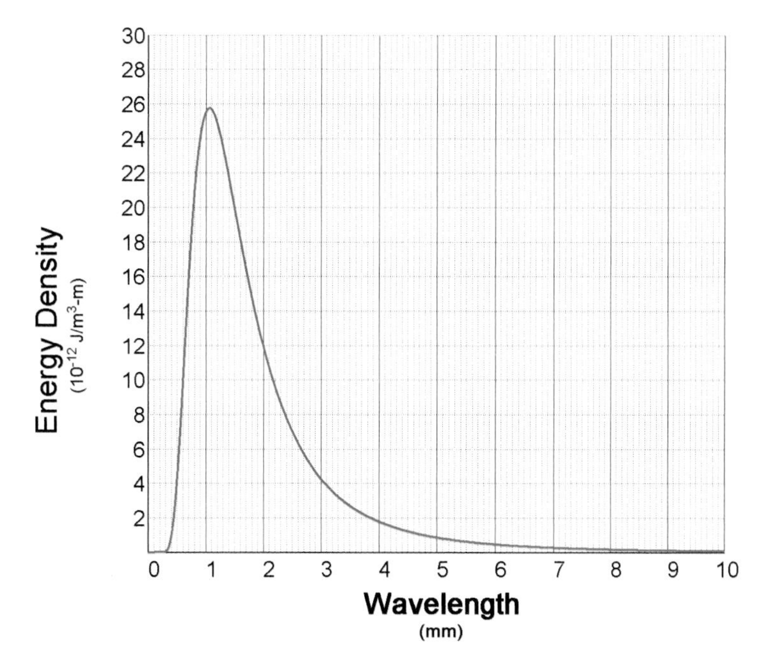

Figure (14.4) CMB energy density as a function of wavelength, T = 2.724 °K

Its photon number density is given by the blackbody expression:⟨8.3⟩

$$n_\gamma = (2.404)8\pi\left(\frac{kT}{hc}\right)^3 \tag{14.4}$$

or 410 million photons/m^3. This is the highest photon population, by far, of any universal background radiation. Blackbody energy density is related to temperature by:

$$\rho_E = \frac{4\sigma T^4}{c} \tag{14.5}$$

where σ is the Stefan-Boltzmann constant. At $4.2(10)^{-14}$ J/m^3, the CMB has significantly more energy per cubic meter than any other bandwidth in space. Energy density and photon number density are the only blackbody parameters used in the development of Null Cosmology, but for more information about blackbody spectra please see Appendix L.

14.2 UNIVERSAL PARAMETERS

Unit hypervolume is the only finite universal constant; it has the same value everywhere across infinite space. Universal *parameters* are averages. They vary from location to location but have a single universal value. Whereas particle, nuclear, and atomic physics are described in terms of unit hypervolume, physical cosmology is the relationship among universal parameters. Those most relevant to our analysis can be found in Appendix C, duplicated here:

Universal Parameters				
Name		Value	Units	Error(%)
H_0	Hubble Constant	60	Hz-km/Mpc	50 ⟨1.2⟩
H_0	Hubble Constant	$1.95(10)^{-18}$	Hz	50 ⟨1.2⟩
ρ_U	Universal Energy Density	$4.0(10)^{-10}$	J/m^3	40 (Chapter 15)
ρ_M	Universal Mass Density	$4.5(10)^{-27}$	kg/m^3	40 (Chapter 15)
j_B	Optical Luminosity Density	$1.4(10)^{-33}$	W/m^3	80 ⟨1.12⟩
j_I	Infrared Luminosity Density	$0.7(10)^{-33}$	W/m^3	80 ⟨1.13⟩
j_ν	Neutrino Luminosity Density	$<5.6(10)^{-35}$	W/m^3	80 (4% of optical) ⟨2.3⟩
j_R	Total Luminosity Density	$2.1(10)^{-33}$	W/m^3	80 ⟨1.12⟩⟨1.13⟩
$\rho_{\gamma\gamma}$	Gamma/Xray Energy Density	$1.5(10)^{-17}$	J/m^3	80 ⟨2.1⟩
$\rho_{B\gamma}$	Optical Energy Density	$1.6(10)^{-15}$	J/m^3	80 ⟨2.1⟩
$\rho_{I\gamma}$	Infrared Energy Density	$1.6(10)^{-14}$	J/m^3	80 ⟨2.1⟩
ρ_{CMB}	CMB Energy Density	$4.165(10)^{-14}$	J/m^3	0.05 ⟨2.2⟩
$\rho_{K\gamma}$	Radio Wave Energy Density	$1.6(10)^{-20}$	J/m^3	80 ⟨2.1⟩
ρ_ν	Neutrino Energy Density	$<6.4(10)^{-17}$	J/m^3	80 (4% of optical) ⟨2.3⟩
ρ_R	Total Radiant Density	$5.9(10)^{-14}$	J/m^3	5 ⟨2.1⟩⟨2.2⟩

Table (C.1) Universal Parameters

Luminous densities (ρ) are the amount of light energy in space, while luminosity densities (j) measure the universe's fusion power output. The term *optical* denotes wavelengths between ultraviolet and infrared, and is often called the integrated starlight. It is also referred to as the *blue band*, hence the B subscripts on the optical parameters. The error (%) is the same for several of the values in Table (C.1) because many are derived from the Hubble constant and are equally sensitive to its accuracy. Unlike universal constants, universal parameters are generally of low resolution, and Table (C.1)'s error margins are conservative. Most cosmological data is by necessity of a statistical nature, and the CMB provides the only information with any significant amount of accuracy. The upper limit of universal neutrino luminosity is estimated at 4% of optical, based on our sun's neutrino/optical luminosity ratio. This fraction varies with star type, so the universe's neutrino luminosity density should be considered a rough estimate. Neutrino energy density is derived in Appendix N, using concepts developed in Chapter 15.

Our goal is not to calculate the universe's lithium abundance. It is to deduce, from the available incontrovertible data, how the infinite machine of reality actually *works* on a cosmically local scale. With this in mind, please note the units for the first Hubble constant in Table (C.1). It is normally listed as km/s/Mpc, owing to its erroneous association with the recession velocities of an expanding universe. Since the universe is not expanding, the Hubble constant will be listed in units of Hz-km/Mpc instead. The numerical value remains the same but the physical significance is more accurate. This constant is also listed in Hz as is more appropriate for some of the applications to follow.

14.3 NULL COSMOLOGICAL MODEL

The null model of the universe is a flat, rectilinear volume, infinite in three spatial dimensions with an infinite history of ultrastasis. It has a finite, universally average energy density that creates a universally average internal curvature. Since its material distribution is limitless, Mach and Einstein's Cosmological Principle applies throughout. Any large sample of the universe looks essentially the same and has the same physics. With the sole exception of the cosmic parity between matter and antimatter, the composition of the distant universe is indistinguishable from the local universe, regardless of the magnitude of the distance. Any astronomical data that allegedly demonstrates universal evolution is a misinterpretation of the signal distortion caused by cosmic distances. *The universe is not evolving or expanding; its global properties are fixed.*

A static universe is certainly not a new idea, and it persisted for some time after the discovery of intergalactic redshift. Edwin Hubble was never convinced the effect he found had anything to do with universal expansion. In 1929, Fritz Zwicky postulated that the observed redshift was due to the gradual loss of a photon's energy over astronomically long distances, an idea referred to as *tired light*.[8.4] Zwicky was entirely correct, but lacked the null principles needed to explain the universe's presence and the underlying *cause* of the redshift effect. So while concepts of infinite, eternal, and quiescent universes have surfaced throughout cosmology's history, no published worldview shares *Null Physics'* basic premise.

COSMOSTASIS

Universal variation ceases at totality, and this in turn defines the equilibrium requirements of the material of which the omnipattern is composed. The interaction between ultrastasis and finite cosmological environments takes the form of *the three laws of cosmostasis*:

Ψ THEOREM 14.1 - THREE LAWS OF COSMOSTASIS {Ψ4.2}

1. *ALL ENERGY FORMS ARE RECYCLABLE, RENEWABLE, AND PART OF AN ETERNAL COSMIC ENERGY CYCLE; THE SUM OF ALL SUCH CYCLES IS THE COSMIC ENGINE*

2. *THE UNIVERSAL DENSITIES OF MATTER, ELECTROMAGNETIC ENERGY, AND LUMINOSITY ARE THE DIRECT RESULT OF UNIVERSAL EQUILIBRIUM, INFINITE IN BOTH TIME AND SPACE*

3. *THE AMOUNT OF ENERGY FLOWING THROUGH ANY STEP OF A GIVEN COSMIC ENERGY CYCLE IS THE SAME AS ANY OTHER STEP*

The reasoning behind each law is as follows.

1. Recyclable and renewable. If an inescapable cul-de-sac existed for energy, such as a gravitational "singularity", energy would flow into this dead end and its universal number density would grow and eventually stop at some final value, with all energy trapped in a single form. The universe can be in a stable equilibrium if and only if each and every energy form within it maintains a constant universal population. This requires cosmic mechanisms that create and destroy all energy forms at equal rates.

2. Energy, mass, and luminosity density. Density represents an equilibrium fulcrum between creation and destruction for any given energy form. When a process converting one form of energy into another reaches equilibrium, it does so because the densities of the *fuel and products* have reached levels where the rate of conversion is equal for both.

3. Uniform energy flow. In an equilibrium system, the density of each form of energy in the cycle remains constant. This means the amount of energy moving from one step to the next is the same *between any two steps*. So even though the total amount of

energy in each step in the cycle may be drastically different, the amount of energy transferred per unit time between the steps is the same for all steps.

These laws presume the inherent *symmetry* of cosmic energy cycles. Unless a cycle is specifically identified as matter-antimatter, it represents the same cycle for both forms of matter. The fusion cycle of hydrogen, for instance, also encompasses the fusion cycle of antihydrogen.

ETERNAL EQUILIBRIUM, BRIEF REVIEW OF THE OPPOSITION

One argument frequently used against an infinitely old, quiescent universe is *gravitational instability*.⟨8.6⟩ Its premise is that matter's natural tendency to aggregate under gravitational attraction would cause irreversible *clumping* throughout space, inevitably leading to a desolate expanse populated with nothing but gravitational singularities. This is an erroneous interpretation for two reasons:

- Black holes are not the irreversible end-states of matter they are currently thought to be.

- Gravitation is not the only long-range universal force that controls matter's cosmic distribution.

As will become apparent in our derivation of galaxies' function in the cosmic engine, electromagnetic forces play a large role, if not larger than gravitation, in the flow of our universe's material.

Another concept commonly cited against a universe in eternal equilibrium is the second law of thermodynamics: *The total entropy of any isolated thermodynamic system tends to increase over time, approaching a maximum value.* Here entropy is defined as a measure of the *unavailable* energy in a closed system. This is used to justify the idea that the universe will grow old and eventually wither away, but is far more applicable to a steam engine than ultrastasis. The universe is infinitely old - *it already exists in a state of maximum entropy!*

14.4 COSMIC ENERGY CYCLES

The universe's material is composed of *electrons* (electrons or positrons) and *protons* (protons or antiprotons). Electrons are ancillary to cosmostasis, as they constitute only 0.05% of the universe's material content. Cosmic energy cycles will therefore be defined in terms of protons. These exist in one of two states, either free as hydrogen, or bound into collections such as helium, carbon, and the rest of the elements. Cosmic equilibrium dictates a constant rate of creation and destruction for both of these proton configurations, so there are two universal energy cycles - the *fusion cycle* and the *annihilation cycle*:

Ψ THEOREM 14.2 - UNIVERSAL ENERGY CYCLES {Ψ14.1}

THERE ARE TWO UNIVERSAL ENERGY CYCLES:

(A) FUSION CYCLE: BOUND COLLECTIONS OF PROTONS

(B) ANNIHILATION CYCLE: FREE PROTONS

Although annihilation and fusion exist in any high-energy environment, such as the interaction between cosmic rays and interstellar material, the cycles noted above represent the large-scale processes responsible for the genesis and overall density of the *majority* of the universe's material. Atomic nuclei containing more than one proton will be referred to as *compound nuclei*.

14.5 FUSION CYCLE

The fusion cycle is the balance between hydrogen and compound nuclei. Our telescopes and space probes have revealed the leading players on the celestial stage, so this cycle's rough outline is already available:

(hydrogen fuses) → (compound nuclei, light) →
(light loses energy by intergalactic redshift, scattering, absorption) ?→
(CMB) ?→ (nuclear dissolution occurs to produce hydrogen) ?

The majority of the luminous energy released by this cycle is from hydrogen fusion, but there is also a small contribution from the fusion of helium and heavier nuclei. The universe is full of galaxies that are in turn composed of stars. They burn hydrogen to produce helium, heavier elements, and light. Fusion is responsible for the universal optical and infrared luminosity densities. The light released by fusion loses its energy through the intergalactic redshift, scattering, and absorption.

Current measurements indicate that photons lose about half of their energy after a journey of ten billion light years through deep space. *This means every ten billion years, half of the universe's entire luminous output is lost to redshift.* Where does all of this energy go? The energy is lost in intergalactic space. The only known energy radiating from deep space is the CMB. Since the CMB is essentially a source of energy emanating from everywhere in deep space it is reasonable to think that it is the direct or indirect by-product of intergalactic redshift.

This is why it is included in the above sequence. The CMB carries energy until such time as it can be used to break down the elements that fusion forms in order to release hydrogen and complete the cosmic process. This is the *fusion cycle*.

Stars have been burning hydrogen forever and require an endless, renewable supply. The amount of energy in the universe does not change over time, so the only possible source of new hydrogen fuel *is the nuclei formed during fusion*. All forms of matter are recyclable - the first law of cosmostasis. This requires *reversibility* in any formative process. The energy in the light released by the fusion of protons must eventually be used to break them apart.

The overwhelming majority of the universe's power output is fusion, predominantly hydrogen fusion. The fusion of helium into carbon, for instance, only produces about 4% as much energy per unit mass as hydrogen fusion. A great deal of energy is released by supernovae, but this is negligible compared to the power generated by main sequence stars over the course of their enormous lifespans. In general, the fusion cycle will refer to all types of fusion, but hydrogen fusion is the first step. Without it there is no helium or carbon fusion:

Ψ THEOREM 14.3 - FUSION CYCLE FUEL {Ψ14.1}

HYDROGEN IS THE FUEL FOR THE FUSION CYCLE

Hydrogen fusion drives the universe's primary energy transfer and will be defined as the fusion cycle's beginning point. The mass fraction of hydrogen fusion, ε, is 0.0073. This is the fraction of mass converted into energy when helium is formed. The term ε_U will be used to denote the energy-averaged mass fraction of all fusion reactions in the universe, from the controlled, slow burning of hydrogen to the catastrophic formation of carbon and iron in supernovae. Since hydrogen fusion produces the lion's share of the cosmic engine's output, ε_U is close to ε. This approximation is more than sufficiently accurate for our calculations in Part IV, given the limited resolution available in the known universal parameters.

§ DEFINITION 14.1 - UNIVERSAL MASS FRACTION

THE UNIVERSALLY AVERAGED MASS FRACTION OF ALL FORMS OF FUSION IS $\varepsilon_U \approx \varepsilon$

The fusion cycle begins with the production of compound nuclei and light. It ends when these recombine and compound nuclei are converted back into hydrogen. This cycle could also be termed the *mass-fraction cycle* or the *binding energy cycle* because its two functions are to bind protons, converting part of their mass into light energy, and then use the energy originally released as light to disassociate nuclei back into individual protons, thereby replacing their lost mass and creating an endless source of hydrogen.

FUSION CYCLE PRODUCTS AND PATHS

The fusion cycle burns hydrogen fuel, leaving two by-products:

Ψ THEOREM 14.4 - FUSION CYCLE PRODUCTS {Ψ14.1}

THE FUSION CYCLE'S PRODUCTS ARE:

(A) LUMINOUS ENERGY

(B) COMPOUND NUCLEI

As is often the case in an equilibrium system, energy flows through a number of intermediate stages. Most of the nuclei the fusion cycle creates are stable and basically on standby until its original luminous output can be captured to blast them apart into hydrogen. The key to understanding the entire process is finding the mechanisms responsible for routing the compound nuclei and luminous energy produced by fusion to the nuclear disassociation that releases the hydrogen needed to perpetuate the fusion cycle.

Compound nuclei have one of two destinies. In smaller stars they generally remain in the stellar interior where they were formed. The star eventually burns out and leaves a cold stellar remnant. In more massive stars, compound nuclei are scattered throughout the galactic disk by ejections and supernovae explosions, creating even heavier nuclei in the process. *In general, compound nuclei remain in the galaxies where they were formed.*

Luminous energy, on the other hand, is released into space and propagates away from the site of fusion at the velocity of light. During the course of its life, the energy an average star emits travels billions of light years away from its source. Thus the fusion cycle splits into two paths. Compound nuclei take one path and light another. The end of this cycle occurs when they are recombined to liberate hydrogen. Fusion's products are too disparate to follow the same path through the cosmic engine:

Ψ THEOREM 14.5 - FUSION CYCLE PATHS {Ψ14.1}

THE FUSION CYCLE DIVERGES INTO TWO PATHS:

(A) COSMIC LUMINOUS PATH (LUMINOUS ENERGY)

(B) COSMIC PROTON PATH (COMPOUND NUCLEI)

These paths separate at fusion and rejoin at the site of nuclear disassociation. The fusion cycle is the universe's largest energy flow and the cosmic distribution of matter is optimized to carry this energy. *It is the reason galaxies are the way they are.*

14.6 ANNIHILATION CYCLE

The annihilation cycle is the balance between matter and light itself:

(matter, antimatter) → (gamma radiation) → (intergalactic red shift, absorption) ?→ (high-energy environment) ?→ (matter, antimatter)

It is not clear whether or not the gamma background radiation seen in space can be directly linked to the cosmic annihilation cycle. Nor is it even apparent that this cycle moves energy on an astronomical scale. Matter and antimatter might be utterly isolated from each other at the level of partial omnielements. Antimatter certainly doesn't survive very long in a neighborhood composed of matter, even one as thin as intergalactic space. That said, the Null Axiom requires a universe of half matter, half antimatter. Regardless of how effectively these might be separated from each other, there will be a certain amount of annihilation, and particles will be lost to radiant energy. An ultrastatic universe makes it necessary to replace these particles.

The general instability of matter and antimatter in close proximity necessitates the existence of containment units large enough so that their ratio of surface area to volume is consistent with either the background gamma flux or the material density of space or both. Since the observed gamma flux is nearly ten orders of magnitude less than fusion's optical flux and since much (if not all) of these gamma rays originate in processes other than cosmic annihilation, the annihilation cycle is far slower and carries quite a bit less energy than the fusion cycle. Annihilation produces about 140 times more energy than fusion. Even a moderate amount of interaction between matter and antimatter leaves an unmistakable signature.

Unlike fusion, matter-antimatter annihilation only has one product: energy. Like fusion, it still requires a path for energy and a path for matter. The only difference is the paths never rejoin per se and therefore are not as intimately related as the fusion paths. Matter and antimatter annihilate to produce energy. This is eventually absorbed to produce particles that in due course annihilate. The energy released in the annihilation cycle is pure gamma radiation. It isn't carried by some other entity and has no recurring connection to the particles it ultimately creates. Annihilation is a cycle in terms of creation and destruction, but there is no *underlying energy buffer.* Fusion, by comparison, doesn't create or destroy elementary particles, so the protons circulating in its cycle provide an enduring and immutable foundation. This creates a sharply defined, contiguous loop that is not available to the annihilation cycle.

There are a number of different gamma ray sources in the heavens, but few are at annihilation energy (940 MEV for protons). Gamma ray bursts typically have energies in the 0.001 to 100 MEV range.[6.2] If this is evidence of the cosmic annihilation cycle, they originate from distances so great that they have been redshifted by a factor of at least $z = 10$.

14.7 THE FOCUS IS FUSION

Fusion accounts for nearly all of the universe's energy flow, so the remainder of Part IV will focus on a detailed description of the paths that energy and protons follow to complete this cycle. For more information about the annihilation cycle and the universe's large-scale antimatter distribution, please refer to Appendix K.

14.8 GENERAL CONCLUSIONS

As an introductory chapter conclusions are few, but the stage is set:

- There are two universal energy cycles, fusion and annihilation, and fusion is by far the more dominant.
- A cosmic mechanism *has to* exist that combines the energy and compound nuclei produced by fusion in order to provide an eternal source of hydrogen.
- The universal rate of fusion's hydrogen consumption is equal to the universal rate of its production by nuclear disassociation.
- The universal density of luminous energy and matter is a product of an infinitely old equilibrium system.

Our goal is to identify and quantify all of the steps in the universe's dominant energy cycle, fusion. Ultimately, this will lead to a detailed description of the role galaxies play as components of the universe's thunderous cosmic engine. Form follows function, but as Part IV unfolds it will become clear that function also provides a great deal of insight about form.

The journey begins with the universe's most spectacular energy loss: *intergalactic redshift*.

15. COSMIC LUMINOUS PATH

15.1 INTERGALACTIC REDSHIFT

Intergalactic redshift is caused by the gradual loss of light energy over great distances, similar to the tired light concept Fritz Zwicky originally postulated. This is why the effect is linear with distance. The farther the source, the more energy is lost, and the greater the redshift. This is also why the structure of the universe looks so uniform; the only real motions in the heavens are the peculiar motions generated by differences in mass distributions. These tend to be no more than a few hundred km/s, not the significant fractions of c required by universal expansion.

The concept of tired light has never had much success in modern cosmology because it fails (at least in its original form) to explain why distant signals are *stretched* as well as redshifted. Supernovae provide the prima facie case for this. Their intensity/duration curves tend to broaden proportionately with distance. An event at a distance corresponding to ($z = 1$) has, on average, twice the duration of a local explosion. If photons are just losing energy by crossing space, why are they also being dispersed along their path? Efforts to resolve tired light's dispersion problem have proven ineffectual and it has been thoroughly trumped by the drama of the Big Bang, pushing cosmology away from the elegant truth Zwicky recognized.

Although Zwicky had the intuition to realize light was losing energy, he lacked the Null Axiom, thereby relegating tired light to just another untestable premise. Our approach suffers no such limitations. It expands the concept significantly beyond Zwicky's original idea, resolves the signal dispersion issues, and is buoyed by the fact that it is the *only possible explanation* for the effect. No changes can occur to the universe as a whole. It doesn't expand, it doesn't contract, and it won't grow old and die. If light acquires a spectral change after crossing billions of light years of deep space, this is the full extent of the phenomenon. It is not a commentary on the dynamics of reality.

Intergalactic redshift has some surprising ramifications that provide the basis for other global phenomena, including the material density of the universe, the CMB field, and dark matter. Even Zwicky would probably have been surprised to learn just how important this process is.

So why does light lose energy when traveling across broad gulfs of space? Why indeed. This phenomenon has some truly curious properties.

ANCIENT LIGHT

Intergalactic redshift is caused by photons' gradual loss of energy over immense distances. Cosmic equilibrium leaves no other alternative. It is a given and is just as certain as the conservation of energy it obeys. But what causes it?

Its characteristics are:

a) Proportional to distance for cosmically short distances. If light travels twice as far, it loses twice the energy.

b) Proportional to photon energy. Visible light and radio waves from the same source lose the same fraction of their energy.

c) No associated scattering. Light reaching us from distant spiral galaxies preserves exquisite details of their disks' structure.

d) Long-range and weak, requiring ~10 Gly to cause a 50% energy loss.

e) Uniform broadening. The duration of a transient signal grows in direct proportion to the increased wavelength of the individual redshifted photons within it.

f) No refractive frequency dispersion, as is typical of an interaction between light and matter. An ancient pulse's red component arrives at the same time as its blue component, although the pulse itself is broadened in accordance with e) above.

One of intergalactic redshift's more intriguing aspects is that light loses energy *without any observable scattering*. If it were caused by a photon-photon or photon-particle interaction, every incremental energy loss would have a corresponding change in direction. Scattering related to a 5% loss is enough to make celestial images unrecognizable, yet our instruments show galaxies in pristine detail even at energy reductions greater than 10%. Also, scattering is generally not linear with incident energy. It is rarely a situation where *energy loss is proportional to the incident energy*. Furthermore, scattering *requires* a change of direction in order to conserve momentum. Preservation of directional integrity with an energy loss proportional to energy has no precedent in known physical interactions.

Intergalactic redshift's other (even more enigmatic) property is its signal dispersion in the absence of refractive dispersion. Not only do photons attain longer wavelengths, the spatial separation between them increases with distance traveled *irrespective of their individual frequency*. The greater an object's range, the greater the duration of its transient signals, *and all frequencies from some remote source experience a comparable amount of broadening*. This means there is far more to this effect than just the energy loss of individual photons. Moreover, the lack of refractive dispersion indicates that *intergalactic redshift is not caused by an interaction between photons and matter*.

Our premise is that intergalactic redshift arises from a *known* entity, because the Cosmological Principle tells us intergalactic space is similar to, albeit more sparse than, local space. The only things found in deep space are:

- Space.
- A weak gravitational field.
- Trace amounts of matter, the equivalent of ~3 hydrogen atoms/meter^3.
- The CMB field.
- Photons of various energies in addition to the CMB.
- Neutrinos.

The dispersion (broadening) found in deep-space signals is the pivotal consideration for evaluating this list. Ancient photons do not lose energy as a result of collisions with particles, other photons, or other entities in deep space. *Collisions would only cause the loss of energy in individual photons; they wouldn't increase the distance between them*. This effectively eliminates the matter, neutrinos, and radiation in intergalactic space, leaving only space and a weak universal gravitational field as possible redshift agents. Space is dimensionally distinct from energy and cannot interact with it, *so a gravitational interaction is the only possible source of intergalactic redshift*.

The universal gravitational background is the one entity in deep space capable of inducing a loss of energy in ancient photons. The only other explanation is photons are intrinsically unstable and spontaneously decay at a certain rate, but this is not viable because it fails to address signal dispersion. Further, all interactions are causal; there is no such thing as "spontaneous" decay. Any transformation photons experience must be facilitated by an external agent and *the only weak mechanism able to act over billions of light years of neutral space is gravity*. This process will be referred to as *lumetic decay*.

Ψ THEOREM 15.1 - LUMETIC DECAY {Ψ4.2}
INTERGALACTIC REDSHIFT IS CAUSED BY THE UNIVERSAL GRAVITATIONAL FIELD

There is no viable alternative, but this phenomenon's underlying dynamic is far from obvious.

15.2 LUMETIC DECAY

Gravitational fields produce a number of different effects on light and matter's interaction with light, all of which have been well documented. Photons appear to lose energy when they move out of a gravitational well; they appear to gain energy upon falling into such a well. They are deflected as they pass near massive objects. What is intriguing is no *known* gravitational effect could possibly be responsible for lumetic decay. The photons traversing deep space move out of as many gravitational wells as they fall in. There should be no net frequency shift. While it is true they are also deflected around superclusters and galaxies, this is an elastic interaction. Even if it were not, gravitational deflection *doesn't cause signal dispersion*. In short, lumetic decay is a new type of photon-gravitational interaction. It occurs between deep space photons and the universal gravitational field through which they pass.

According to the General Theory of Relativity, the universe's energy content, composed of elementary particles, photons, and neutrinos, gives space an average radius of curvature of the form:[17.1]

$$R_G = \frac{c^2}{\sqrt{4\pi G \rho_U}} \tag{15.1}$$

where ρ_U is universally average energy density in units of (energy/volume), and R_G will be called the *gravitational radius* of space.

General Relativity relies heavily on the mathematical equivalence between free space acceleration and the acceleration induced by a gravitational field. Space has no physical reality in this theory. As such, Equation (15.1) represents the *effect* of internal spatial distortion, not the spatial distortion itself. Since space's *net* internal distortion necessarily sums to zero as a consequence of its overview, R_G is the result of its *average* spatial distortion. Think of space as a sheet of rubber that has been unevenly stretched across a perfectly flat floor. Its gravitational radius is the averaged effect of its internal nonlinearity.

Even though General Relativity does not address space's physical geometry, it is a brilliant and accurate generalization of the measurement of space and time, and can be used to calculate the magnitude of lumetic decay. The easiest way to understand the curvature of Equation (15.1) is to take the case of two photons moving down the same axis in curved as compared to Euclidean (rectilinear) space, departing two seconds apart:

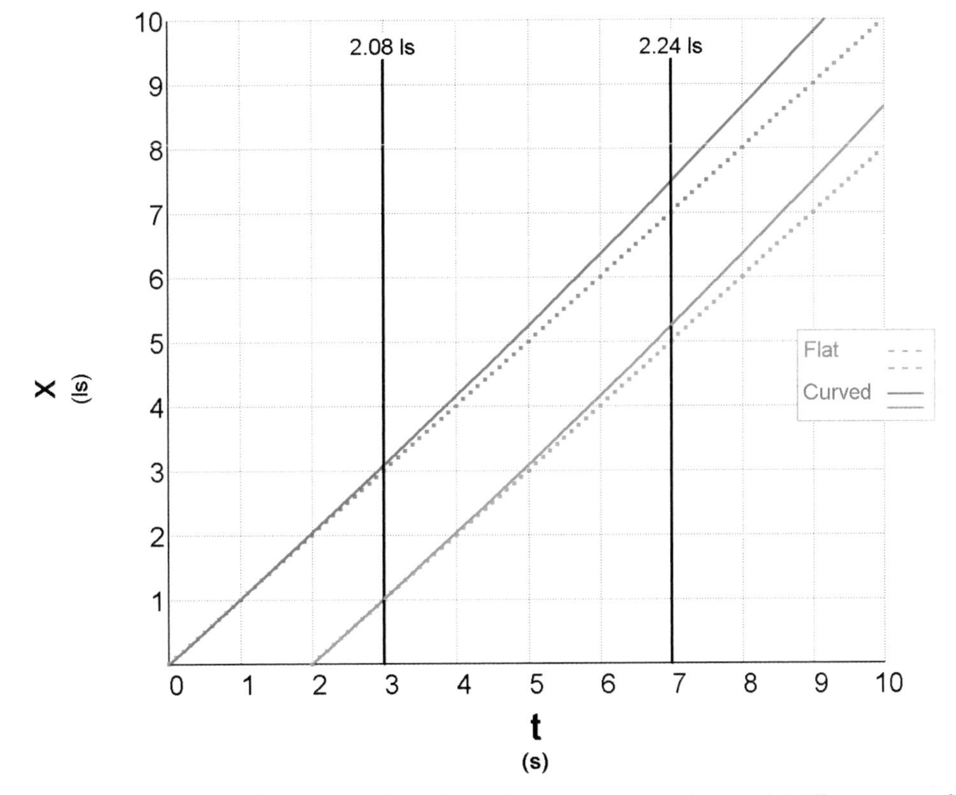

Figure (15.1) Photons moving along an axis through rectilinear versus curved space, initially two seconds apart

Time is the horizontal axis, in seconds, and distance is the vertical axis, in light-seconds (ls). This graph is a physically accurate representation of motion through space because time, in relation to motion, is the internal, fourth-dimensional difference of space. It is an additional dimension that does not increase space's universal extent: *space-time*.

If space were perfectly rectilinear, the two photons shown in Figure (15.1) would follow the dotted traces, always separated by a distance of two light-seconds. However, due to its energy content, space exhibits, on average, a slight fourth-dimensional curvature. This has been markedly exaggerated (by a factor of $\sim 10^{16}$) in Figure (15.1) to make its effect visible. The photons' paths through this artificially curved space-time corresponds to the solid traces. Three seconds after the first (red) photon departs, the two photons are separated by a distance of 2.08 ls. Four seconds later, they are separated by 2.24 ls. The space of our universe exhibits such a small average curvature that this same deviation would require $\sim$1.2

billion years. *Curved space simulates spatial expansion*. Ironically, whereas Special Relativity shows us that it is not possible to directly measure a reference frame's movement, the only way to measure spatial curvature is to move through it.

Gravity's universal curvature creates a differential velocity difference per unit distance. *This is what the Hubble constant represents*:

$$\frac{dv}{dx} = \frac{c}{R_G} = H_0 \tag{15.2}$$

A value of 60 Hz-km/Mpc exposes light to a tiny differential velocity gradient of $1.95(10)^{-18}$ meters per second per meter. Although H_0 is usually evaluated in terms of megaparsecs of distance, *space is continuous, so it is equally applicable to the space within a photon's topology*.

Energy *acquires* the differential velocity of Equation (15.2) through propagation. Since this occurs along straight line trajectories through the third dimension, the Hubble constant represents a tiny positive acceleration, and the energy distributed throughout a photon's substructure accrues differential velocity at a steady pace. A photon's leading edge has traveled farther than its trailing edge by its wavelength λ, inducing an internal velocity gradient of the form:

$$dv = \frac{d\lambda}{dt} = \frac{c}{R_G}\lambda = H_0\lambda \tag{15.3}$$

Differential velocity generated by interaction with universal curvature stretches photons, causing a continuous expansion throughout their energy distributions. *The universe isn't expanding; intergalactic photons are*. Moreover, just as it acts within a single photon's structure, spatial curvature also creates differential velocity between photons, causing a general broadening of ancient signals in direct proportion to their redshift. From a local perspective, photons, regardless of their age, appear to be moving at c, but on a larger scale their effective speed along the path they have taken depends on their departure time relative to other photons.

Although the universe is ultrastatic, it does have a relationship between time as change and time as dimension. R_G manifests as spatial expansion because it induces a *nonlinear* relationship between distance and time. Even though space's size is invariant, its phenomenological curvature represents a differential velocity field - the same field that would exist if it were actually expanding. The difference between a billion light years now and a billion light years a billion years ago *is not linear*, and gives the false impression of universal expansion. Scientists have grossly misinterpreted the physical reality of gravitation in both of its most extreme environments - black holes and deep space.

All photons move at *c*, so it certainly seems like it should take the same amount of time to cross the same amount of space. But this is only true if a photon's motion had a purely linear relationship to time, and this is not possible when they move through space infused with fourth-dimensional curvature.

In summary, universal curvature causes photons to expand on their epic journey across deep space. This results in an isotropic redshift of all signals from distant sources, the magnitude of which is roughly proportional to their distance from Earth. *The intergalactic redshift of photons is nothing more than a direct measurement of the average universal curvature predicted by the General Theory of Relativity.*

Ψ THEOREM 15.2 - LUMETIC DECAY EXPANSION {Ψ15.1}

RADIANT ENERGY DECAYS AS A RESULT OF THE INTERNAL ISOTROPIC EXPANSION OF ITS STRUCTURE CAUSED BY THE DIFFERENTIAL VELOCITY FIELD OF UNIVERSAL CURVATURE

Universal expansion is not only wrong - it is unnecessary and redundant. The universe's average spatial curvature is already known to exist. It ought to have *some* effect, and the fact that it emulates universal expansion is certainly no coincidence. Nor is the fact that the magnitude of its effect (redshift) is consistent with the universe's average energy density.

Lumetic decay is not limited to a photon's wavelength. The universe's differential velocity field is isotropic, and so therefore is the expansion it causes, increasing the size of a photon's spatial footprint along all three of its extents. Let's quantify the energy loss that this expansion causes.

LUMETIC DECAY'S ENERGY LOSS

Simplify Equation (15.3):

$$\frac{d\lambda}{dt} = H_0\lambda \tag{15.4}$$

Solve for wavelength:

$$\lambda = \lambda_0 e^{H_0 t} \tag{15.5}$$

As a photon expands, the differential velocity at its boundaries slowly increases, producing an exponential decay.

Convert Equation (15.5) to energy and simplify:

$$\frac{1}{E}=\frac{1}{E_0}e^{H_0t}\rightarrow E=\frac{E_0}{e^{H_0t}}\rightarrow E=E_0e^{-H_0t} \tag{15.6}$$

When a photon expands over time, it has to release energy in order to satisfy both the unit hypervolume relationship and conservation, so it is forced to slowly decay:

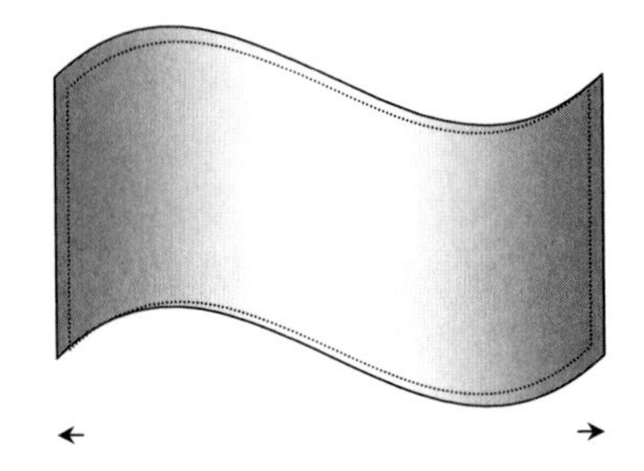

Figure (15.2) Universal curvature stretches photon wavelength, inducing a gradual energy loss

From Equation (15.6), a photon's energy loss rate is given by the Hubble constant as:

$$E=E_0e^{-H_0t}\rightarrow\frac{dE}{dt}=H_0E_0e^{-H_0t} \tag{15.7}$$

Energy loss rate *per unit energy* is the Hubble constant:

$$\left(\frac{1}{E}\right)\left(\frac{dE}{dt}\right)=\frac{H_0E_0e^{-H_0t}}{E_0e^{-H_0t}}=H_0 \tag{15.8}$$

Decay rate is not a function of gravitational potential, but rather the average spatial *distortion* caused by energy's presence. Thus it is relatively uniform even though the density of matter and other forms of energy varies throughout the universe.

All photons lose the same fraction of energy per unit time, so the decay rate of any frequency band is simply the product of its total energy density and the Hubble constant:

$$\frac{d\rho_E}{dt}=H_0\rho_E \qquad \left\{\rho_E=\int\frac{\rho_E(\lambda)}{\lambda}d\lambda\right\} \tag{15.9}$$

Distant objects appear to lose surface brightness far more rapidly than an exponential decay would suggest, but this is only because the optical band is relatively narrow and has virtually no energy density above it. Please see Appendix M for a detailed analysis of the relationship between surface brightness loss and redshift.

The relationship between photon energy, distance traveled, and redshift follows from Equation (15.6):

$$x = \left(\frac{c}{H_0}\right)\ln\left(\frac{E_0}{E}\right) = \left(\frac{c}{H_0}\right)\ln(z+1) \tag{15.10}$$

On object with a redshift of ($z = 6$), for instance, is ~30 billion light years from Earth.

15.3 DECAY VERSUS RECESSION

Before delving more deeply into lumetic decay's numerous and profound ramifications, let's briefly compare its energy loss profile with the expanding universe concept in accelerating and non-accelerating formulations. Simplified versions of both are presented below. They do not contain many of the Big Bang's ad hoc revisions, such as inflation, but are sufficiently representative for the purposes of this section.

The relationship between a deep-space photon's energy loss and the Hubble constant for a universe with uniform expansion has the form:

$$E = E_0\sqrt{\frac{c - xH_0}{c + xH_0}} \tag{15.11}$$

This restricts the observable universe's size, since a photon's energy goes to zero at any distance x greater than (c/H_0). Cosmologists have recently (at least as compared to the introduction of the original expanding universe concept) embraced the idea of a universal expansion whose rate changes with time.

In the most straightforward rendition of the *accelerating* expanding universe, the Hubble constant is a linear function of distance (and therefore time) of the form:

$$H_0^*(x) = (1 - a_U x)H_0 \tag{15.12}$$

where a_U is the acceleration of the universe's expansion in units of distance^{-1}. a_U is positive for an accelerating universe because the Hubble constant needs to be smaller in the past if the rate of universal expansion were actually increasing. The energy loss associated with this acceleration is somewhat more complex than for the uniform expansion of Equation (15.11):

$$E = E_0\sqrt{\frac{c - (x - a_U x^2)H_0}{c + (x - a_U x^2)H_0}} \tag{15.13}$$

Compare lumetic decay to the two simplified expanding universe models:

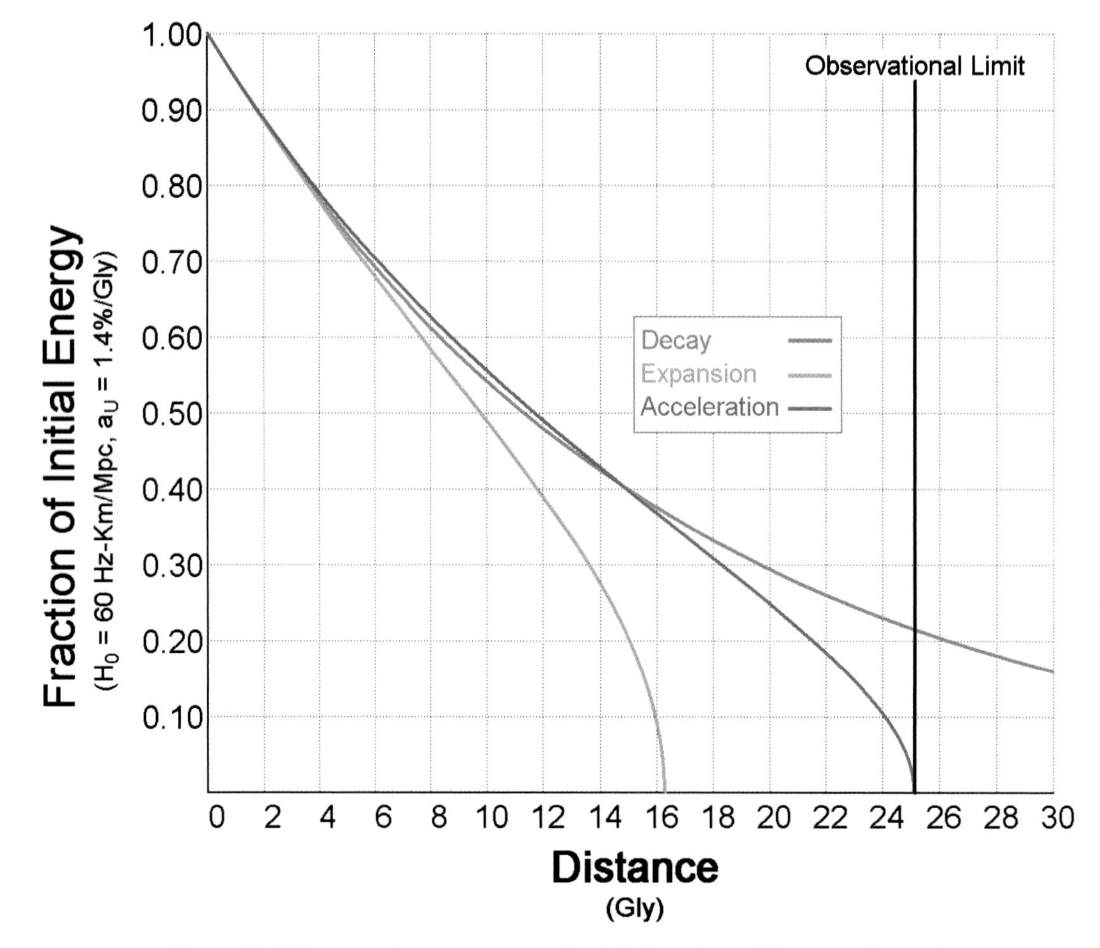

Figure (15.3) Lumetic decay versus two simplified versions of the expanding universe

The orange trace ending at ~16 Gly is of the constant expansion model. The red trace ending at 25 Gly is the accelerating expansion model of Equation (15.13) with an a_U of 1.4% per Gly. The green trace extending past 30 Gly is lumetic decay. As shown in the graph, it became necessary for the universe's expansion to accelerate in the Big Bang model because long-distance measurements of supernovae (>4 Gly) are clearly more consistent with lumetic decay than with a fixed rate of expansion.[37] As long as the Big Bang is the dominant cosmology, its expansion/acceleration profile will have to be repeatedly adjusted as new technology allows astronomers to penetrate more deeply into space. Indeed, timing is everything in science. If Einstein would have predicted intergalactic redshift based on the universal curvature required by his own theory of gravitation, modern cosmology would have a very different conceptual landscape.

According to the plot of lumetic decay, an energy loss of 50% occurs over a distance of about 11 Gly. Photon age is given by solving Equation (15.6) for time:

$$t = -\left(\frac{1}{H_0}\right)\ln\left(\frac{E}{E_0}\right) \tag{15.14}$$

Lumetic half-life is the special case where the energy ratio in Equation (15.14) is 0.5:

$$\tau_\gamma = \frac{-\ln(0.5)}{H_0} \tag{15.15}$$

This is not the average time it takes for a decay to occur. The smoothness of the redshift distribution indicates at least a hundred decays occur prior to reaching a 50% energy loss. Equation (15.15)'s half-life is the time required for a photon to lose half its energy.

A Hubble constant of 60 Hz-km/Mpc corresponds to a photon half-life of 11.3 Gyr. The Hubble constant is thought to lie somewhere between 50 and 85 Hz-km/Mpc.[1.2] This puts lumetic half-life in the range {8 Gyr $< \tau_\gamma <$ 13.6 Gyr}.

15.4 UNIVERSAL DENSITY

The Hubble constant measures the net effect of the strained geometry of space-time, as presented earlier by Equation (15.2):

$$H_0 = \frac{c}{R_G}$$

Substituting Equation (15.1) for R_G yields the Hubble constant as a function of the universe's average energy density:

$$H_0 = \frac{\sqrt{4\pi G\rho_U}}{c} \tag{15.16}$$

Solve for energy density:

$$\rho_U = \frac{H_0^2 c^2}{4\pi G} \tag{15.17}$$

or $4.0(10)^{-10}$ J/m^3 at a Hubble constant of 60 Hz-km/Mpc.

The universe's average energy density is related to the Hubble constant as follows:

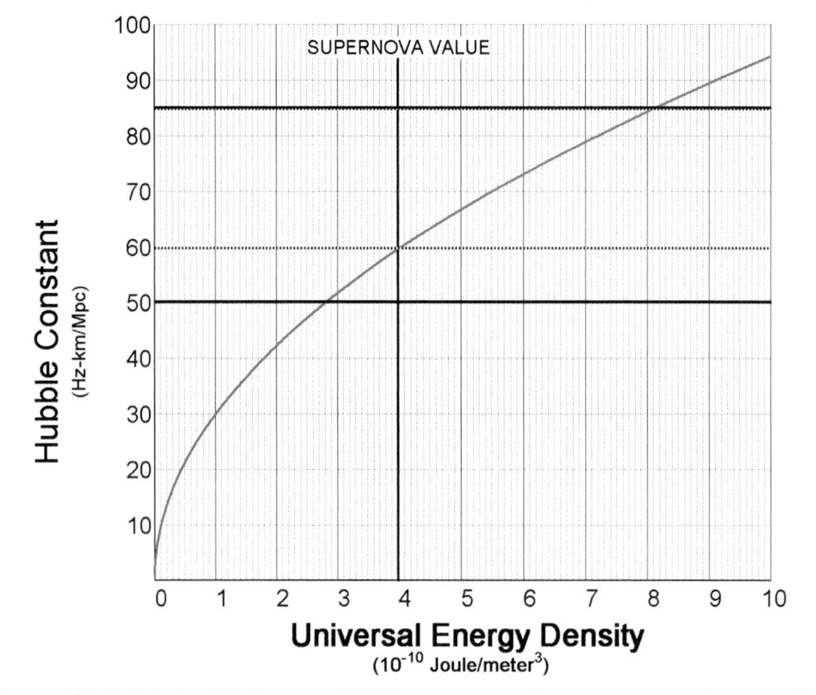

Figure (15.4) Relationship between Hubble constant and average universal energy density

The vertical line indicates the best current estimate of the Hubble constant using supernovae as standard candles. It puts the universe's average energy density at about 2.7 hydrogen atoms per cubic meter. This is consistent with the estimates of the total amount of matter in space and supports the idea that the majority of it is nonluminous. The horizontal lines reflect the amount of error thought to currently exist in the Hubble constant, from 50 to 85 Hz-km/Mpc. This corresponds to a fairly wide (~3x) range of universal energy density, $\{2.8(10)^{-10}\ \mathrm{J/m^3} < \rho_U < 8.1(10)^{-10}\ \mathrm{J/m^3}\}$. Luminous matter, whose energy density is estimated to be on the order of $\sim 10^{-11}\ \mathrm{J/m^3}$,[3.1] constitutes only a small fraction ~(1%–3%) of the universe's average energy density.

UNIVERSAL COMPOSITION

In accordance with nature's causal singularity, radiant energy ultimately originates from matter. This means every photon or neutrino throughout space at this very moment has a direct relationship to binding energy at either an atomic or nuclear level. Our universe's chemical binding energy is negligible in comparison to its nuclear, so *the universe's radiant energy density is the difference between the mass density of free protons and the mass density of compound nuclei.* Although annihilation is also a source of radiant energy, the gamma flux in space is so small that its mass density equivalent is insignificant.

The cosmic balance between compound nuclear matter and electromagnetic energy density will be referred to as *correspondence*:

Ψ THEOREM 15.3 - MATTER-ENERGY CORRESPONDENCE {Ψ4.9}

THERE IS A ONE-TO-ONE CORRESPONDENCE BETWEEN MATTER FIELD POTENTIAL AND PHOTON/NEUTRINO/KINETIC ENERGY

This occurs exclusively in terms of energy's *quantity* and has no relationship to quantal number. The binding energy between two nucleons, for instance, might eventually take the form of millions of CMB photons, but the primary consideration is underlying magnitude, not quantal configuration.

Matter-energy correspondence can be expressed mathematically as:

$$\rho_U = \rho_\wedge m_\wedge c^2 = \rho_M c^2 \tag{15.18}$$

where ρ_M is the product of the numerical density of the universe's elementary particles, $\rho_\wedge$, and their free-space rest masses, $m_\wedge$. In other words, if all of the universe's particles were unbound and at no potential, gravitational or otherwise, the sum of their rest energies would be precisely equal to the universe's total energy density.

In essence, the temporal closure (Ψ4.9) between matter and (photons/neutrinos) requires the sum total of the universe's kinetic, luminous, and neutrino energy to be balanced by and equal to the negative potential of all of its particles, primarily protons. Thus:

$$\rho_R + \rho_{KE} \cong \varepsilon_U \rho_{bp} c^2 + \rho_{\Phi_g} \tag{15.19}$$

where ρ_{bp} is the universally average concentration of bound protons, ρ_R is the average density of all radiant energy, and ρ_{Φ_g} is matter's average gravitational potential. Equation (15.19) is shown as a close approximation because the universe's electron content, whose negative potential contribution is negligible, has been excluded. Solve for the concentration of bound protons:

$$\rho_{bp} \cong \left(\frac{\rho_R + \rho_{KE} - \rho_{\Phi_g}}{\varepsilon_U c^2} \right) \tag{15.20}$$

Most of the universe's mass is dark, so it has low average kinetic energy content. Also, compact objects tend to emit radio and high-energy radiation. Since the cosmic density of such sources is small, most of the universe's dark mass exists at a relatively weak average gravitational potential.

Applying these provisions to Equation (15.20) yields:

$$\rho_{bp} \cong \frac{\rho_R}{\varepsilon_U c^2} \tag{15.21}$$

NEUTRINO DENSITY

Given neutrinos' remarkably small absorption cross-sections,[2.4] one might think that their equilibrium number density in an eternal universe is extraordinarily high. This is not the case. Neutrinos move at the speed of light, so they decay just like electromagnetic radiation, and the only neutrino sources with significant luminosity are transient - supernovae or compact objects experiencing a change of state. Thus their universal density is fairly small. Matter-energy correspondence is consistent with this, as it also requires a low universal neutrino density. Neutrinos are only released by the binding of free particles. Even if all of the universe's matter were confined into compact objects, the *complementary* cosmic neutrino energy density would be orders of magnitude less than the electromagnetic radiation released from the mass fraction of all of these bound particles. For additional information about neutrinos, please refer to Appendix N.

When neutrino and other low-density radiant bands, such as gamma and radio wave, are omitted from Equation (15.21), the result is the universe's average bound proton concentration as a function of the most prominent luminous backgrounds in space:

$$\rho_{bp} \cong \frac{\rho_R}{\varepsilon_U c^2} \cong \frac{\rho_{B\gamma} + \rho_{I\gamma} + \rho_{CMR}}{\varepsilon_U c^2} \tag{15.22}$$

or $9(10)^{-29}$ kg/m^3 using the values listed in Appendix C.

DARK HYDROGEN

Measurement of *luminous* matter density doesn't reveal much about cosmic composition since it originates from a mixture of free (hydrogen) and bound protons. Equation (15.22) does, however, in concert with Figure (15.4), indicate that the universe is composed of a large fraction of nonluminous material. Although modern cosmology is full of speculation about a host of exotic dark matter candidates, such as gravitons and HIGGS and Z bosons, there is really only one viable contender for the missing energy density - *hydrogen.*

Electrons constitute only ~0.05% of the universe's mass, so its average material density is close to the sum of its free and bound proton concentrations:

$$\rho_M \cong \rho_{fp} + \rho_{bp} \tag{15.23}$$

where ρ_{fp} is the density of free protons (hydrogen).

The universally average *fraction* of bound protons is given by the ratio of Equations (15.22) and (15.17):

$$f_{bp} \cong \frac{4\pi G(\rho_{B\gamma} + \rho_{I\gamma} + \rho_{CMB})}{H_0^2 \varepsilon_U c^2} \tag{15.24}$$

or 2% at a Hubble constant of 60 Hz-km/Mpc. The universe is about 98% hydrogen (and antihydrogen). This fraction is inversely proportional to the square of the Hubble constant so is fairly sensitive to its value. If the Hubble constant is 50 Hz-km/Mpc, the universe is 97.3% hydrogen; at 85 Hz-km/Mpc, it is 99%. Astronomers have estimated the composition of luminous matter to be about ~70% hydrogen, ~25% helium, and perhaps ~5% other elements.⟨1.1⟩ Equation (15.24) indicates that dark matter is overwhelmingly hydrogen. For more information about its cosmic distribution, please refer to Appendix N.

FUSION ENDURANCE

Dark matter represents an extraordinary buffer for cosmic activity. In the absence of some form of replenishment, the amount of time it would take the universe to consume all of its available fuel is given by the ratio of its fuel density to total luminosity density. This will be called its *fusion endurance*, τ_ε:

$$\tau_\varepsilon = \frac{(1 - f_{bp})\varepsilon_U \rho_U}{j_R} \tag{15.25}$$

At (f_{bp} = 0.02), with the values given in Appendix C, and in the absence of some form of hydrogen renewal, it would require ~45 trillion years for the universe to exhaust its fuel. *A star only burns the hydrogen in its core region, which accounts for about 10% of its mass. Since luminous material represents at most ~2% of all mass, less than 0.2% of the universe is on fire, and this fire is exceptionally slow-burning.*

The value of luminosity density j_R used in Equation (15.25) was averaged over the entire universe to include extensive stretches of virtually empty intergalactic space. A similar calculation can be done for individual galaxies to show that their fuel consumption is appreciably higher. *Galactic* fusion endurance is given by a slight modification of Equation (15.25), with the simplifying assumption that the bound proton fraction is the same:

$$\tau_{\varepsilon} = \frac{\left(1 - f_{bp}\right)\varepsilon_U M_g c^2}{L_g} \quad (15.26)$$

where M_g and L_g are a galaxy's mass and luminosity, respectively. The luminous portion of the Milky Way, for instance, has an estimated power output of $1.4(10)^{37}$ W⟨1.14⟩ and mass of $8(10)^{41}$ kg.⟨1.16⟩ Its total fuel reserve (at a composition of 98% hydrogen) is about 2% of the universal average, 1.1 trillion years.

15.5 LUMINOUS LIMIT

The original version of Olbers' paradox was:⟨6.6⟩

> "*If the universe is infinite, then a line extending in any direction from Earth would eventually intersect a star's surface. Why isn't the night sky white? Proposed solution: Because the universe is finite.*"

Once more was learned about the breadth and low density of the heavens, it became clear that light would probably be scattered long before it traveled the distance necessary to make Olbers' evenings white. But its energy would still remain, so the paradox was modified to a more modern, thermodynamic form:⟨6.7⟩

> "*If the universe is infinite then the luminous output of stars would quickly build to the point where its accumulated heat would burn nonluminous objects, including planets. Proposed solution: This doesn't happen because the universe is expanding, lowering its energy density, cooling it down.*"

Neither of these interpretations is even marginally close to the mark. There is a far more interesting reason why space is a chilly 2.7 °K in an infinite, nonexpanding universe.

Fusion is by far the universe's largest power output, and lumetic decay is by far its largest *power loss*. Even though this effect is weak and requires billions of years to cause a significant energy deficit in individual photons, its universal consequence is staggering. *Since the light given off by all luminous objects decays over time, the cumulative energy in space associated with any luminous object is limited. After a given length of time, the loss due to the lumetic decay of prior luminous output will balance an object's current luminous output.*

Let L be the luminosity of some celestial object such as a star or galaxy. The energy it radiates in some small time interval is initially given by:

$$dE = Ldt \tag{15.27}$$

Since luminous energy decays according to the Hubble constant, the energy in this differential is actually a function of time:

$$dE(t) = Le^{-H_0 t} dt \tag{15.28}$$

The total energy in space associated with an object is given by the aged sum of its radiant output from some initial time ($t = 0$) to its current age, τ:

$$E = \int_{t=0}^{t=\tau} Le^{-H_0 t} dt \tag{15.29}$$

which evaluates to:

$$E = \frac{L}{H_0}\left(1 - e^{-H_0 \tau}\right) \tag{15.30}$$

This is assuming the physical size of celestial objects is small in comparison to the onset of lumetic decay, as is the case. The luminous portion of most galaxies is less than 100 Kly in radius, and lumetic decay doesn't have much of an effect until after a few hundred million light years. Equation (15.30) also assumes the rate of absorption of luminous energy in intergalactic space is small in relation to lumetic decay, which is also the case. Decay is weak, but deep space absorption is orders of magnitude weaker.⟨10.2⟩

When τ is large, Equation (15.30) goes to the limit:

$$E_L = \frac{L}{H_0} \tag{15.31}$$

Lumetic decay reduces the luminous legacy of celestial objects so effectively that there is a limit to the amount of energy they can maintain in space. The energy of Equation (15.31) will be referred to as the *luminous limit*.

Ψ THEOREM 15.4 - LUMINOUS LIMIT {Ψ15.2}

THE LUMETIC DECAY RATE OF ANY CELESTIAL OBJECT'S PRIOR LUMINOUS OUTPUT WILL EVENTUALLY BALANCE ITS CURRENT LUMINOUS OUTPUT

The luminous limit for our sun, with an output of $3.8(10)^{26}$ W,⟨1.20⟩ is $2(10)^{44}$ J. This is the amount of energy it emits in 16.3 billion years, which is longer than its estimated lifespan. The luminous limit for the Milky Way, with an estimated power output of $1.4(10)^{37}$ W,⟨1.14⟩ is $7(10)^{54}$ J. This also represents its output for 16.3 billion years since it is defined by the same Hubble constant, but unlike our sun it will last long enough to achieve it (barring collisions with other galaxies, such as the impending disaster with Andromeda).

15.6 LUMINOUS BALANCE

The universe has existed forever, so it represents a *universal luminous limit*. When Equation (15.31) is evaluated on a per-volume basis, the Hubble constant relates universal *luminous energy* density with universal *luminosity* density:

$$\rho_{B\gamma} = \frac{j_B}{H_0} \qquad (15.32)$$

where $\rho_{B\gamma}$ is the energy density of the optical band. Energy falling into the optical spectrum from the decay of higher-energy bands enhances j_B slightly, but the content above the optical band is much smaller than $\rho_{B\gamma}$. Substituting the optical luminosity density and Hubble constant of Appendix C into Equation (15.32) results in a universally average optical energy density of $\rho_{B\gamma} = 7(10)^{-16}$ J/m^3, within the estimated range of the observed value.

The Hubble constant is the relationship between the energy and luminosity densities of integrated starlight. Or to express this in terms of universal equilibrium, *the output of luminosity density balances the decay loss of luminous energy density*:

$$j_B = H_0 \rho_{B\gamma} \qquad (15.33)$$

This is the *luminous balance*:

Ψ THEOREM 15.5 - LUMINOUS BALANCE {Ψ15.4}

THE ENERGY LOST IN THE DECAY OF THE UNIVERSE'S OPTICAL ENERGY DENSITY IS REPLENISHED BY ITS LUMINOSITY DENSITY

The luminosity density needed to balance optical decay represents nearly all of the universe's power output. The only bandwidths with significant cosmic luminosity density are optical and infrared. X-ray and radio wave sources are sparsely distributed throughout space but their output is several orders of magnitude less than the glow and afterglow of stars.⟨2.1⟩

DECAY IMMUNITY

The Hubble constant defines the geometric expansion of any photon over distance, so it is reasonable to think that Equation (15.33) ought to hold for the entire radiant spectrum. It does not. The balance between the universal density and luminosity of electromagnetic radiation has the form:

$$j_R = \beta_R H_0 \rho_R \tag{15.34}$$

where j_R is the universal luminosity density of all photons and neutrinos, ρ_R is their average total energy density in space, and β_R will be called the *lumetic decay fraction*. Solving for the decay fraction in Equation (15.34) yields:

$$\beta_R = \frac{j_R}{H_0 \rho_R} \tag{15.35}$$

For the values listed in Appendix C, this amounts to ~0.02. *The universe's total luminosity can only support ~2% of the lumetic decay of its average electromagnetic energy density.*

Since the CMB represents nearly all of this density, it follows that it is *immune* from lumetic decay:

Ψ THEOREM 15.6 - CMB DECAY IMMUNITY {Ψ15.5}

THE CMB IS IMMUNE FROM THE ENERGY LOSS OF LUMETIC DECAY

This is consistent with the CMB's appearance. It contains, as shown in Appendix L, no observable redshift components (assuming its nominal state is thermal). Beyond that, the conspicuous disparity between the CMB's luminous density and galaxies' virtually nonexistent microwave luminosity provides additional confirmation. Luminosity, at any frequency, is ultimately responsible for replacing the luminous density lost to lumetic decay. The CMB's energy density in space is over twenty times that of integrated starlight, yet microwave radiation amounts to only parts per million of a galaxy's total output. Just below the CMB, in the universe's radio wave background, there is virtually no energy density.

How is decay immunity possible? Decay is induced by the motion of photons through curved space, a process to which all photons, including those of the CMB, are exposed. But before the CMB's decay immunity can be fully understood, the first mystery to solve is how, exactly, photons lose energy during intergalactic redshift.

15.7 DECAY MECHANISM

Intergalactic photons decay, and the energy they lose has to go *somewhere*. There are only three energy transfer scenarios that could conceivably support this phenomenon:

1. Photon-photon upscattering. **Premise**: Decaying photons transfer their energy loss directly to the CMB via a photon-photon interaction. The CMB avoids decay because it transfers its decay energy directly to itself. **Discussion**: If the CMB upscattered itself directly with its own decay energy, then it would also upscatter radio waves as well, preventing their lumetic decay. Observations of 21 cm radio signals with redshifts in excess of ($z = 4$) contradict this assertion. **Determination**: Unviable.

2. Photon-matter-photon upscattering. **Premise**: Decaying photons transfer their decay energy directly to the matter of deep space (IGM), which in turn transfers it back to the CMB through the processes that maintain its thermal equilibrium. **Discussion**: If decaying photons upscattered matter directly, they would incur either angular or frequency dispersion or both, and intergalactic redshift has neither. **Determination**: Unviable.

3. Radiative emission. **Premise**: Expanding photons release energy by the emission of *decay photons*. **Discussion**: The only way a photon can expand in response to a differential velocity field is if its trailing edge moves slightly slower than its leading edge. As such, the average velocity of its energy distribution is less than c, and it is for all intents and purposes an unstable, relativistic system. Just as relativistic particles emit photons in response to external fields, a photon can emit photons in order to preserve its Planck/energy relationship as its wavelength is distended by curved space. **Determination**: Viable.

Light decays by emitting light of lower energy:

Ψ THEOREM 15.7 - QUANTIZED DECAY {Ψ15.6}

EXPANDING PHOTONS RELEASE ENERGY IN DISCRETE STEPS BY EMITTING DECAY PHOTONS PARALLEL TO THEIR DIRECTION OF MOTION

In accordance with the conservation of momentum and energy, decay photons coexist with their source, moving along the same trajectory in a closely linked ensemble. An expanding photon can shed its retinue of decay products when it passes through any material opaque to its low-energy companions. Quantized decay is not constrained by photon number conservation because it produces no scattering, refractive or otherwise.

DECAY QUANTIZATION

An expanding photon is a relativistic particle moving very close to the speed of light. The slight slowing it receives due to its internal expansion causes internal change, and this promotes a gradual decay on its journey across deep space. The rate of this decay is governed by the rate that change can occur within the photon's structure, which is in turn controlled by the speed of its expansion. From Equation (15.3) and the inherent linearity of a photon's topology, the *average* differential velocity in the energy distributed along an expanding photon's wavelength is half of the differential speed between its leading and trailing edges:

$$\overline{dv} = \frac{H_0 \lambda_s}{2} \tag{15.36}$$

where λ_s is the wavelength of the expanding (source) photon.

Equation (15.36) is not, however, an accurate portrayal of the average internal motion a photon experiences. Internal motion is, as shown in Appendix E, perpendicular to a moving object's trajectory, and a photon's expansion is isotropic (Ψ15.2). This means that a photon's internal motion is its expansion *normal* to its velocity, as given by:

$$\overline{dv_i} = v_i = \frac{H_0 h_\gamma}{2} \tag{15.37}$$

where h_γ is a photon's height, its maximum extent perpendicular to its trajectory as defined earlier by Equation (8.18).

Let:

$$\beta_\gamma = \frac{h_\gamma}{\lambda_s} \tag{15.38}$$

where β_γ is the ratio between a photon's wavelength and height, introduced earlier by Equation (8.20) as its *scale*. This, along with a photon's profile and wavelength, fully define its three-dimensional spatial footprint.

Like its profile, a photon's scale has a fixed ratio to its wavelength. Substitute Equation (15.38) into Equation (15.37):

$$v_i = \frac{H_0 \beta_\gamma \lambda_s}{2} \tag{15.39}$$

From Appendix E, the magnitude of a moving system's *internal change* is the ratio of its internal motion, v_i, to c. Time dilation, Θ, is the inverse of this, increasing as internal change slows. Thus:

$$\Theta = \frac{c}{v_i} = \frac{2c}{H_0 \beta_\gamma \lambda_s} = \left(\frac{2}{H_0 \beta_\gamma h} \right) E_s \tag{15.40}$$

where E_s is the energy of an expanding photon of wavelength λ_s. *A decaying photon's internal time dilation is proportional to its energy.* The greater the photon's energy, the smaller its internal change, and the greater its time dilation.

The amount of energy released in a decay photon is defined by the Hubble constant, the decay interval, and the source photon's energy. From Equation (15.7):

$$dE = E = H_0 E_0 e^{-H_0 t} dt = H_0 E_s \tau \tag{15.41}$$

where E is the energy of the decay photon, τ is the time interval between decays (*decay period*), and E_s is the energy of the source photon. The release of any photon, including those of lumetic decay, is governed by the Planck relation. This is dilated by the source photon's relativistic speed as:

$$E\left(\frac{\tau}{\Theta} \right) = h \tag{15.42}$$

where τ is the decay period and Θ is given by Equation (15.40).

Combining Equations (15.40), (15.41), and (15.42) and solving for the decay period yields:

$$\tau = \sqrt{\frac{2}{\beta_\gamma}} \left(\frac{1}{H_0} \right) \tag{15.43}$$

Lumetic decay's period is a function of two things - the geometry of photons and the geometry of the space they traverse, and *it is constant*.

Ω HYPOTHESIS 15.1 - LUMETIC DECAY PERIOD {Ψ15.6}

ALL PHOTONS HAVE THE SAME LUMETIC DECAY PERIOD

This will be listed as a hypothesis since so little is currently known about a photon's actual spatial footprint. Polarization suggests that a photon's profile exceeds 10^6 [Equation (8.18)]. If its width and wavelength are comparable (or in fact the same), its scale exceeds 10^6 as well. Substituting a *minimum* photon scale of 10^6 and Hubble constant of 60 Hz-km/Mpc into Equation (15.43) yields a *maximum* decay period of ~20 million years. As it turns out, however, the actual decay period of ancient photons is substantially shorter.

DECAY PERIODOCITY

If lumetic decay's period is sufficiently long, intergalactic redshift ought to exhibit a certain amount of granularity or quantization. As it turns out, it does. Discovered first by Tifft for optical photons⟨40⟩ and verified by Napier⟨41⟩ for the 21 cm radio band, the energy loss of ancient light is *quantized*. Since this effect is monumentally inconsistent with the expanding universe of the Big Bang model, it has been ignored by mainstream cosmology even though the measurements have been duplicated and verified. *Intergalactic redshift is caused by a series of discrete events, not the expansion of space.*

Tifft discovered two distinct redshift quantizations, ($z = 1.2(10)^{-4}$) for the optical and 21 cm bands and ($z = 2.4(10)^{-4}$) for the optical band. These correspond to time periods, in accordance with a slight modification of Equation (14.2):

$$\tau = \frac{z}{H_0} \tag{15.44}$$

of 2 and 4 million years, respectively (at a Hubble constant of 60 Hz-km/Mpc). The smaller redshift quantization was confirmed in the 21 cm radio band by Napier. Also note that the observed quantizations are multiples of each other, suggestive of trace amounts of decay immunity in the radio and optical bands. Decay immunity is virtually complete in the CMB, but in the other bands it only delays their inevitable energy loss.

Equation (15.41) gives the energy of the decay photons released by ancient light:

$$E = H_0 E_s \tau$$

Rewrite this in terms of wavelength and solve for decay wavelength:

$$\lambda = \left(\frac{1}{H_0 \tau}\right) \lambda_s \tag{15.45}$$

where λ_s is the source wavelength as before. This can be used to calculate the *decay spectrum* that integrated starlight produces.

The energy density in any spectrum is the product of the number density of photons and their individual energies. Since lumetic decay's period is constant, the number of decay photons produced by any given wavelength of integrated starlight is proportional to the numerical photon density of said wavelength. Further, the energy of decay photons is proportional to the energy of their source photon. Thus the energy density of integrated starlight's decay spectrum is directly proportional to its own energy density, in accordance with the wavelength reduction (transformation) shown at Equation (15.45).

When integrated starlight is approximated as an attenuated 10,000 °K blackbody and its decay spectra, at the observed redshift quantizations, are superimposed on the CMB, the result is startling:

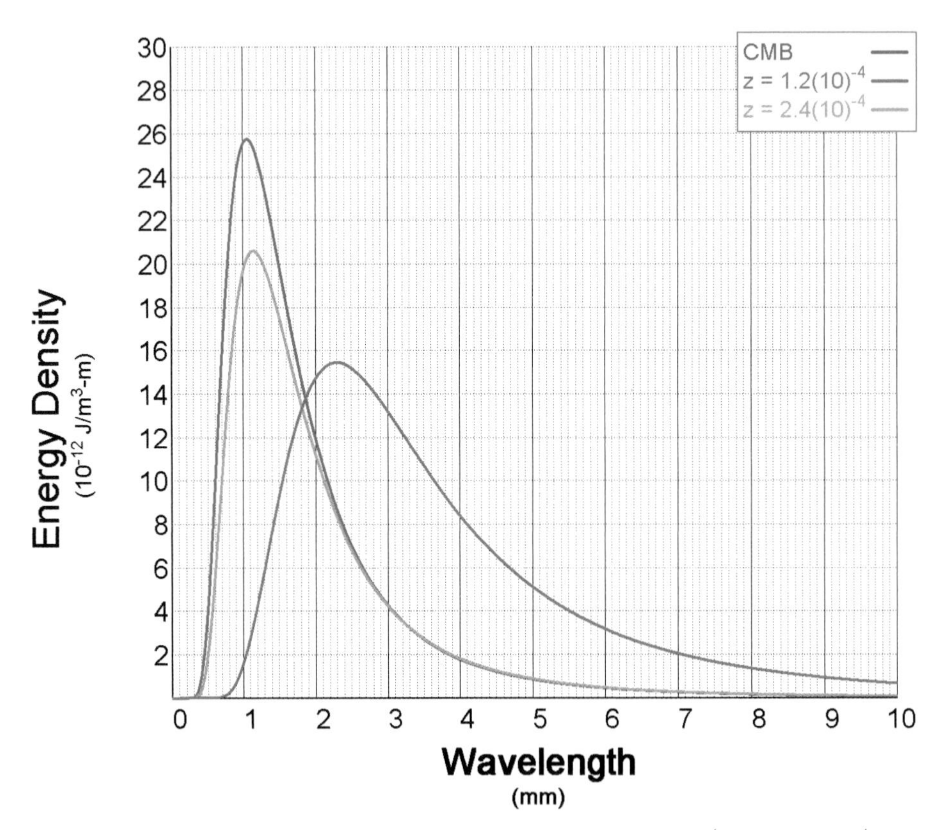

Figure (15.5) Optical decay energy in the CMB spectrum, $z = 1.2(10)^{-4}$(red), $2.4(10)^{-4}$(orange)

Integrated starlight's ($z = 2.4(10)^{-4}$) decay spectrum is shown in orange, and it coincides almost exactly with that of the CMB. *The energy the universe's optical band loses to intergalactic redshift is emitted directly into the CMB.* Integrated starlight, as shown in Appendix I, has about 4300 optical photons in every cubic meter of the universe. These decay, on average, once every 4 million years, and release microwave radiation into deep space. The power output this represents is, as shown previously by Equation (15.33), equal to the universe's total fusion luminosity.

Galaxies illuminate the universe, and the deep space between them returns the favor by showering them with microwaves.

Ω HYPOTHESIS 15.2 - LUMETIC DECAY OUTPUT {Ψ15.6}

DECAYING STARLIGHT RELEASES MICROWAVES DIRECTLY INTO THE CMB BAND

The universe's fusion cycle is a forced, unidirectional flow of energy. It is not a case of optical redshift energy passively heating the IGM, which, because of its temperature, radiates in the microwave band. It is a case of optical energy being pumped into deep space by fusion, being forced to decay by space's curvature, then filling the void with microwave radiation at a rate capable of driving energy back into the galactic environment. The optical band's microwave decay output is not a thermal spectrum, but it is close. Since it would take 600 billion years for decaying starlight to fill deep space to CMB energy density, there is more than enough time for decay energy to be properly thermalized by the IGM. Although the universe is infinitely old, the movement of energy through its fusion cycle can still be understood in terms of an equilibrium perspective.

A photon's gravitational expansion is a continuous by-product of spatial geometry, and the release of its decay energy occurs through discrete emissions, so old photons spend a considerable amount of time in an expanded state, where their wavelength is not consistent with their energy as defined by the Planck relation. Such photons will be referred to as *gravid*, and their gestation period is a multiple of ~2 million years. Even so, decay photons actually begin their own gravitationally induced expansion long before they are eventually released from their source. The only case where source and decay photons have the correct Planck energy/wavelength configuration is at the moment of decay emission.

DECAY ECHOES

Integrated starlight decays into the CMB, and although the CMB has strong decay immunity, no process is 100% efficient, and a few of its photons (0.3 to 10 mm) still manage to decay into radio waves with wavelengths from 2 to 80 m. This is consistent with the universe's weak extragalactic radio wave background.[1.17] These photons decay into longer radio waves, and so on. Each band echoes the last, the ratio of their photon energies defined by Equation (15.45). Integrated starlight and the CMB both have a spectral width of at least two orders of magnitude of wavelength, so it is difficult to establish a direct connection to their decay products. There is, however, a universal and precise frequency that might be used for this purpose - the 21 cm radio wave band. Here is a single, ubiquitous frequency, pervading all of space, and decaying into 1700 m radio waves. Unfortunately, lumetic decay and the velocity dispersion of radio sources tend to blur this precise signature.

All photons decay into progressively longer photons, so a limit is soon reached where a decay photon has so little energy that the IGM becomes perfectly opaque to its passage. There is a geometric limitation as well - *the differential velocity within a photon's topology can never exceed the speed of light*. Hence the maximum wavelength of any photon is governed by Hubble's constant, not Planck's. When:

$$\lambda = \frac{2c}{H_0 \beta_\gamma} \tag{15.46}$$

a photon can no longer expand. The factor of 2 arises because a photon's transverse speed is zero at its longitudinal axis. At a lumetic decay period of ~2 million years, Equation (15.43) gives photon scale β_γ as $\sim 1.26(10)^8$ at $H_0 = 60$ Hz-km/Mpc. This corresponds, by Equation (15.46), to a maximum photon wavelength of ~250 light years.

15.8 THE CMB

The fact that photons decay by radiative emission provides an important clue to the nature of the CMB's curious decay immunity. It is a direct result of one or both of the following two mechanisms:

- Decay opacity. **Premise**: The IGM is opaque to CMB decay photons and the CMB is upscattered through thermalization faster than it decays. **Discussion**: The wavelength peak of the CMB's energy density is near 1 mm, corresponding to decay photons ~8 m long. There is a cosmic radio background at this wavelength, but its energy density is negligible. Since radio telescopes demonstrate the phenomenal mean free path that radio waves have in space, it is safe to say that the IGM is not terribly opaque at this frequency. Or at least it is not so opaque as to maintain a radio background a million times less energetic than the CMB. **Determination**: Unviable.

- Decay inactivation. **Premise**: The thermalization between the IGM and CMB is so active that it interrupts the lumetic decay process. **Discussion**: The CMB is the only truly thermal band in space, and is also the only band that avoids lumetic decay. The time dilation associated with lumetic decay slows the entire decay emission down, from beginning to end. It is not a case of a microwave releasing a radio wave every ~2 million years - it is a case of it takes ~2 million years for a microwave's radio decay *to be emitted*. If the microwave gets scattered by even the smallest amount during this time frame, its momentum or energy will change, and the emission of its decay product cannot proceed. **Determination**: Viable.

The CMB's thermalization is responsible for its lumetic decay immunity, as confirmed by the *redshift of radio signals* from distant objects. Why, after all, should such feeble photons lose energy across deep space while the much more powerful CMB maintains a thermal equilibrium that upscatters its lowest energy components? *Scattering and decay are intimately linked.* If radio waves were in thermal equilibrium with the IGM, they could take advantage of the gentle heating it receives from lumetic decay energy, but they are not. As such, they are redshifted along with all bandwidths *except the CMB*.

The CMB is a universal reservoir of antique photons, the only photons with a number density sufficient to process the energy of gravitationally induced decay:

Figure (15.6) WMAP without galactic emissions or Earth's motion dipole
(Courtesy NASA/WMAP Science Team, shown previously as Figure (6.7))

The IGM is so thin and cold that it has virtually no energy content, yet needs to be able to process the CMB's energy *at least* once every 2 million years to prevent its decay. In thermal equilibrium, the average kinetic energy of a particle in deep space (as calculated in Appendix P), is $3kT/2$. At 2.7 °K this is only $5.6(10)^{-23}$ J/particle. Since the IGM's average density amounts to three electrons and protons per cubic meter, deep space electrons have an average kinetic energy density of $1.7(10)^{-22} f_e$ J/m^3, where f_e is the fraction that are free. Most of the CMB's energy is thermalized by the IGM's (much lighter) electrons, so in order to process the CMB's energy at a rate sufficient to defeat lumetic decay (once every 2 million years), the IGM needs to cycle the kinetic energy of its electrons, on average, once every $3 f_e$ days.

By cosmic standards, the CMB's thermalization occurs at a frenzied pace. This is possible because the IGM is literally *immersed* in CMB photons. But there is more at work here than a random thermal interaction between the CMB and IGM. As will soon become apparent, the IGM's electrons are *tuned* to the CMB, for without this connection the comic luminous path couldn't carry its energy to its final destination.

OLBERS' PARADOX, REVISITED

The CMB receives a huge influx of energy from the lumetic decay of optical energy density. Since spectral backgrounds below the CMB's frequency have negligible energy density and the CMB has virtually no decay, the only way it can maintain its constant energy density is through a loss *other than lumetic decay*. Absorption is the only possible agent for this loss. The absorption of CMB radiation is the true solution to Olbers' paradox:

Ψ THEOREM 15.8 - CMB INFLUX/ABSORPTION BALANCE {Ψ14.1}

SPACE MAINTAINS CONSTANT TEMPERATURE BECAUSE THE CMB HAS A STRONG DECAY IMMUNITY AND THE ENERGY IT RECEIVES FROM LUMETIC DECAY IS BALANCED BY THE UNIVERSE'S RATE OF MICROWAVE ABSORPTION

Lumetic decay cools luminous radiation, but microwave absorption cools the CMB. *The CMB is a cosmic buffer for the universe's luminous output.* At equilibrium, luminous energy density is constant in space, so universal luminosity must balance universal absorption.

Ψ THEOREM 15.9 - PHOTONIC EQUILIBRIUM {Ψ14.1}

UNIVERSAL LUMINOSITY = UNIVERSAL ABSORPTION

Luminous energy of all wavelengths is emitted into space, decaying with distance. Optical energy becomes infrared, infrared becomes microwave, and radio waves become longer radio waves. Most of this energy eventually takes the form of microwaves, but this is a remarkably slow process. As noted earlier, it would take the universe over 600 billion years to generate the CMB's energy density with its current fusion output.

15.9 COSMIC MICROWAVE ANTENNAS

One of the most telling clues to the CMB's true nature is that *celestial objects emit only trace amounts of microwave radiation*. The natural inclination is to think, since the present level of galactic microwave emission is far too low to account for the CMB, that it originated from some cataclysm in the distant past. This is clearly not the case, as has been demonstrated by a preponderance of evidence and logic. *Microwaves didn't come from the Big Bang - they are produced throughout space in teeming numbers by the decay of optical photons*. Indeed, the CMB's functionality is not subtle. Here are two unambiguous pieces of evidence:

- The CMB is the band whose energy *lies immediately below the active production of photons*. Universal luminosity consists of ~70% optical energy; ~30% infrared; and trace amounts of radio, X-ray, and gamma. There are isolated and powerful radio wave and high-energy sources, such as neutron stars and active galactic nuclei, but in terms of sheer power output, the universe's luminosity ends abruptly with infrared. The CMB is the band *directly beneath it.* Our sun, for instance, produces microwave energy at only ~10 ppm above the cosmic background level.

- The CMB contains the greatest reservoir of electromagnetic energy, and just below it, in the radio band, there is virtually no energy. The energy density difference between these two backgrounds is a factor of a million. Energy flowing into the CMB from the lumetic decay of optical and infrared bands *is not converted into a different form of light; it is converted into a different form of energy.*

Why does electromagnetic energy density essentially end with the CMB? *Because the next step in the cosmic luminous path is not photonic.* Microwave energy takes on a new form in order to continue along this path.

The CMB must lose energy to maintain its equilibrium temperature, and the only possible destination for this energy loss is away from deep space, back into the galactic environment. Galaxies emit luminous radiation that decays, pumping energy into the CMB. In order to complete the circuit, CMB energy must return to the galactic fold. But in what form? Microwaves can immediately be ruled out. Galaxies all produce trace levels of microwave *emission.* Indeed, the red region in our galaxy's COBE profile, at about 4° mK warmer than background, is incontrovertible evidence that galaxies do not capture CMB energy *as microwaves*:

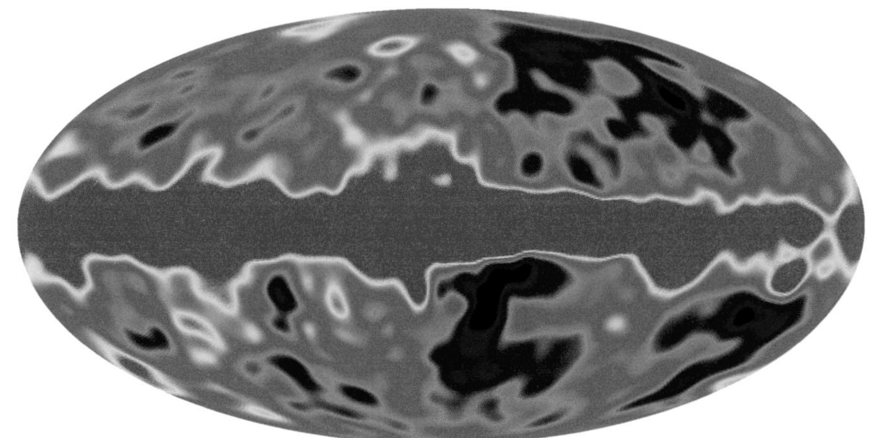

Figure (15.7) COBE field with slightly warmer galactic region (red)
(Courtesy NASA/WMAP Science Team)

There are a limited number of mechanisms available to transfer energy from the CMB back to the galactic environment. The COBE image shown in Figure (15.7) eliminates the possibility of thermal absorption. Discounting gravitational effects, scattering, neutrino interaction, and a host of other phantasmagorical agents, only one carrier remains: *electrical current*. Galaxies absorb CMB energy using broad, ethereal streams of charged particles moving through deep space. Hannes Alfvén recognized the plasma nature of the large-scale universe; he just failed to understand the *necessity* of it. Galaxies flood deep space with their luminous roar. The only way to complete the cosmic loop for universal energy flow is to carry energy back to galaxies *from* deep space.

Electrical current is the only agent available for this transfer, and galactic halos are the only structures capable of capturing microwave energy and converting it into electrical current.

Ψ THEOREM 15.10 - GALACTIC POWER RETURN {Ψ15.9}

CMB ENERGY IS CAPTURED BY GALACTIC HALOS AND TRANSFERRED TO THEIR DISKS AS ELECTRICAL CURRENT

Lumetic decay heats the CMB; cosmic currents cool it down. *A galaxy's halo acts like a gigantic microwave antenna, capturing energy to balance the power it loses as light.* This is why the CMB has a thermal spectrum, and also why it avoids lumetic decay. Deep-space electrical currents are tuned to receive CMB energy on a scale consistent with the distribution of matter in intergalactic space. This distribution, which is on average a few million light years between galaxies, defines the CMB's decay immunity.

CMB POWER GRID

When it was first measured, the CMB's uniformity baffled theorists. If it were in fact the by-product of the universe's birth, its original state would have had to be impossibly smooth. Much to the relief of cosmologists the world over, slight CMB temperature fluctuations were found by the ultra-sensitive COBE satellite. According to the current Big Bang paradigm, these tiny deviations arise from vanishingly small imbalances present in the primordial fireball. Nothing could be further from the truth. CMB temperature fluctuations are direct evidence of the transfer of energy from deep space to galactic systems - an explicit demonstration of cosmic equilibrium. The CMB is heated by luminosity and cooled by electricity. *The only way to transfer power from one to the other is through CMB temperature differentials.* Energy cannot flow without a corresponding thermal gradient. The CMB is isotropic, but if it were perfectly uniform it would be incapable of transferring energy to a galaxy's dark halo.

Consider a universally average galaxy S in the center of a universally average spherical intergalactic spatial volume, V_a. V_a is cosmologically small, so lumetic decay has virtually no effect on the light that S releases within its interior. At equilibrium, the amount of galactic luminosity that S pumps through V_a's surface must equal the microwave radiation that this surface absorbs through the flow of charged particles. This corresponds to an average temperature difference across V_a's entire spherical surface. Regardless of how large V_a is, the microwave temperature of its surface must be slightly different from ambient CMB in order to transfer energy. V_a is a representative sample of the universe. The ratio of its luminosity, L_a, to volume is equal to:

$$j_R = \frac{L_a}{V_a} \tag{15.47}$$

where j_R is the universally average luminosity density (creation of new photons/neutrinos) throughout space. This is dominated by optical and infrared radiation.

The relationship between small changes in CMB luminosity/temperature across V_a's surface is given by the CMB's radiancy:

$$\frac{\Delta L_{CMB}}{\Delta T_{CMB}} = 4A_a \sigma T_{CMB}^3 \tag{15.48}$$

where A_a is V_a's spherical area. At equilibrium, the CMB luminosity difference at V_a's surface is equal to the luminous output of the universally average galaxy within its volume, L_a:

$$\Delta L_{CMB} = L_a \tag{15.49}$$

Substitute this into Equation (15.48) and solve for the temperature difference:

$$\Delta T_{CMB} = \frac{L_a}{4A_a \sigma T_{CMB}^3} \tag{15.50}$$

Since the total luminosity in V_a increases as the cube of radius while its area increases as the square, the required temperature differential for equilibrium increases indefinitely with scale. However, galaxies do not exchange energy with the CMB across some gargantuan surface spanning billions of light years; they exchange energy with the CMB in the intergalactic space of their immediate surroundings. Use Equation (15.47) to define the area of V_a's surface in terms of luminosity density and luminosity:

$$\begin{aligned} V_a &= \frac{L_a}{j_R} = \frac{4\pi}{3} R_a^3 \rightarrow R_a = \left(\frac{3}{4\pi}\right)^{\frac{1}{3}} \left(\frac{L_a}{j_R}\right)^{\frac{1}{3}} \\ A_a &= 4\pi R_a^2 = (36\pi)^{\frac{1}{3}} \left(\frac{L_a}{j_R}\right)^{\frac{2}{3}} \end{aligned} \tag{15.51}$$

Substitute this area into Equation (15.50):

$$\Delta T_{CMB} = \frac{L_a}{4(36\pi)^{\frac{1}{3}}\left(\frac{L_a}{j_R}\right)^{\frac{2}{3}}\sigma T_{CMB}^3} \tag{15.52}$$

Simplify:

$$\Delta T_{CMB} = \left(\frac{1}{4\sigma T_{CMB}^3}\right)\left(\frac{L_a j_R^2}{36\pi}\right)^{\frac{1}{3}} \tag{15.53}$$

At the total universal luminosity density listed in Appendix C and Milky Way luminosity of $1.4(10)^{37}$ W,⟨1.14⟩ Equation (15.53) gives a deep space CMB temperature fluctuation of $1.8(10)^{-5}$ °K, or 18 μK. This is within the error limits of the perturbations evident in the CMB (18 ± 1.6 μK).⟨38⟩

The CMB's tiny temperature ripples have nothing to do with a primordial explosion; they are the signature of the power transfer that galaxies receive from deep space.

Ψ THEOREM 15.11 - CMB SMALL-SCALE ANISOTROPY {Ψ15.10}

SMALL-SCALE ANISOTROPIES IN THE CMB ARE ARTIFACTS OF THE GALACTIC POWER RETURN

Electrical currents carry energy captured from microwaves in intergalactic space to galactic space, while photons carry optical and infrared energy in the opposite direction. Perhaps one of the reasons galactic halos are such efficient microwave antennas is because their electrical currents are slightly cooler than the CMB; they might be most appropriately viewed as supercooled receivers in this context. Indeed, the temperature of deep space is 2.7 °K, so galactic halo material, while sparse, has the temperature of a superconducting medium.

HOT RICH CLUSTERS

Galaxies are bound into groups, groups into clusters, and clusters into superclusters. In many cases their spacing is close enough to heat the intergalactic gas throughout their entire neighborhood to blistering temperatures. As a case in point, measurements indicate the hydrogen dispersed throughout the Virgo cluster has a temperature of 28 million degrees.⟨1.4⟩ Yet even this is cool for typical clusters, whose temperatures average about 75 million degrees.

A correlation has been found between rich galactic clusters and the CMB. The intensity of the CMB they emit is cooler than the average background by about one part in 10^4. This is known as the Sedouski-Zeldelov (S-Z) effect and is thought to be caused by the up-scattering of CMB photons by high-energy plasma particles.⟨1.5⟩ This is not the case. *The S-Z effect is direct evidence of galactic CMB energy absorption.*

Rich clusters are common throughout the universe. They are blazingly luminous, and this is why their S-Z effect is particularly evident - *they must absorb enough microwave energy to power their entire system.* On the smaller galactic scale, CMB anisotropies are not obviously *cool* spots because there is no absolute baseline for CMB temperature. This makes some spots in the CMB appear hotter than others. In the S-Z effect, the CMB baseline temperature is defined by the remainder of the sky, so in relation to this an unambiguous cooling correlation is evident.

The S-Z effect produces a negative image that loses intensity and contrast with distance. It is a shadow of the effect of microwave absorption currents, not the up-scattering of the CMB by high-energy particles:

Figure (15.8) Rich clusters are large galactic groups embedded in hydrogen plasma, pulling energy from the CMB (Abell 2218, courtesy NASA/Hubble Heritage Project)

15.10 GALACTIC CURRENT

Galactic halos convert microwave energy into electrical energy, which is then carried inward to the galactic environment by massive currents. Given the magnitude of the electrical current necessary to support the power output of an entire galaxy, there ought to be unambiguous evidence of its presence. There is. Galactic currents reveal themselves in the most striking way - *the banding of spiral galaxies.*

The arms of a spiral galaxy have the following general characteristics, all of which arise from the flow of a remarkable amount of electrical current:

a) An average material density about 10%-20% higher than the rest of the galactic disk.⟨6.8⟩ **Cause**: The magnetic pinch effect of current through plasma.

b) Accelerated stellar ignition. **Cause**: Increased material density due to magnetic pinch and increased energy available to promote material condensation and ignition.

c) Radio wave emission.⟨16.2⟩ **Cause**: Acceleration of the galactic current's charged particles.

d) Rotation slower than a galaxy's disk material. At the Earth's current location (R_0) of 28 Kly⟨3.4⟩ from the center of our galaxy, the Milky Way's arms rotate about its center once every ~460 million years, whereas its disk material completes an orbit in ~240 million years.⟨3.5⟩ **Cause**: Spiral bands rotate differently than a galaxy's disk because they are electrically induced pressure waves, not material structures.

The torrential flow of electrical current adorns the disk of every spiral galaxy:

Ψ THEOREM 15.12 - GALACTIC CURRENT {Ψ15.10}

A SPIRAL GALAXY'S BANDING MARKS THE FLOW OF ELECTRICAL CURRENT FROM ITS CMB POWER RETURN THROUGH ITS DISK

This effect is most apparent in spiral galaxies, but ellipticals and lenticulars also have complex filaments betraying the electrical current moving through their interior regions.⟨15.2⟩

At equilibrium, the amount of electrical power moving through a galaxy's disk balances its luminosity. *The fusion cycle is completed when the luminous energy a galaxy releases is recombined with the compound nuclei it creates.* The only environment in the universe powerful enough to disassociate compound nuclei on a grand scale is a galaxy's core, and the energy to accomplish this is supplied by the currents generated when microwave energy is captured from deep space.

A galaxy is *an electric furnace*.

Ψ THEOREM 15.13 - ELECTRIC GALACTIC FURNACE {Ψ15.12}

GALACTIC CURRENTS TRANSPORT ENERGY TO A GALAXY'S CORE TO FACILITATE THE DISASSOCIATION OF THE COMPOUND NUCLEI ITS STARS PRODUCE BY FUSION

This is why jets of hydrogen have been observed escaping from the cores of so many galaxies. There is, after all, no other possible source for new hydrogen. *The luminous banding of a spiral galaxy is the unequivocal signature of the galactic currents flowing through its disk, betraying an enormous power return in the direction of its core*:

Figure (15.9) NGC 3310
(Courtesy NASA/Hubble Heritage Project)

Thus completes the cosmic luminous path:

1. Hydrogen fuses, producing compound nuclei and light.

2. Light loses energy to the CMB through microwave emission, driven by gravitationally induced photon expansion.

3. Galactic halos capture CMB energy via deep-space electrical currents.

4. Electrical currents carry their captured energy through galactic disks to galactic cores.

5. Electrical energy supplied to galactic cores disassociates compound nuclei to produce hydrogen.

This is also why there is a predominance of H II (ionized hydrogen) in galactic arms.⟨3.4⟩⟨14.2⟩ Its principle function is to *provide a charged path to carry electrical current into a galaxy's core while transporting hydrogen outward.* As further support for the idea that galactic halos are actually gigantic microwave antennas, a galaxy's banding activity begins in the material that lies just beyond its luminous rim, thereby confirming the extragalactic source of its electrical current. *Galactic output illuminates the heavens; galactic input is dark but just as unmistakable.*

BARRED SPIRALS

Spiral galaxies' bands curve with decreasing radius in their glowing disks, but tend to have a bar-shaped structure of varying prominence in their central region. When the barring is particularly evident, a galaxy is referred to as a *barred spiral*:

Figure (15.10) Barred spiral, NGC 1300
(Courtesy NASA/Hubble Heritage Project)

The Milky Way has a barred region near its center, but it is too subtle to be considered a barred spiral.⟨6.9⟩

The reason a typical spiral galaxy's bands are twisted is because they are pressure waves, and their underlying disk material rotates at angular velocities that decrease as distance from the galaxy's core increases. This is not the case in the bars of barred spirals. They are not pressure waves - the stars of which they are composed move in lockstep with the bar's overall rotation.⟨14.5⟩ Although its resident stars gyrate in complex, noncircular orbits, a galactic bar spins like a solid object, so electrical current takes a straight line directly through it to the galaxy's core. If a spiral galaxy's entire disk had the same rotation profile as a galactic bar, it would look like a pinwheel.

GALACTIC EFFICIENCY

There are a variety of different galactic morphologies. All are solutions to the cosmic power grid, and all live longer than we could possibly imagine. This does not, however, indicate guaranteed perpetuity for any given type. Perhaps ellipticals, with their fine, uniform structure, are the most stable galaxies and begin their life as spirals. Or perhaps a galaxy's morphology is purely a function of the ambient material composition in which it forms. In any event, the power relationship between a galaxy and its intergalactic neighborhood will be referred to as its *galactic efficiency*:

$$Q_g = \frac{P_g}{L_g} \tag{15.54}$$

A galaxy's efficiency is the ratio of its electrical power capture, P_g, to luminous output, L_g.

Cosmostasis requires that the universe's fusion is perfectly balanced by its hydrogen production, so the universally average value of galactic efficiency is unity:

Ψ THEOREM 15.14 - UNIVERSAL GALACTIC EFFICIENCY {Ψ14.1}

GALACTIC EFFICIENCY'S UNIVERSALLY AVERAGE VALUE IS UNITY

Different types of galaxies process energy differently. Ellipticals burn clean and slow; spirals burn hot and fast. Certain types of galaxies are supremely efficient with ($Q_g > 1$), and are offset by their less efficient brethren, ($Q_g < 1$). Elliptical galaxies, with copious amounts of hydrogen in their vicinity, stable galactic cores, and clean fusion, are likely to be the most efficient morphology. That said, our analysis will continue to focus on spirals.

The electrical current carried by a spiral galaxy depends on its power output, electron energy transfer, and efficiency:

$$i_g = \frac{Q_g L_g q}{E_e - E_{ex}} \tag{15.55}$$

where E_e is the average kinetic energy of electrons moving inward along galactic bands and E_{ex} is *exit energy*, the average kinetic energy of electrons upon completing the *galactic circuit*. There are only two paths available for closing this circuit, either normal to a galaxy's plane or out through its disk. The symmetry of most galaxies' banding and the flow of hydrogen outbound along these bands suggest that the roaring current that sustains a galaxy migrates uniformly *inward* through its disk. Hence the galactic circuit is in all likelihood closed normal to the galactic plane. This is, after all, the shortest route back to intergalactic space.

GALACTIC AMPERAGE

The angle and rotation speed of a spiral galaxy's bands can be used to estimate the lower limit of its galactic current. *Pitch* is the average angle between a galaxy's banding and a tangent to its circular disk:[14.1]

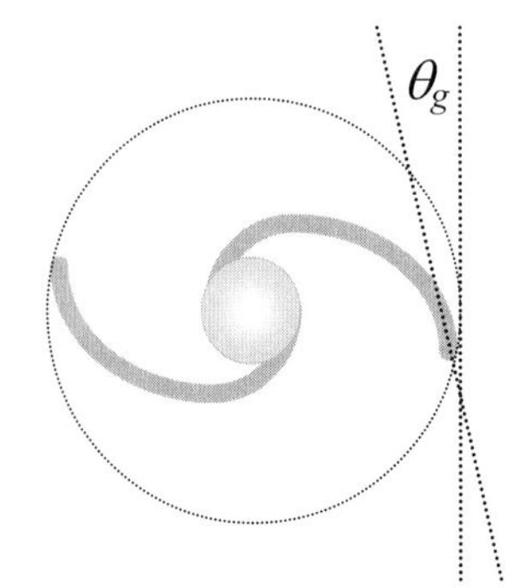

Figure (15.11) Spiral pitch angle, θ_g

Spiral galaxies with prominent bulges, a tightly wound structure, and smooth stellar distributions were first classified by Edwin Hubble as type Sa, whereas loosely wound galaxies with negligible bulges were denoted as type Sc. The pitch angle of type Sa galaxies averages ~6° while that of type Sc is three times as great, at ~18°.[16.1]

As noted earlier, a spiral galaxy's bands move independently of its disk material. The net motion of its inbound electrical current is radial, toward its center. This is spread across its disk by the rotation of the underlying material, but the effective speed of the current in a galaxy's bands is the radial component of its differential disk-band motion. *This is equal to the tangent of the pitch angle.*

The electron drift speed of a spiral galaxy's disk currents is given by:

$$v_e = (v_c - v_b)\tan(\theta_g) \qquad (15.56)$$

where v_c is the average rotational speed of its disk, v_b is the average speed of its bands, and θ_g is its pitch angle. If a spiral galaxy's disk were motionless or rotated like the center of a barred spiral, its inbound currents would move in straight lines directly toward its core, yielding no information about its electron drift speed [$v_c = v_b$ and tan(90°) undefined]. This is not the case, and the ratio between the average rotation speeds of a spiral galaxy's bands and disk will be referred to as its *galactic rotation ratio,* η_g:

$$\eta_g = \frac{v_b}{v_c} \qquad (15.57)$$

The Milky Way's rotation ratio in our local stellar neighborhood is ~0.5[3.5] - its bands moving about half as fast as its disk material. This will be assumed to be close to the universal galactic average for the estimates in this section. Rewrite Equation (15.56) in terms of the galactic rotation ratio:

$$v_e = v_c(1-\eta_g)\tan(\theta_g) \quad (15.58)$$

Generally, the more luminous a galaxy, the faster its rotation. Take the case of galaxies of absolute magnitude (**M_B**) −22, for instance. Tightly wound (6°) type Sa galaxies of this magnitude rotate (on average) at 320 km/s, more loosely (12°) wound type Sb rotate at 245 km/s, and the most loosely wound (18°) type Sc are the slowest at 190 km/s.[6.21] If galactic bands rotate, on average, half as fast as their disks, *magnitude -22 type Sa galaxies have an average electron drift speed of 17 km/s, increasing to 31 km/s in type Sc*. Note that more tightly wound Sa galaxies have a greater banding density than Sc galaxies, so their total electrical power can be comparable even if electron drift speed differs.

The differential motion between a spiral galaxy's bands and disk material defines the drift speed of its inbound galactic currents, but this difference might not be limited to its disk's rotational speed. Recent measurements suggest that the banding of NGC 4622 *actually moves in the opposite direction of its disk rotation*:

Figure (15.12) Counter-rotating spiral, NGC 4622
(Courtesy NASA/Hubble Heritage Project)

This provides additional evidence that galactic banding is electrical, not gravitational, since it opposes the motion of the disk's material.

A galaxy's absolute magnitude $\mathbf{M_B}$ can be converted to watts by:

$$L = 3.845(10)^{26}\left(10^{\left(\frac{4.76-M_B}{2.5}\right)}\right) \tag{15.59}$$

An $\mathbf{M_B}$ of −22 corresponds to $1.9(10)^{37}$ W. The current required to carry this much power at a given drift speed is given by substituting the kinetic energy associated with E_e in Equation (15.58) into Equation (15.55):

$$i_g = \frac{2Q_g L_g q}{m_e v_c^2 (1-\eta_g)^2 \tan^2(\theta_g) - E_{ex}} \tag{15.60}$$

The greater a galactic circuit's exit energy, the more current it requires to transfer a given amount of power. An exit energy of zero corresponds to minimum current. The Milky Way, with a rotation speed of 220 km/s, is thought to be a type Sb galaxy. At its estimated luminosity of $1.4(10)^{37}$ W,[⟨1.14⟩] it carries a *minimum* current of $1.3(10)^{40}$ Amps at 100% efficiency, ~10° pitch,[⟨15.1⟩] and 0.5 rotation ratio.[⟨3.5⟩]

Anyone who believes that the dynamics of a galactic disk are governed solely by the gravitational force of its mass distribution need look no further than the warped disk of ESO510 G13:

Figure (15.13) The warped disk of ESO510 G13
(Courtesy NASA/Hubble Heritage Project)

Gravitation is a spherically symmetric force, governed purely by material distribution. Could the rings around any of the planets in our solar system ever look like Figure (15.13)?

15.11 THE WINDING PROBLEM

The banding of spiral galaxies poses quite a dilemma for astrophysicists because in the absence of galactic currents it makes absolutely no sense. Imagine viewing a galaxy only as a gravitationally bound structure. Where do its bands come from? Even more incongruous, why do they persist as a galaxy's disk rotates? This last question has proven quite intractable, and is known as the *winding problem*.[6.19]

Galactic disks have a peculiar rotational profile. The angular velocity of their material varies with its orbital radius, but not in the way expected of a gravitationally bound object. This anomalous motion is derived in the next chapter, but for now let's consider the net effect of *angular velocity varying with radius*, as this is the source of the winding problem. In essence, radial variance of angular velocity means that stars at various radii of a galactic disk move relative to each other. If stars at a radius R_1 move a certain number of degrees per unit time, and stars exterior to this at R_2 move at a different rate ($R_2 > R_1$), they will shear away from each other. A galactic band stretching through a disk with this kind of underlying motion ought to be torn apart. Yet this is not what happens. Galactic bands somehow persevere, giving a spiral galaxy its spectacular pattern.

In the following sequence, the winding problem is simulated by exposing a thin radial of material to the motion of a typical galactic disk. Its dispersal over time is shown below:

Figure (15.14) Material disperses when subjected to angular velocity that varies with radial distance

Figure (15.14)'s disk is scaled to a radius of 80,000 ly, comparable to that of the Milky Way, and the radial distance between any two adjacent marks represents 500 light years. As is evident from the above, after a few orbits the initially linear material becomes progressively more randomly distributed throughout the disk. For reference, a small barred region (10000 ly in diameter) has been included at the galaxy's center.

When banding is simulated by injecting electric current at opposite sides of a galaxy's rim, the result is a stable spiral, as shown at left after 50 rotations:

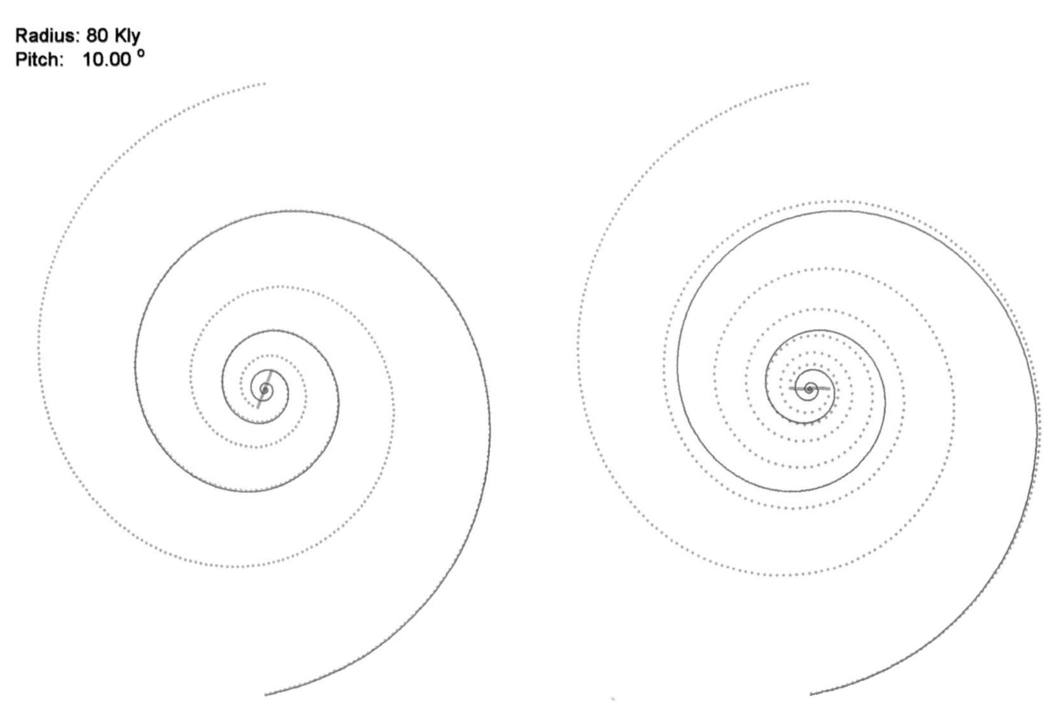

Figure (15.15) Spiral banding, $\eta_g = 0$ (left) and $\eta_g = 0.5$ (right)

The violet fitting trace has the form:

$$R = Ae^{(\tan\theta_g)\theta} \tag{15.61}$$

where θ_g is pitch angle.

If the spiral pattern is given a global rotation, such as half the speed of the underlying disk material ($\eta_g = 0.5$), the shape of the resulting bands is stable, but has a slight deviation from a perfect exponential spiral, as shown at right in Figure (15.15) after 50 rotations. This is the same pattern speed that was used in our galactic amperage calculations, and it represents the rotation of galactic halo current sources around a galaxy's rim. *Galaxies have complex kinematics, but the only way to understand their basic nature is to recognize the magnitude of their electrical component.* The source code for the winding calculations presented in this section can be found at *www:nullphysics.com*.

15.12 GENERAL CONCLUSIONS

The fate of luminous energy in a cosmostatic universe is inevitable - what goes out must someday return.

- Intergalactic redshift is not caused by the expansion of the universe; it is caused by the expansion of photons.

- The universe's average energy density and composition are directly related to the Hubble constant. At H_0 = 60 Hz-km/Mpc its energy density is ~4.0(10)$^{-10}$ J/m^3, of which ~98% is hydrogen. About 98% of this material is nonluminous.

- All electromagnetic radiation moving through deep space, with the exception of the CMB, decays in periods that are multiples of ~2 million years. The CMB's decay is deactivated by its thermalization with intergalactic material.

- Optical photons reconcile the energy they lose from intergalactic redshift by emitting microwaves directly into the CMB band.

- A galaxy's halo pulls power out of the CMB with huge, deep space electrical currents that leave slight temperature ripples in their wake. These currents pass through the galaxy's disk inward to its core, providing power for the disassociation of the compound nuclei created by galactic fusion. This liberates hydrogen, which flows up the galaxy's arms and is ejected from its core directly into space.

- The small-scale temperature ripples in the CMB are an artifact of electrical power capture in our galactic neighborhood; they are not a relic of a primordial universal explosion.

- The S-Z effect is not the up-scattering of CMB photons by high-energy plasma particles; it is the shadow of the large-scale microwave capture necessary to power massive galactic superclusters.

- The Milky Way, assuming it is in steady-state with its microwave power capture, carries a minimum galactic current of ~1.3(10)^{40} Amps.

- Galactic currents provide an eloquent and effective solution to the winding problem.

Any cosmological theory that is stymied by finite universal age will fail to detect the cosmic fusion cycle - eternity's heartbeat.

16. COSMIC PROTON PATH

16.1 GALACTIC VORTEX

Our analysis of the cosmic flow of material will focus on spiral galaxies because they constitute a large fraction of the universe's galactic population and their relationship between form and function is far more apparent than in other types. Whenever the term *galaxy* is used with no modifier it will mean *spiral galaxy*. Spiral galaxies have an intricate structure, but their predominant energy transport is unambiguously emblazoned across their disks.

The Null Axiom requires a universe of eternal equilibrium. Stars burn hydrogen and release light during the formation of compound nuclei, and this energy must eventually recombine with compound nuclei to produce hydrogen. *This is the fusion cycle.* Stellar luminosity decays, slowly transferring its energy into the microwave background. This is converted into immense electrical currents by the dark halos of *galaxies*. The glowing bands woven through galactic disks are an indisputable demonstration of this, but the even more manifest reality is that dark halos are the only available option for large-scale microwave capture. *There is nothing else capable of absorbing the torrents of energy required to power galaxies.*

The energy released by fusion has to be transferred to an environment capable of breaking nuclei apart in order to complete the fusion cycle's proton path. Just as there is a severely limited number of candidates for the destination of absorbed CMB energy (read galaxies), the list of places where compound nuclei can be torn apart at a pace rivaling universal fusion is of similar brevity.

- Cosmic rays. Unviable. Their measured flux is orders of magnitude too low to undo the work of the universe's luminosity density, even if it is assumed a collision is a guaranteed disassociation. In many cases a cosmic ray impact creates a heavier radioactive nucleus, not nuclear debris.

- Antimatter. Not even close. The observed gamma flux is far too low to support the existence of significant annihilation reactions. Annihilation releases about 140 times more energy than fusion. If it were responsible for universal nuclear dissolution, the gamma flux in space would be far in excess of the luminous flux, not orders of magnitude less.⟨39⟩

- Galactic core region. This is the only viable location. Nuclei condense into complex forms inside stellar cores; the site of their disassociation needs to be far more powerful. *The only known environment with this kind of energy density is the galactic core, and the evidence is as blatant as it is inescapable.* Many galaxies have jets of high-speed hydrogen erupting from their cores, streaking thousands of light years across space. Galactic cores also emit hard X-ray and gamma radiation, further evidence that they are significantly more energetic than stars.⟨1.6⟩

The region where universal nuclear disassociation occurs has to be directly linked with the cosmic microwave absorption agent, and our investigation of the fusion cycle is inexorably drawn toward a single conclusion. A galaxy is a vast microwave antenna. It transfers captured CMB energy into its core in the form of electric current for the purpose of burning compound nuclei back into hydrogen. No other scenario is possible. *Galaxies are the cogs of the universal engine.*

Three things are necessary to undo the work of nuclear fusion:

- Energy.

- Compound nuclei.

- A region where energy is applied to compound nuclei to break them apart.

A galactic core provides this environment, and there is only one way to transport the compound nuclei found in a galaxy's disk to its core. It arrives there by virtue of a steady inflowing movement. *Galaxies are vortices, and the material of their disks falls inward towards their cores.* Form follows function:

Ψ THEOREM 16.1 - GALACTIC VORTEX {Ψ14.1}

(A) A GALAXY IS A MICROWAVE ANTENNA THAT TRANSFERS CMB ENERGY DIRECTLY FROM ITS DARK HALO TO ITS CORE IN THE FORM OF ELECTRICAL CURRENTS, MARKING ITS DISK REGION WITH LUMINOUS BANDS OF ENHANCED STELLAR IGNITION AND PRESSURE WAVES INDUCED BY MAGNETIC PINCH

(B) *DISK MATERIAL FALLS SLOWLY INWARD TOWARD A GALAXY'S CORE REGION, WHERE IT IS RECOMBINED WITH THE ELECTRICAL ENERGY ARRIVING ALONG ITS LUMINOUS BANDS*

(C) *COMPOUND NUCLEI ARE DISASSOCIATED IN THE EXTREME CONDITIONS PRESENT IN A GALAXY'S CORE, RELEASING ATOMIC HYDROGEN INTO SPACE; THIS HYDROGEN RAINS DOWN ONTO THE GALACTIC DISK AND IS TRANSPORTED OUTWARD ALONG THE SAME BANDS THAT CARRY ELECTRICAL ENERGY INWARD; UNDER CERTAIN CONDITIONS A GALAXY'S HYDROGEN PRODUCTION IS ALSO LAUNCHED INTO SPACE AS A VIOLENT STREAMING JET*

(D) *THE HYDROGEN RELEASED BY A GALAXY'S CORE IS THE FUEL USED FOR ITS FUTURE STELLAR FORMATION*

This derivation explains each and every one of a galaxy's basic characteristics as described in the sections to follow. Hydrogen is the universe's fuel. As the material of a galactic disk percolates, bathed in the glow of the currents driven by CMB radiation, *hydrogen completes the proton path while free electrons complete the luminous path*. Galaxies are cosmic whirlpools of matter and energy:

Figure (16.1) M74, an archetypical spiral galaxy with supernovae (lower right)
(Courtesy NASA/Hubble Heritage Project)

No galaxy achieves a *perfect* balance between the energy it captures and the energy it radiates, so the calculations to follow use a galactic efficiency of Q_g, as defined previously in Equation (15.54):

$$Q_g = \frac{P_g}{L_g} \tag{16.1}$$

where P_g is a galaxy's captured power input and L_g is its luminous output.

16.2 GALACTIC ROTATION CURVE

Galactic rotation is one of the great mysteries of modern astronomy. The material of a spiral galaxy's disk orbits a common center, but not in any way expected from Newtonian mechanics. This *circular velocity profile* has the following form, shown in comparison to a Newtonian profile:

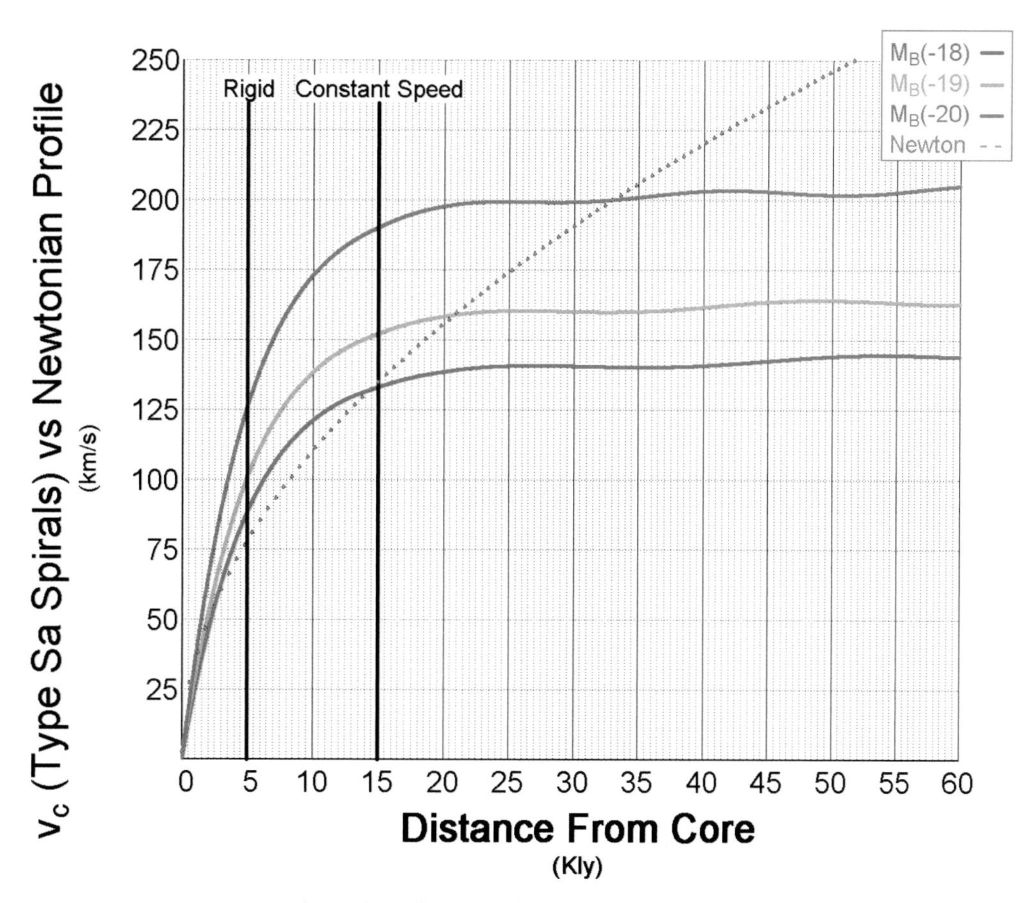

Figure (16.2) Typical circular velocity profiles for Sa spirals versus a Newtonian profile

The central region of a spiral galaxy, from its center to a radius of about 5 Kly, exhibits a great deal of *rigid-body rotation*, moving as if it were a solid object.⟨6.22⟩ Here circular velocity is roughly proportional to radius and angular velocity is constant. Beyond this is a transitional region, where its material departs markedly from a rigid profile. Exterior to this, at about 15 Kly, a galaxy's disk moves with nearly constant circular velocity. This is called its *constant speed region*.⟨6.21⟩ As noted in the previous chapter, a galaxy's speed increases with increasing luminosity (as denoted by a more negative absolute magnitude, $\mathbf{M_B}$).

Galactic rotation profiles have baffled astrophysicists since they were first discovered. If a galaxy was a distribution of gravitationally bound objects like our solar system, circular velocity would vary with radial distance according to:

$$\mathrm{v}_c(r)=\sqrt{\frac{GM}{r}} \tag{16.2}$$

This is shown as the dotted trace in Figure (16.2). Pluto's circular speed, for instance, is far slower than Earth's, and Mercury's is much faster. A galaxy's motion is entirely different from this. So different in fact, that some scientists have even considered the possibility that galactic dynamics might be an exception to Newton's gravitational laws.⟨8.5⟩ This is because constant circular speed makes absolutely no sense in a universe that is only 13.7 billion years old. As it turns out, however, the galactic motion shown in Figure (16.2) makes perfect sense in an eternal universe.

A stable galaxy maintains a constant mass distribution. This means its inflow is balanced at any given radius by a comparable hydrogen emission. *Material falls into a galactic disk to balance the mass loss caused by the ejection of disassociated protons.* The motion of galactic material is in fact governed by the Newtonian constraint of Equation (16.2). Its central mass M at any given radius is controlled (by hydrogen emission) to stay slightly larger (~0.7%, as calculated in Appendix P) than what would be required to maintain a perpetual circular motion for all of its material. This *vortical mass driver* causes the stars, dust, and gas in its disk to slowly spiral inward. Galactic inflow is a direct, balanced response to the mass loss associated with hydrogen discharge - the galactic vortex.

A galaxy's vortex causes the material density of its disk to vary with radius in a precise way. If most of a galaxy's mass resides within a disk of thickness h_g, its circular velocity is related to material density by Equation (16.2):

$$\mathrm{v}_c^2=\frac{GM}{r}=\frac{G\bar{\rho}V}{r}=\frac{G\bar{\rho}(r)h_g\pi r^2}{r} \tag{16.3}$$

Since circular velocity is constant, average density varies inversely with radius:

$$\bar{\rho}(r) = \frac{v_c^2}{Gh_g \pi r} \tag{16.4}$$

In a flat disk, the instantaneous density for a $(1/r)$ profile at some radius r is half of the average density within it:

$$\rho(r) = \frac{\bar{\rho}(r)}{2} \tag{16.5}$$

Substitute this into Equation (16.4):

$$\rho(r) = \frac{v_c^2}{2Gh_g \pi r} \tag{16.6}$$

The reason a galaxy's circular velocity is constant is because it conforms to Newton's laws, not because it violates them. As its disk material moves towards its core into progressively smaller volume, its density necessarily increases as $(1/r)$. The Milky Way's disk is about 5 Kly thick and its flat orbital profile extends past ~80 Kly, so its density is ~$3.5(10)^{-21}$ kg/m^3 at R = 80 Kly, increasing to ~$1(10)^{-20}$ kg/m^3 at R = 30 Kly.[1.7]

VORTICAL FLUX

Galaxies have a steady luminous output, creating compound nuclei at an even rate. These nuclei are disassociated in their galactic cores at a comparable rate, so the flow of material through their vortices is uniform. This material always follows the same path through the disk to the core, so the amount moving across any given radius is constant. In a disk of uniform thickness, the total material inflow flux at any radius r is given by:

$$\frac{dM}{dt} = (2\pi r h_g)\rho_i(r) v_r(r) \tag{16.7}$$

where h_g and $\rho_i(r)$ are a galaxy's disk thickness and *inflow density*, respectively, and $\mathbf{v}_r(r)$ is the *radial inward speed* of this material, not to be confused with its circular orbital speed, $\mathbf{v}_c$.

Infalling material is an electrically neutral mix of stars, dust, and planets. It is captured by the slight mass density surplus distributed throughout the galactic vortex. The only way to balance this inward flux is to drag mass outward, a process accomplished by electric fields acting on ionized hydrogen. At any given location in a galaxy's disk, the inward flow of material is offset by the exodus of mass in the opposite direction.

Ejected gas moves far faster than the stately vortical inflow, so:

$$\rho_i(r) \cong \rho(r) \tag{16.8}$$

Most of a galaxy's mass is moving slowly towards its core. This inward dM/dt is balanced by a small density of rapidly outbound hydrogen. To find a galactic vortex's radial velocity, substitute Equations (16.6) and (16.8) into the expression for the radial mass transfer of Equation (16.7):

$$\frac{dM}{dt} = (2\pi r h_g)\left(\frac{\mathrm{v}_c^2}{2Gh_g \pi r}\right)\mathrm{v}_r(r) \tag{16.9}$$

Simplify and solve for radial velocity:

$$\mathrm{v}_r = \frac{G}{\mathrm{v}_c^2}\left(\frac{dM}{dt}\right) \tag{16.10}$$

This is the speed that material falls through a galactic disk into its core. It is constant - a function of circular velocity and the rate that material passes through any radius.

RADIAL VELOCITY

The amount of compound nuclei a galaxy creates is given by its luminosity:

$$\frac{dn}{dt} = \frac{L_g}{\varepsilon_U c^2} \tag{16.11}$$

where ε_U is the power-weighted average mass fraction of all types of fusion, and L_g is galactic luminosity, including the light given off by the burning of helium and heavier elements. The averaged mass fraction ε_U is about 0.007. The Milky Way, with an estimated power output of $1.4(10)^{37}$ W,[1.14] produces compound nuclei at a rate of $2.2(10)^{22}$ kg/s. Although this might seem prodigious by Earthly standards, on the galactic scale it is miniscule. It takes our entire galaxy ~3 years to produce one solar mass equivalent of compound nuclei.

The amount of compound nuclei that a galactic core consumes is given by:

$$\frac{dn}{dt} = \frac{Q_g L_g}{\varepsilon_U c^2} \tag{16.12}$$

This differs from Equation (16.11) by a factor of galactic efficiency. Whereas the amount of nuclei a galaxy creates is driven purely by its luminosity, the amount it consumes is a function of its efficiency. Moreover, all of the material flowing through the Milky Way's disk also flows through the disassociation environment in its core - the core *drives* the galactic vortex. Any material not transiting the core is not a part of the vortex. But how much material must pass through this vortex to achieve a given efficiency?

A galaxy's vortical inflow is governed by the total amount of compound nuclei exposed to its core. Its core's primary function is to dissolve these nuclei at a certain rate. *It has no effect on hydrogen.* Equation (15.24), based on matter-energy correspondence (Ψ15.3), tells us that the universe has an *average* composition of about 2% bound protons. If the material entering the Milky Way's core is 2% compound nuclei, for instance, and it leaves as pure hydrogen with 0% compound nuclei, then *mass moves through its vortex fifty times faster than the rate at which it creates compound nuclei.* The hydrogen entering the core leaves as hydrogen with no net effect, except to dilute the compound nuclei the core burns. This necessitates the transfer of a far greater amount of material for a given amount of compound nuclei.

At equilibrium, a galaxy's vortical inflow is given by Equation (16.12) and its material composition:

$$\frac{dM}{dt} = \left(\frac{dn}{dt}\right)\frac{1}{f_{cn}} = \frac{Q_g L_g}{f_{cn}\varepsilon_U c^2} \tag{16.13}$$

where f_{cn} is the fraction of compound nuclei passing into a galactic core.

Material falls from intergalactic space into the galactic vortex, reaching the density required to initiate fusion and its attendant nucleosynthesis at the galactic rim. As it descends towards the galaxy's center, its production of compound nuclei slowly increases. Finally, this ancient disk material is shunted through the galactic core, *as a balanced reversal of a galaxy's fusion activity.*

Combining Equations (16.10) and (16.13) results in an expression for the rate that a galaxy's disk material is falling towards its inevitable disintegration:

$$\mathrm{v}_r = \frac{G}{\mathrm{v}_c^2}\left(\frac{Q_g L_g}{f_{cn}\varepsilon_U c^2}\right) \tag{16.14}$$

For a given luminosity, the faster the radial velocity $\mathbf{v}_r$, the less compound nuclei a galaxy has a chance to create in its disk and destroy in its core.

Substituting values for universal constants and the Milky Way's luminosity yields:

$$v_r \cong 30\left(\frac{Q_g}{f_{cn}}\right) \qquad (16.15)$$

in units of meters per second.

Assuming:

a) The Milky Way is a steady-state system with a galactic efficiency Q_g of 1.0 (100%). **Justification**: The universally average galactic efficiency is unity (Ψ15.14), and spiral galaxies have an intricate structure that suggests long-term stability, which in turn requires close to break-even efficiency.

b) The fraction of compound nuclei moving through the Milky Way's core is close to the universal fraction of bound protons, or $f_{cn} \cong f_{bp}$. **Justification**: A galaxy is a self-sustaining, perpetual engine, so the hydrogen it burns in fusion has to be continually replaced, regardless of where this fusion occurs. This is the function of the galaxy's core. It creates hydrogen at the same rate that it burns compound nuclei, and transfers it, in various quantities, throughout a galaxy's disk. This is why the relative elemental composition of any location on this disk is similar to any other. There is a high density of dust and other material in a galaxy's central region, but it also has a corresponding abundance of hydrogen. So although the material density and rate of fusion in a galaxy's disk increases on its way to the core, its compound nuclear fraction does not. The proliferation of stellar activity near a galaxy's core is a function of material and energy density, not composition.

Equation (15.24) gives f_{bp} as 0.02 (2%) at a Hubble constant of 60 Hz-km/Mpc. Applying an f_{cn} of 0.02 along with the efficiency (1.0) cited above yields a vortical inflow rate for the Milky Way of 1.5 km/s.

Our local stellar neighborhood is falling toward the galactic core far faster than this, at a rate of u = 9 km/s, but this is not an accurate reflection of the actual rate for the entire galactic disk. Nor does it represent our average infall rate, when averaged over billions of years. These same stars also have a motion component *perpendicular to the galactic plane* of ~7 km/s.[6.5] Galactic dynamics are complex. The variability of intragalactic mass distribution causes a great deal of turbulence among star populations throughout the disk region. Even averaging hundreds of local stars may not give an accurate assessment of the net inward motion of the billions scattered throughout the galaxy. Fortunately, there is a great deal of independent evidence of our galaxy's inflow rate.

WHITE DWARFS IN MOTION

Using the detailed work of Sion and McCook, a compilation of the temperatures of white dwarfs was made, resulting in a dataset of 1262 stars. Figure (16.3) shows a distribution of their temperature versus number density:

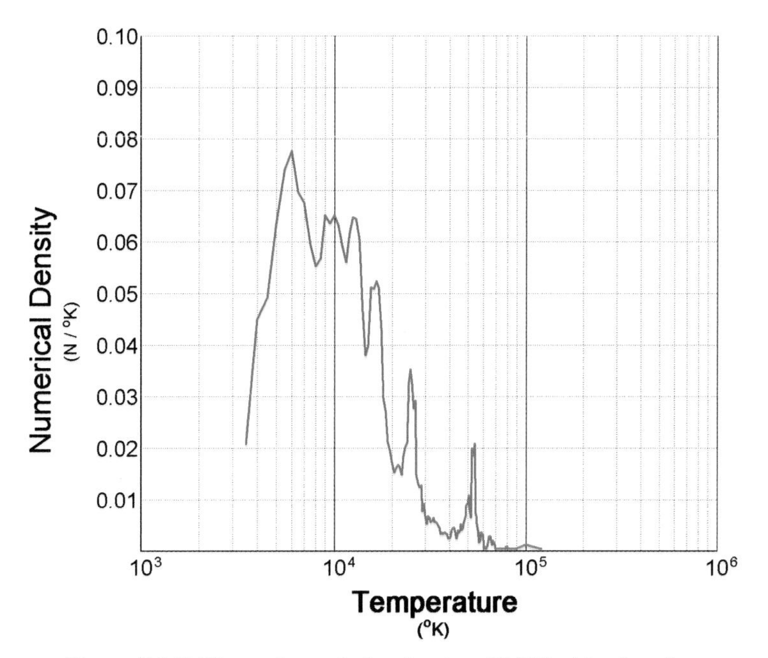

Figure (16.3) Thermal population density of 1262 white dwarf stars

White dwarfs in our local stellar neighborhood have a temperature distribution beginning near 120,000 °K and decreasing down to about 3600 °K, where there is an abrupt cutoff, as is evident in the graph. This isn't because the universe has a finite age; it is because our local neighborhood does. Its coolest white dwarfs are the remnants of stars that ignited on the Milky Way's rim billions of years ago, long before our solar system existed.

The radial drift of a galaxy's disk material is constant, so the distance that stars move inward from its rim over time is given by:

$$D_r = \mathrm{v}_r \tau \qquad (16.16)$$

where τ is the age of a stellar neighborhood. Solve for radial drift:

$$\mathrm{v}_r = \frac{D_r}{\tau} \qquad (16.17)$$

The oldest white dwarfs in our local neighborhood are thought to be ~10 Gyr, including the limited time they spent in the main sequence (about 300 million years for the massive stars responsible for most of these remnants).[6.10] Our stellar neighborhood lies about 55 Kly from the galactic rim,[6.11] so it has been moving toward the Milky Way's core at an average rate of 1.6 km/s for billions of years, consistent with our earlier estimate using Equation (16.15):

$$v_r \cong 30\left(\frac{Q_g}{f_{cn}}\right)$$

For additional information about the white dwarf distribution of our local stellar neighborhood, please refer to Appendix J.

16.3 GALACTIC LUMINOSITY PROFILE

White dwarf age is a welcome correlation, but as it turns out, a galaxy's vortex leaves a far more indelible mark on its disk. The surface brightness of a spiral galaxy's constant speed region adheres to an exponential relationship of the form:[14.3]

$$I \approx I_0 e^{-\left(\frac{r}{h_R}\right)} \tag{16.18}$$

where r is the distance from the galaxy's center and the decay constant h_R is known as the galactic *disk scale length*.[14.3] Disk scale length defines the rate that a galaxy's surface brightness attenuates with distance from its center:

Figure (16.4) Galactic surface brightness falls off exponentially with distance from core (M74, Courtesy NASA/Hubble Heritage Project)

A short disk scale represents a rapid brightness loss. The surface brightness of a galaxy's central portion more closely resembles:

$$I \approx I_0 e^{-7.67\left(\left(\frac{r}{R_e}\right)^{\frac{1}{4}}-1\right)} \tag{16.19}$$

where R_e is known as a galaxy's *effective radius*. Gérard de Vaucouleurs identified this empirical relationship in 1948 and it is known as de Vaucouleurs $R^{1/4}$ law.⟨6.12⟩ As was the case with disk scale length, a small effective radius denotes a precipitous loss of brightness.

The exponential brightness profiles of Equations (16.18) and (16.19) have proven particularly inexplicable to astrophysicists because they don't correspond to any known material density profile, gravitational or otherwise. As it turns out, however, a galaxy's vortical nature explains both, and they can be used to measure its radial inflow speed. Let's begin by deriving galactic surface brightness directly from stellar population density.

STELLAR LUMINOSITY DENSITY

The number density of stars in any volume of space is related to their lifespan and rate of formation by:

$$\rho_s = \tau_s q_s \tag{16.20}$$

where q_s is stellar formation rate in (stars/volume/second) and τ_s is *average stellar lifespan*, which will be defined as the average duration of a star's fusion light production. Equation (16.20) applies to all stars with a defined main-sequence lifespan. It does not include white dwarfs, neutron stars, and other objects that lack specific termination scenarios.

Stars produce, through their solar winds, flares, and explosive deaths, the nuclear by-products necessary for the formation of more stars. This means their formation rate is proportional to stellar number density:

$$\frac{d\rho_s}{dt} \propto q_s = \frac{\rho_s}{\tau_s} \tag{16.21}$$

The hundreds of white dwarfs that provided the basis for Figure (16.3) were all products of explosions that scattered heavy nuclei throughout the Milky Way's disk region - seeds for the next generation of new stars. *A galactic disk's stellar population growth operates much like an ultra-slow nuclear chain reaction*. This process is not limited by the available fuel. A galaxy's core provides all the fuel it might need. And even if a core's hydrogen production fell short, the estimate done with Equation (15.26) demonstrates that the fusion endurance of a typical galaxy like the Milky Way is over a trillion years. *Stellar number density is governed purely by the rate that stars are able to coalesce and ignite, not by a lack of fuel.*

In an environment lacking fuel or volume limitations, stellar number density doubles every $0.7\tau_s$ ad infinitum. So if there are no other selection criteria, the spatial density of stars of shorter lifespans grows faster than those with longer lifespans. The dominance of younger, brighter stars near the center of a galaxy is a natural consequence of their brief lifespan combined with a vortex that makes this region the oldest and most energetically dense part of a galactic system.

Stars descend slowly toward the galactic core as they age. The differential relationship between their coreward transit time t and radial position r is given by:

$$dt = -\left(\frac{dr}{\mathrm{v}_r}\right) \tag{16.22}$$

Substitute Equation (16.22) into Equation (16.21):

$$\frac{d\rho_s}{dr} \propto -\frac{\rho_s}{\mathrm{v}_r \tau_s} \tag{16.23}$$

This has an exponential solution of the form:

$$\rho_s = \rho_0 e^{-\left(\frac{r}{\mathrm{v}_r \tau_s}\right)} \tag{16.24}$$

where ρ_0 is peak stellar number density.

Equation (16.24) represents star density growth due to formation rate, but this isn't the only density consideration. As stars fall inward in a galactic vortex, they are packed into decreasing space, so their volume density varies with radius as:

$$\frac{\rho_{s_1}}{\rho_{s_2}} = \frac{r_2}{r_1} \tag{16.25}$$

Apply Equation (16.25) to (16.24):

$$\rho_s = \rho_{s_c}\left(\frac{r_c}{r}\right) e^{-\left(\frac{r}{\mathrm{v}_r \tau_s}\right)} \qquad \{r \geq r_c\} \tag{16.26}$$

where ρ_{s_c} is the peak stellar number density in a galaxy's most central region ($r \le r_c$), and r is defined down to r_c, where a galaxy's surface brightness plateaus.

VORTICAL BRIGHTNESS FUNCTION

The final step in our derivation is to convert Equation (16.26) into an expression for galactic surface brightness. Given a roughly homogenous stellar population and galactic disk of uniform thickness, the surface brightness at any location is proportional to the number of stars per volume. Average lifespan, however, takes on an entirely different meaning. Red dwarfs live for tens of billions of years, but their luminous output is insignificant. Supergiants have roaring luminance but a brief life. Most of a spiral galaxy's light comes from hot, young stars about twice as massive as our sun. *Luminosity lifespan*, τ_L, will be defined as the lifespan of the stars that provide the majority of a galaxy's light.

Write Equation (16.26) in terms of surface brightness and τ_L, resulting in what will be referred to as the *vortical brightness function*:

$$I = I_0\left(\frac{r_c}{r}\right)e^{-\left(\frac{r}{\mathrm{v}_r\tau_L}\right)} \qquad \{r \ge r_c\} \qquad (16.27)$$

where I_0 is the peak surface brightness throughout a galaxy's inner central region ($r \le r_c$).

DISK SCALE LENGTH

Compare the vortical brightness function to Equation (16.18):

$$I \approx I_0 e^{-\left(\frac{r}{h_R}\right)}$$

The term ($\mathrm{v}_r\tau_L$) is immediately recognizable as disk scale length, h_R. A galaxy's disk scale length as originally presented in Equation (16.18) is in reality the *product of its vortical radial velocity and average luminosity lifespan*:

$$h_R = \mathrm{v}_r\tau_L \qquad (16.28)$$

Disk scale length is the ancient luminous trail left by stellar number density evolution, painted on a galaxy's disk as it spirals toward its core.

Ψ THEOREM 16.2 - GALACTIC SCALE LENGTH {Ψ16.1}

THE SCALE LENGTH OF A GALAXY IS THE PRODUCT OF ITS RADIAL FLOW VELOCITY AND AVERAGE LUMINOSITY LIFESPAN

The rate of stellar population density evolution is driven by the average lifespan of its hot young stars, but the distribution of this evolution across a galactic disk is governed by vortical inflow velocity.

The following depicts a galaxy's surface brightness as a function of radial distance from its center:

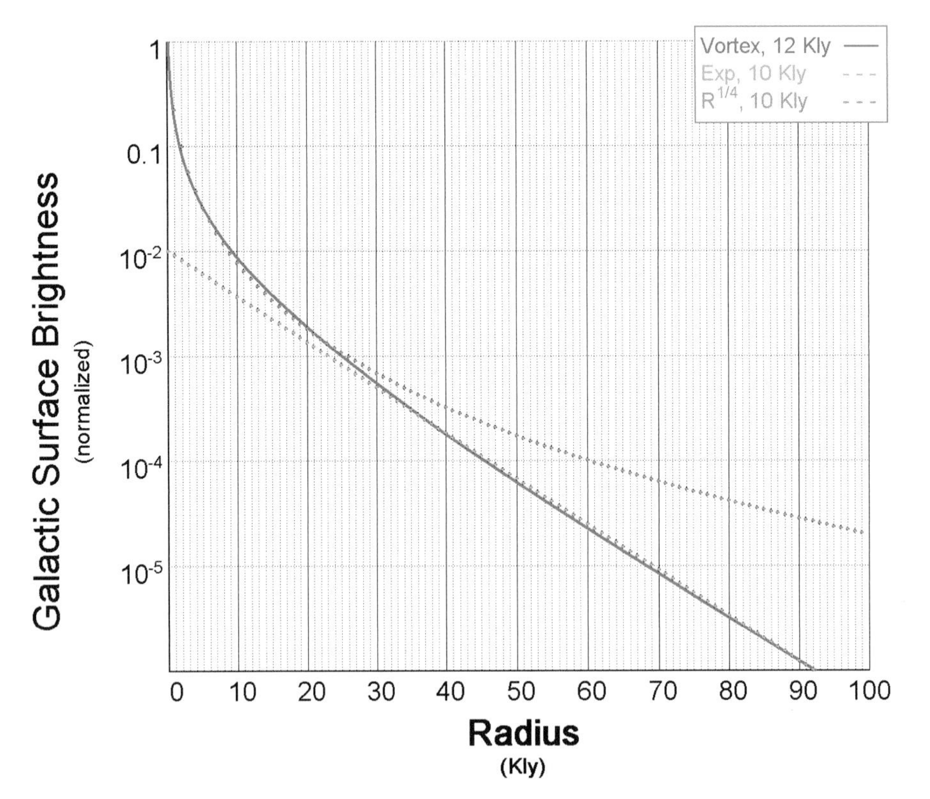

Figure (16.5) Spiral galaxy surface brightness, vortical disk scale length 12 Kly, r_c = 1 Kly

The vortical brightness function of Equation (16.27) with a disk scale length of 12 Kly is shown as the green trace. The curved violet trace is de Vaucouleurs $R^{1/4}$ law at an effective radius of 10 Kly, and the orange trace is the simple exponential of Equation (16.18) with a scale length of 10 Kly. *The vortical brightness function provides a unified representation of the surface brightness of a spiral galaxy's entire disk.* Note, however, that vortical inflow serves to understate a galaxy's actual disk scale length by ~20% (a 12 Kly scale length will appear as 10 Kly).

16.4 LUMINOSITY LIFESPAN

The Milky Way's disk scale length has been estimated by astronomers, so it is possible to use Equation (16.28):

$$h_R = \mathrm{v}_r \tau_L$$

to derive its radial inflow speed if its average luminosity lifespan is known. The first step in this calculation is to derive the general relationship between a star's luminosity and lifespan. This is defined by the amount of its available fuel and the rate that it is consumed, both related to the star's mass.

The mass and luminosity of stars whose mass is greater than 0.2 M_{sun} are related as:⟨1.21⟩

$$\frac{L}{L_{sun}} = 10^{0.08}\left(\frac{M}{M_{sun}}\right)^{3.8} \qquad (16.29)$$

Due to temperature gradients and other physical limitations, stars only burn the fuel in their core regions during their main-sequence lifespans. This amounts to ~12% of their total mass.⟨3.3⟩ Combining this available fuel fraction of 0.12 with the fusion mass fraction ε yields what will be called the *burn fraction*, ε_*. It is equal to 0.00088. A star's main-sequence lifespan is given by the ratio of its available fuel energy and the rate that it releases it:

$$\tau_s = \frac{\varepsilon_* c^2 M}{L} \qquad (16.30)$$

This amounts to 12.8 billion years for our sun, consistent with the commonly used ZAMS theoretical model. Since mass is a function of luminosity by Equation (16.29), the lifespan of Equation (16.30) can be written solely in terms of luminosity. Solve Equation (16.29) for M and substitute into Equation (16.30):

$$\tau_s(L) = \left(\frac{\varepsilon_* c^2 M_{sun}}{10^{\left(\frac{0.08}{3.8}\right)} L_{sun}^{\left(\frac{1}{3.8}\right)}}\right) L^{\left(\frac{1}{3.8}-1\right)} = 1.5(10)^{37} L^{-0.7368} \qquad (16.31)$$

This has units of seconds when L is in watts.

The following shows, as a function of luminosity, the numerical space density (green), luminosity-weighted space density (orange), and calculated lifespan (violet) of the stellar population of the Milky Way's disk:⟨1.8⟩

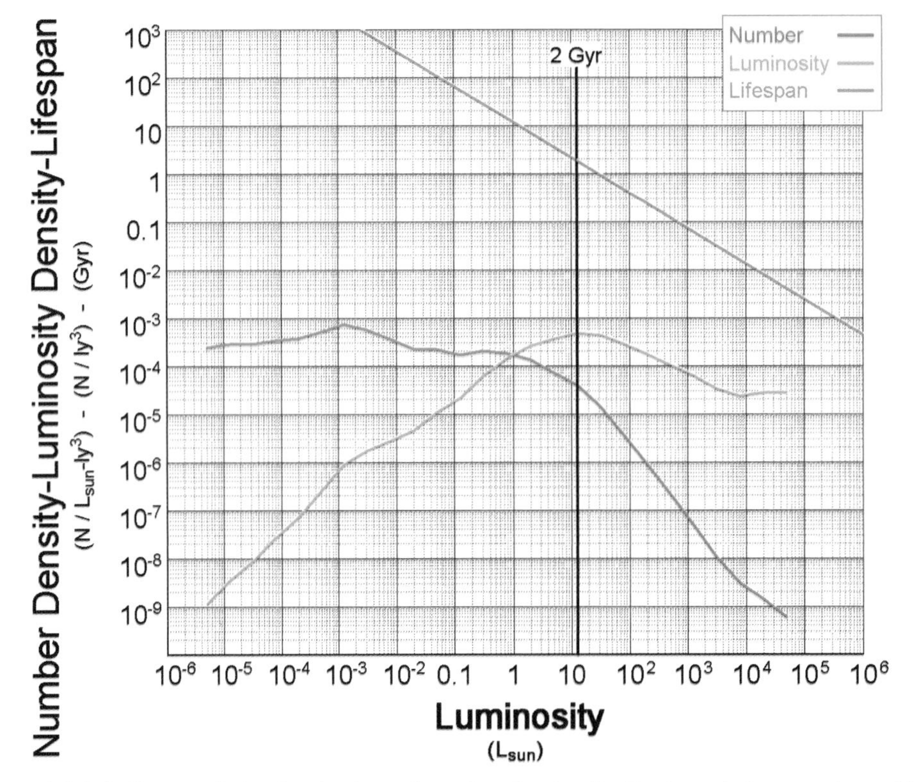

Figure (16.6) Stellar population luminosity and number density distribution relative to estimated lifespan

The luminosity distribution of the Milky Way's star population is fairly consistent throughout its disk. The green trace shown in Figure (16.6) is the number density of stars. Its maximum near 0.001 L_{sun} represents the vast numbers of red dwarfs scattered throughout our galaxy. The orange trace peaking near 10 L_{sun} is luminosity density - the product of number density and luminosity. This shows that most of the Milky Way's light originates from hot young stars with over ten times the power output of our sun.

The violet trace in Figure (16.6) is estimated stellar lifespan as a function of luminosity, Equation (16.31). At the peak of luminosity density, 12 L_{sun}, it has a value of ~2 Gyr. *This is the lifespan of the stars most visible to our instruments,* τ_L.

$$\tau_L \cong \tau_s(12L_{sun}) \tag{16.32}$$

The scale length for the Milky Way's disk has been estimated to be in the range of 6.5 - 13 Kly.⟨14.4⟩ Since this is derived using a simple exponential in the absence of a vortical inflow component, it is understated by about 20%, so its actual range ought to be closer to 8 - 16

Kly. From Figure (16.6), the lifespan of its most visible stars is about 2 Gyr. Rewrite Equation (16.28) and substitute these values:

$$\mathrm{v}_r = \frac{h_R}{\tau_L} \qquad (16.33)$$

A luminosity lifespan of 2 Gyr at the Milky Way's observed disk scale lengths represents an inward radial velocity range of 1.2 to 2.4 km/s, bracketing the values (1.5 km/s and 1.6 km/s) derived earlier using Equation (16.15) and the age of our stellar neighborhood.

THE HIDDEN VORTEX

The vortical nature of galaxies has evaded detection until now because it is a subtle effect buried by a myriad of competing dynamics. It is so small in relation to most galactic motions, such as circular velocity, that it can only be found by knowing in advance what to look for. Equation (16.13) gives the total material flow through a galactic vortex:

$$\frac{dM}{dt} = \left(\frac{dn}{dt}\right)\frac{1}{f_{cn}} = \frac{Q_g L_g}{f_{cn}\varepsilon_U c^2}$$

To put this into perspective, the Milky Way's total vortical flow, at f_{cn} = 0.02 and 100% efficiency, is a vanishingly small ~18 M_{sun}/yr. This amounts to less than 40 parts per trillion of its total mass *per year*. Indeed, our galaxy's vortical inflow provides an ideal test for Null Cosmology because (a) it is small, and (b) there is no reason, from the perspective of the Big Bang at least, to measure it.

16.5 GALACTIC TRANSIT TIME

The galactic vortex is responsible for why our local neighborhood appears about ten billion years old, and it is also the reason why globular clusters often seem older than the galaxy they orbit.[1.3] Their trajectories lie immediately outside the galactic vortex, so they are not pulled into its recycling engine with the same regularity as its disk material. The net effect of the galactic vortex on the rest of the universe's appearance, however, is more profound.

It took about ten billion years for our solar system to reach its current distance from the galactic rim. How long until it falls into the core? The time required to flow across our galaxy's entire luminous disk will be called the *galactic transit time*. The close correspondence between the vortical brightness function of Equation (16.27) and the luminosity profiles of galaxies suggests that their radial inflow speed is constant throughout their entire structure, to include their central region of rigid-body rotation. This means galactic transit time is simply the ratio of radius to radial velocity:

$$t_g = \frac{R_g}{v_r} \tag{16.34}$$

where R_g is the radius of the galactic rim. In the case of our galaxy, with a luminous radius of ~82 Kly⟨6.11⟩ and radial velocity at ~1.5 km/s, the time required to traverse its entire luminous structure is ~16 billion years.

All galaxies are vortices similar to ours, so there is very little material in the universe that appears much older than the galactic transit time of the Milky Way. *This is why the entire universe appears to have a finite age.* Reality's building blocks are quite ancient but are frequently reshuffled by galactic cores.

Ψ THEOREM 16.3 - APPARENT UNIVERSAL AGE {Ψ16.1}

THE UNIVERSE'S MATERIAL APPEARS TO BE ON THE ORDER OF 12 – 18 BILLION YEARS OLD BECAUSE THIS IS THE TIME REQUIRED FOR GALAXIES TO RECYCLE IT THROUGH THEIR CORES

The aptly named Whirlpool Galaxy, M51, was the first galaxy observed to have internal structure. This was in 1851, and is the height of irony. Even when given a name *that tells us exactly what it is*, the true nature of galactic systems has eluded scientists for over 150 years.

Figure (16.7) The Whirlpool galaxy, M51
(Courtesy NASA/Hubble Heritage Project)

The time it takes to fall through the luminous portion of our galaxy's disk is about 16 billion years. Any star born on the Milky Way's rim with a mass smaller than that of our sun will still be burning when it falls into our galaxy's core. Indeed, our own sun has enough fuel to burn at its current luminosity for another four or five billion years. At its current rate of descent, our solar system will be in the Milky Way's core region *in less than four billion years.* This means our sun might still be burning when it is consumed by the beast living in the heart of our galaxy.

There are two unforgiving time constraints for the evolution and long-term survival of life in the universe. It must advance from bacteria to full-scale space transport before its sun fails or falls into the core of its galaxy, whichever comes first. Earth life had plenty of time to spare on both accounts. Once our planet was first capable of supporting life, it took about 3 Gyr to reach our current level of biological complexity, a total run of about 4.5 Gyr after our sun's ignition. If Sol had started burning closer than ~25 Kly from the Milky Way's core, our civilization would simply not exist. *A significant portion of any galaxy is too close to its core to allow the emergence of life in newly formed solar systems, at least at the pace it occurred on Earth.*

16.6 ENIGMATIC GALACTIC CORE REGION

The spectral characteristics and velocity dispersion of the material in galaxies' central regions indicate the presence of extremely energetic processes near enormous black holes. In our own galaxy, the motions of supergiant stars close to its center betray the presence of a black hole about three million times as massive as our sun.[1.9] Let's quickly review black hole dynamics prior to investigating their function as galactic cores.

The gravitational potential energy near a nonrotating spherical mass is given by the Schwarzschild solution to Einstein's field equations. From Equation (5.19):

$$\Phi_g = m_0 c^2 \left(\sqrt{1 - \frac{2GM}{c^2 r}} - 1 \right)$$

When the radius is equal to the Schwarzschild value:

$$R_S = \frac{2GM}{c^2} \qquad (16.35)$$

the potential of Equation (5.19) becomes $-m_0c^2$, the energy of an object's rest mass. This is consistent with the Maxfield Theorem (Ψ5.5). It is not possible for an object to have a gravitational potential energy in excess of its rest mass because its rest mass is the ultimate source of this potential energy. The unification of particle cores and gravitation presented in Part III also demonstrated that matter's limited compressibility makes it impossible for compact objects to actually achieve a Schwarzschild radius.

Gravitational potential energy is irrespective of kinetic energy. At any given location in space, a relativistic proton moving at 0.999 c has the same gravitational potential as that of a motionless proton. This means that *the kinetic energy a particle requires to exit a black hole is close to, but less than, its rest energy*. Photons are redshifted by the expanded particles of a black hole's gravitational veneer. Particles, on the other hand, only lose an amount of energy equal to its veneer potential. Black holes are a superbly efficient filter between matter and light, not the unbreachable gravitational containers they are currently thought to be. The escape velocity for light is c in a black hole because this is the speed of photons of any energy; the escape velocity is less than c for matter *because it has the capability of carrying a kinetic energy greater than its rest mass.*

Although it will be some time before the detailed aspects of the galactic core environment are revealed, this much is known with a great amount of certainty:

- The velocity necessary for any particle to escape a black hole's veneer is at most $0.86c$.

- A galaxy's core absorbs electrical energy from its disk and uses this energy to disassociate the compound nuclei it captures through its vortical action, thereby producing hydrogen.

Galactic cores are electrical storms of inconceivable proportions, capable of breaking any material back into its component protons, *to include neutron stars and stellar black holes*. The first law of cosmostasis is *all of the universe's material is recyclable*. If supernovae produce stellar black holes at a certain rate then the galactic core possesses an environment capable of destroying them at the same rate.

NUCLEAR STRETCH

A black hole's interior is composed of a degenerate neutron superfluid with a gravitational potential strong enough to lower its own density through the hyper-expansion of particle cores. Otherwise it is similar to a neutron star's composition, which is in turn similar to that of atomic nuclei.

Strong potential is inversely proportional to the core radius of particles, as given by Equation (12.9):

$$\Phi_s(d) = \int_{r=d}^{\infty} F_s(r)dr$$

The greater the gravitational expansion, the lower the nuclear potential. Binding energy scales with nuclear potential because both are governed by energy density. This means that the extraordinary particle core expansion present in galactic cores drastically reduces both nuclear potential and inter-nucleon binding energy. The binding energy of heavy nuclei averages about 7 MEV/nucleon in free space, and they begin to break apart at temperatures of a few billion degrees.[9.2] When these same nuclei are gravitationally expanded by a factor of several thousand, *their binding energy falls accordingly. A galaxy's core doesn't just burn nuclei; it first forces them apart with its gravitational potential.*

The universe is about balance. If nuclei are formed in heat they must be dissolved in relative cold. A galaxy is an engine and heat is a waste product. If it produces compound nuclei in the cores of stars, at temperatures of 10 to 100 MK, burning them in its galactic core at a temperature of billions of degrees is not a complementary or efficient process. *At a billon degrees, a thermal object ten kilometers in diameter would radiate more energy than the entire Milky Way.* Clearly, a galaxy's core is not a normal thermal object. It is an environment where electrical energy is applied to degenerate matter, producing hydrogen and *virtually no radiant energy.* At first glance it might appear that a galactic core's electromagnetic filtering capability would allow it to maintain a temperature of billions of degrees, but at this temperature it would lose far too much high-energy matter. *The gravitational expansion required to disassociate nuclei at a relatively low temperature is why galaxies need massive black holes.*

The only place where nuclei could possibly be dissolved in the sweeping manner necessary to balance a galaxy's fusion is in its core. Imagine an object with a gravitational field strong enough to *inflate* atomic nuclei, blacker than the deepest space. *Universal equilibrium demands the existence of black holes.* It is an unavoidable consequence of cosmic renewal. Astrophysicists may argue about the face they show to the rest of the universe, but the reality of their existence is a foregone conclusion before the discussion even begins. There is quite literally no other way to disassociate compound nuclei in the virtual absence of radiant emission.

Ψ THEOREM 16.4 - GALACTIC CORE {Ψ5.6, Ψ16.1}

COSMIC EQUILIBRIUM REQUIRES MASSIVE GALACTIC CORES; THEY ARE GRAVITATIONAL FILTERS THAT SEPARATE LIGHT FROM HOT, ULTRA-EXPANDED MATTER, AND THEY FACILITATE NUCLEAR DISASSOCIATION BY PROVIDING AN ENVIRONMENT THAT RADICALLY REDUCES NUCLEAR BINDING ENERGY

A galaxy's core might not be at a temperature of billions of degrees, but it still needs to be hot enough to break the bonds between gravitationally expanded protons. Its mass varies with temperature. Too hot and too much matter escapes; too cold and its mass increases. *Electrical current heats the core; the absorption of compound nuclei cools it.* Breaking the bonds between protons converts kinetic energy into mass fraction. Essential to this process is the (relatively) low-energy transport of hydrogen away from the core itself.

The Chandra X-ray observatory reveals some of the amazing processes occurring near our galaxy's core:

Figure (16.8) Core region of the Milky Way in long X-ray
(Courtesy NASA/CXC/MIT/F.K. Baganoff et al.)

The Milky Way's core is known as Sagittarius A^* (Sgr A^*). It is centered though not visible in Figure (16.8). The galactic plane runs top left to bottom right. Note the clouds of hydrogen on either side of the core, separated by the Milky Way's disk.

16.7 A QUANTUM CORE (ADIABATIC APPROXIMATION)

A galactic core is an immense superfluid sea, the degenerate superposition of two fundamental gases - bound protons and bound electrons. Its energy distribution can be accurately described by a Fermi expression of the form:

$$n(E) = \left(e^{\left(\frac{E-E_F}{kT}\right)} + 1 \right)^{-1} \qquad (16.36)$$

where n is number density, k is the Boltzmann constant, T is temperature in degrees Kelvin, and the term E_F is known as the *Fermi energy level*.

FERMI ENERGY LEVEL

The Fermi level is the maximum energy state that a particle can achieve in a system at a temperature of absolute zero. Particles must attain kinetic energy (temperature) in order to fill energy states above this. When ($kT << E_F$), a material is said to be predominantly *degenerate*, though not in terms of the elementary core volume limitations discussed in Part III. The valance electrons of a metal at room temperature are said to be degenerate because virtually all have energy less than E_F. Conversely, when ($kT >> E_F$), a large fraction of particles have energy in excess of the Fermi level. A metal at ~100,000 °K is a good example of a nondegenerate electron gas. The difference between a galactic core and a bar of copper is one of degree, not substance. When copper is hot enough it emits valance electrons; when a galaxy's core is hot enough it emits protons and electrons.

The Fermi level of a system is related to the density and mass of its component particles, and lies below the total potential to which they are exposed. Valance electrons of silver, for instance, have a Fermi level of 5.5 EV and an average Coulomb potential of 10.2 EV. The difference between the two is known as the *work function*. Similarly, the nucleons of a typical atomic nucleus have a Fermi level of ~43 MEV and a total Strong potential of ~50 MEV, for a difference of ~7 MEV.

The difference between a particle's total potential and Fermi level is referred to as *binding energy*. The average 7 MEV binding energy of nucleons in atomic nuclei corresponds to their mass fraction ε; (940 MEV rest energy)(0.0073) $\cong$ 7 MEV.

In the adiabatic case, the total potential of a black hole's protons is defined by Maxfield as their rest mass. *The difference between this and their Fermi level is their average binding energy*:

$$m_p c^2 - E_F = E_{s\vee} - E_{q\vee} \tag{16.37}$$

where $E_{s\vee}$ is the *Strong veneer binder* and $E_{q\vee}$ is the *Coulomb veneer binder*. $E_{q\vee}$ varies with a black hole's net charge - it is not the Coulomb interaction between adjacent particles. When a black hole's charge is positive it repels protons and lowers their total binding energy.

Gravitation is uniquely different from the Strong interaction because it doesn't involve field cancellation or core excision. It is a pure conversion of field energy to kinetic energy, and its binding energy is virtually nonexistent. Black holes are a combination of the Strong, Weak, Coulomb, and gravitational potentials. Even though protons must relinquish a small amount of nuclear binding energy to settle into such remarkable density, most of their field energy is preserved as kinetic energy. Bound electrons move at relativistic speeds in an atomic nucleus. The same is true of the protons in a black hole.

The Strong veneer binder $E_{s\vee}$ varies inversely with proton core size and is significantly less than the average ~7 MEV binding energy of atomic nuclei for two reasons:

a) A neutron superfluid has a higher concentration of bound electrons than atomic nuclei and is therefore a less efficient environment for the Strong force.

b) Particle cores are gravitationally extended by the near-unity surface potential of a black hole, significantly lowering their energy density.

A particle's residual rest energy *fraction* in a massive black hole is the absolute magnitude of the difference between its veneer potential and negative unity. From Equation (13.39):

$$\frac{E_{\triangle}}{m_0 c^2} = \left(\frac{\Phi_{\vee}}{m_0 c^2} + 1\right) = \left(\frac{3R_p^3 c^6}{8\pi G^3 m_p M^2}\right)^{\frac{1}{3}} \quad \{M >> M_{sun}\} \tag{16.38}$$

The Strong veneer binder is nuclear binding energy scaled by this fraction:

$$E_{s\vee} = \left(\frac{\Phi_{\vee}}{m_0 c^2} + 1\right) E_{nb} = \left(\frac{3R_p^3 c^6}{8\pi G^3 m_p M^2}\right)^{\frac{1}{3}} E_{nb} \quad \{M >> M_{sun}\} \tag{16.39}$$

where E_{nb} is the average binding energy of protons in an electrically neutral nuclear matrix at zero gravitational potential.

E_{nb} differs from E_b, the average nuclear binding of atomic nuclei, because an electrically neutral matrix has a higher concentration of bound electrons than atomic nuclei. The relationship between E_{nb} and E_b can be expressed as:

$$E_{nb} = E_b - \frac{3E_{ec}}{8} \tag{16.40}$$

where E_{ec} is the amount of energy needed to compress an electron to nuclear dimensions, as derived earlier by Equation (12.11). In atomic nuclei there are, on average, 5 bound electrons for every 8 bound protons. This means a black hole's nuclear composition has an additional 3 bound electrons for every 8 protons, which reduces nuclear binding energy with electron compression energy. This is what the 3/8th term represents. Equation (12.11) indicates that the amount of energy required to compress an electron to nuclear dimensions is 2.8 MEV. Substituting this value into Equation (16.40) puts E_{nb} at 6 MEV. *The binding energy of a nuclear matrix is slightly lower when it is electrically neutral.*

The Milky Way's core represents a particle core expansion of ~6300x. The Strong force is purely a function of geometry and scales inversely with core radius, so a core expansion of 6300x reduces electrically neutral nuclear binding energy from 6 MEV to about 950 EV. In a typical nuclear environment ~7 MEV is required to control protons moving at near-relativistic speeds, but 950 EV provides more than adequate containment in the intense gravitational potential of a black hole, where protons' rest masses have been dramatically reduced. Also note that while the Strong veneer binder attenuates with increasing gravitational potential, the Coulomb veneer binder is unaffected by gravitation. It is a direct consequence of the unit polarvolume distribution, not particle core size.

The Strong veneer binder is only ~950 EV, and the Coulomb binder has to be less than this for a black hole's degenerate material to remain stable. This means that the Fermi level of a galactic core's protons is exceptionally close to their rest energy. Rewriting Equation (16.37):

$$E_F = m_p c^2 - E_{s\vee} + E_{q\vee} \tag{16.41}$$

Black holes are *gravitationally degenerate*. Nuclear binding energy is a small fraction of total nuclear potential, but the Strong veneer binder is an even smaller fraction of rest energy.

ADIABATIC FERMI-MAXFIELD DISTRIBUTION

Substituting the Fermi level of Equation (16.41) into Equation (16.36) yields the energy distribution of the protons in a gravitational veneer, as a function of temperature and binders:

$$n_{\text{F-M}}(E) = \left(e^{\left(\frac{E - m_p c^2 + E_{sv} - E_{qv}}{kT} \right)} + 1 \right)^{-1} \tag{16.42}$$

This expression works for a gravitational veneer, but does not accurately describe the state of protons deep in a black hole's interior, as it contains no factor to account for the tremendous pressure in this region. However, since a black hole communicates with the rest of the universe only through its veneer, the energy distribution of its central material has no tangible effect on cosmic equilibrium.

Equation (16.42) is nature's maximal expression of degeneracy, the *Fermi-Maxfield distribution*. It is shown below for the protons of an electrically neutral galactic veneer at temperatures of 0.2 MK (green) and 2 MK (orange):

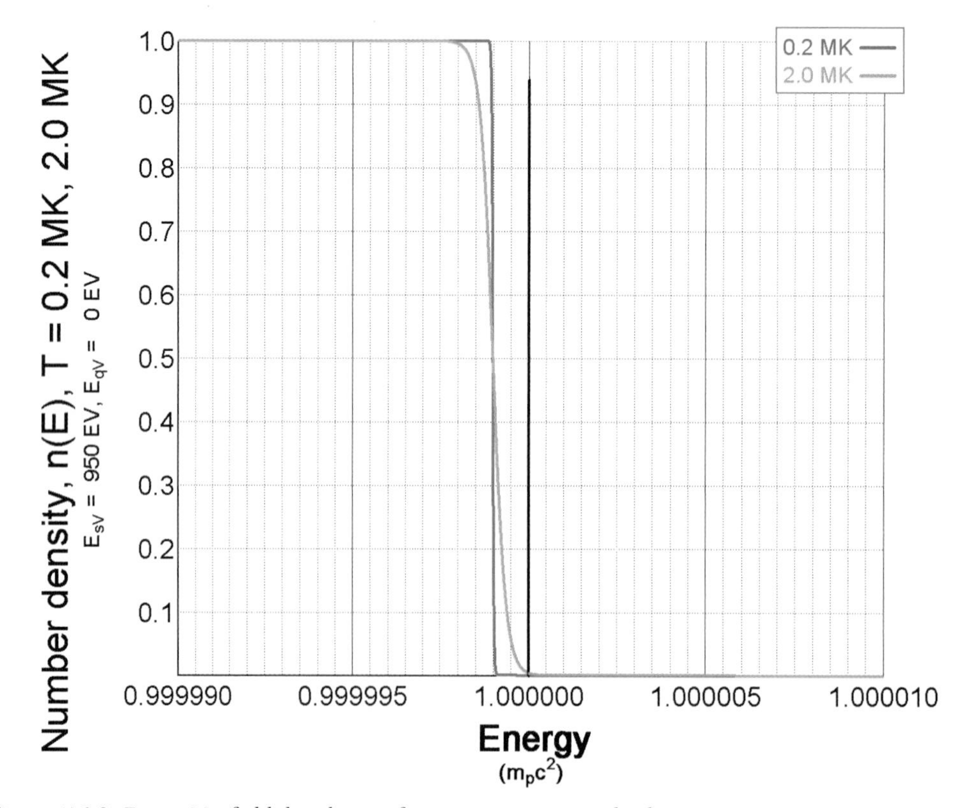

Figure (16.9) Fermi-Maxfield distribution for protons in a neutral galactic veneer at $T = 0.2$ MK, 2 MK

At a number density of 0.5, both traces coincide with the Fermi energy level, ~950 EV less than a proton's rest energy. Note that the trace for 2 MK has a far more conspicuous high-energy tail then the one for 0.2 MK. This is the way a galaxy's core regulates itself. If its temperature becomes too great, its high-energy tail sheds heat as relativistic protons.

The fraction of protons in the Fermi-Maxfield distribution greater than a veneer's escape energy is given by:

$$f_{\vee} = \left(\frac{1}{E_F}\right) \int_{m_p c^2}^{\infty} n_{\text{F-M}}(E)\,dE = \frac{\displaystyle\int_{m_p c^2}^{\infty} \left(e^{\left(\frac{E - m_p c^2 + E_{s\vee} - E_{q\vee}}{kT}\right)} + 1 \right)^{-1} dE}{m_p c^2 - E_{s\vee} + E_{q\vee}} \tag{16.43}$$

where $f_{\vee}$ will be called the *veneer ejection fraction*. Integrating yields:

$$f_{\vee} = \frac{\left(kT \ln\left(1 + e^{\left(\frac{E_{s\vee} - E_{q\vee}}{kT}\right)} \right) - E_{s\vee} + E_{q\vee} \right)}{m_p c^2 - E_{s\vee} + E_{q\vee}} \tag{16.44}$$

Equation (16.44)'s numerator is the correct analytic solution for the integral in Equation (16.43), but its unfortunate combination of logarithmic and exponential functions obliterates the accuracy of floating point arithmetic for certain combinations of temperature and binder values. Equation (16.43) should be integrated numerically for the best results in these cases. It indicates a fairly small veneer ejection fraction of $7.4(10)^{-10}$ for an electrically neutral core at T = 2 MK with a Strong binder of 950 EV. At 0.2 MK this fraction plummets to $2.2(10)^{-32}$. Let's use galactic dynamics to estimate the temperature of the Milky Way's veneer.

GALACTIC CORE FLUX

The Fermi-Maxfield distribution gives us the fraction of protons with enough energy to escape a galactic core, but they must also have a clear route. If they lie deep in its interior, their energy level is irrelevant. The total material flux of particles available at the surface of a massive black hole, introduced earlier as *veneer capacity*, is given by Equation (13.45):

$$C_{\vee} = \frac{\overline{\mathrm{V}} c^2}{8G}$$

In a predominantly degenerate environment, average energy is equal to the Fermi level because a Fermi distribution is symmetric about a number density of 0.5. Average velocity, however, is not the speed of a particle of average energy. Velocity does not vary linearly with energy in the Fermi or any other statistical distribution.

The average velocity of a fully degenerate Fermi-Maxfield distribution, which will be referred to as the *Fermi-Maxfield speed*, $v_{\text{F-M}}$, is given by the weighted average of relativistic kinetic energy:

$$v_{\text{F-M}} = \left(\frac{c}{m_0c^2}\right)\int_{E=0}^{m_0c^2}\sqrt{1-\left(\frac{1}{\left(\frac{E}{m_0c^2}+1\right)^2}\right)}\,dE = \left(\frac{c}{m_0c^2}\right)\int_{E=0}^{m_0c^2}\frac{\sqrt{\left(\frac{E}{m_0c^2}\right)^2+2\left(\frac{E}{m_0c^2}\right)}}{\left(\frac{E}{m_0c^2}+1\right)}\,dE \qquad (16.45)$$

$$= c\int_{x=0}^{1}\frac{\sqrt{x^2+2x}}{(x+1)}\,dx$$

Numerical evaluation of this integral yields 68.49% of the speed of light.

Thus:

$$v_{\text{F-M}} \cong 0.6849c \qquad (16.46)$$

This is a blistering speed, but negative field potential allows bound particles to have a low temperature even when moving close to the speed of light. Hydrogen's ground-state electron, for instance, moves at $2(10)^6$ m/s with its small negative Coulomb potential, even at a temperature near absolute zero. If it were moving this fast in an electron gas in free space, its temperature would be over 100,000 °K. *Bound states conceal kinetic energy*.

Combining Equation (16.46) with Equation (13.45) yields the material flux capacity of a galactic core's veneer:

$$C_v = \frac{v_{\text{F-M}}c^2}{8G} \qquad (16.47)$$

Capacity is independent of temperature until $(kT>>m_0c^2)$, where our degeneracy approximation breaks down. The calculations to follow are for cores in an equilibrium state with their host galaxy, so Equation (16.47) is appropriate, and we are to find that a galactic core's temperature is many orders of magnitude cooler than the minimum required to maintain its degenerate state.

The rate that protons are discharged from a galactic veneer is given by the fraction of veneer capacity with energy greater than a proton's rest mass. Applying Equation (16.44) to Equation (16.47) yields a galaxy's *veneer flux* in terms of temperature and binders:

$$\Delta_{\vee} = C_{\vee} f_{\vee} = \frac{\left(\frac{\mathrm{v}_{\text{F-M}} c^2}{8G}\right)\left(kT \ln\left(1 + e^{\left(\frac{E_{s\vee} - E_{q\vee}}{kT}\right)}\right) - E_{s\vee} + E_{q\vee}\right)}{m_p c^2 - E_{s\vee} + E_{q\vee}} \quad (16.48)$$

Proportional to Equation (16.44), this represents a flow of $2.6(10)^{25}$ kg/s in an electrically neutral core at 2 MK with a 950 EV Strong veneer binder. A lower temperature reduces this output precipitously, to 750 kg/s at 200 KK.

A core's thermal proton current, in amperes, is given by a slight modification of Equation (16.48):

$$i_p = \frac{\Delta_{\vee} q}{m_p} = \frac{\left(\frac{\mathrm{v}_{\text{F-M}} c^2 q}{8G m_p}\right)\left(kT \ln\left(1 + e^{\left(\frac{E_{s\vee} - E_{q\vee}}{kT}\right)}\right) - E_{s\vee} + E_{q\vee}\right)}{m_p c^2 - E_{s\vee} + E_{q\vee}} \quad (16.49)$$

Even at 2 MK, Equation (16.49) only amounts to $2.5(10)^{33}$ Amperes, tiny in comparison to a typical galactic current.

THERMAL CORE CURRENTS

As noted earlier, black holes are the superposition of two degenerate gases - protons and bound electrons. The preceding calculations have assumed a zero Coulomb binder, but this is not an accurate representation of a galactic core's charge. The disparity between proton and electron rest masses, coupled with the fact that electrons are not subject to the Strong interaction, means a galactic core's thermal electron flux will exceed its proton flux for any given temperature. *This in turn dictates a net positive charge.*

Ψ THEOREM 16.5 - GALACTIC CORE> ELECTRICAL CHARGE {Ψ16.4}

GALACTIC CORES HAVE A NET POSITIVE ELECTRICAL CHARGE

In an equilibrium state, the electrical charge of a galactic core is constant. This means its thermal proton flux equals its thermal electron flux. Whereas thermal proton flux is governed by the Strong and Coulomb potentials, *thermal electron flux is driven by the Weak and Coulomb potentials.*

Weak binding energy, E_{wb}, is the amount of energy expended to compress a free electron to nuclear size. Unlike nuclear binding energy, Weak binding energy is endothermic. Its simplest case is the neutron, which requires 0.78 MEV to compress its electron to a radius of ~1.7 F. Bound electrons in medium to heavy nuclei are smaller than this, comparable to the size of bound protons, at ~0.75 F. The Weak binding energy required for nuclear electrons is given by the electron compression energy of Equation (12.11), $E_{wb} = E_{ec}$. This amounts to ~2.8 MEV per bound electron.

The *Weak veneer binder* is Weak binding energy scaled by the residual rest energy of the particles on a galactic core's surface:

$$E_{w\vee} = \left(\frac{\Phi_{\vee}}{m_0 c^2} + 1\right) E_{wb} = \left(\frac{3R_p^3 c^6}{8\pi G^3 m_p M^2}\right)^{\frac{1}{3}} E_{wb} \quad \{M >> M_{sun}\} \quad (16.50)$$

where E_{wb} is the average Weak binding energy in heavy nuclei at zero gravitational potential, 2.8 MEV. The Weak binder's effect is opposite that of the Strong binder, and its value amounts to 440 EV for the Milky Way's core.

The Fermi level of a black hole's bound electron gas is given by:

$$E_F = m_e c^2 - E_{q\vee} + E_{w\vee} \quad (16.51)$$

where the Coulomb veneer binder changes sign from the Strong case since electrons are negatively charged. The Weak binder also changes sign as it represents an endothermic process.

A galactic core's electron flux can be calculated from the parity between the sizes of bound electrons and protons in a nuclear matrix. Even in a positively charged black hole, bound electrons have virtually the same number density as protons. As a Fermi-Maxfield distribution, their average energy is close to their rest mass, so their average speed is also comparable to protons. The key difference between the two degenerate gases is electrons are not subject to the Strong force, so their flux is governed by the Weak and Coulomb binders. A galactic core's thermal electron current is given by:

$$i_e = \frac{\left(\frac{v_{\text{F-M}} c^2 q}{8Gm_p}\right)\left(kT \ln\left(1 + e^{\left(\frac{E_{q\vee} - E_{w\vee}}{kT}\right)}\right) - E_{q\vee} + E_{w\vee}\right)}{m_e c^2 - E_{q\vee} + E_{w\vee}} \quad (16.52)$$

Even though Equation (16.52) is a thermal *electron* flux, m_p still governs its number density since protons represent most of a black hole's mass, which in turn defines its area and volume.

A galactic core's thermal electron and thermal proton currents are equal when:

$$\frac{\left(kT\ln\left(1+e^{\left(\frac{E_{sv}-E_{qv}}{kT}\right)}\right)-E_{sv}+E_{qv}\right)}{m_pc^2-E_{sv}+E_{qv}}=\frac{\left(kT\ln\left(1+e^{\left(\frac{E_{qv}-E_{wv}}{kT}\right)}\right)-E_{qv}+E_{wv}\right)}{m_ec^2-E_{qv}+E_{wv}} \qquad (16.53)$$

This balance requires the cooperation of all four universal forces, and it defines a one-to-one correspondence between the Coulomb veneer binder and temperature. When temperature goes to zero, the Coulomb binder goes to the limit $E_{qv} = (E_{wv} + E_{sv})/2$. Although current at low temperature is exceptionally small, it will eventually charge the veneer to an electrical binder intermediate between the Strong and Weak binders.

The relationship between veneer temperature and the Coulomb veneer binder is shown below:

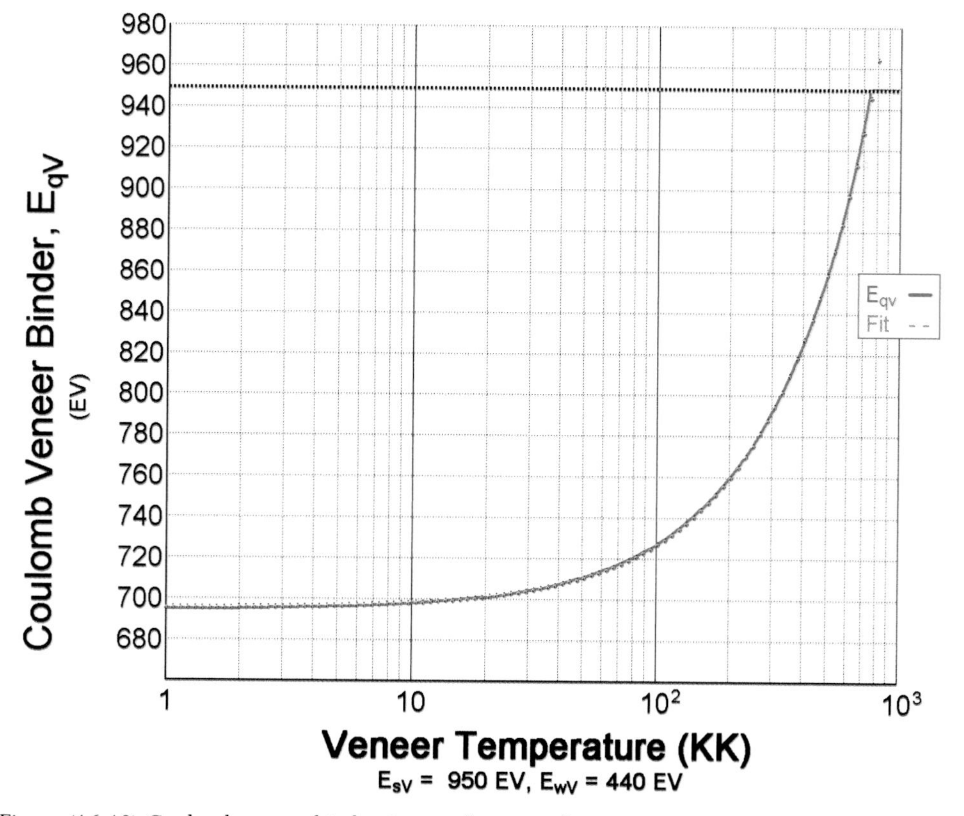

Figure (16.10) Coulomb veneer binder E_{qv} as a function of temperature, E_{sv}= 950 EV, E_{wv}= 440 EV

The greater the Coulomb binder (positive veneer charge), the greater the temperature where the thermal fluxes of electrons and protons are equal. The Coulomb binder's upper limit is the Strong binder because nuclear binding energy is required to keep a black hole's material in place. This is shown as the horizontal line at 950 EV. *It follows that, at least for the estimated binder values shown above, a galactic veneer's maximum temperature is less than 1,000,000 °K.* The Coulomb binder's lower limit, based on our estimates of the Strong and Weak binders, is shown as ~695 EV.

The violet fitting trace of Figure (16.10) is a power function of the form:

$$E_{q\vee(T)} = E_{q\vee(0)} + aT^b = \left(\frac{E_{s\vee} + E_{w\vee}}{2} \right) + 2.2(10)^{-4}\, T^{1.03} \tag{16.54}$$

where the Coulomb binder goes to the average of the other binders at low temperature. Since the masses of electrons and protons are so dissimilar, the only way their thermal fluxes can balance in a galaxy's core is if it is positively charged. *A galactic core is a gigantic celestial cathode.*

VENEER TEMPERATURE

A galactic core's primary function is to undo billions of years of fusion in a galactic disk. In order to accomplish this, the entire disk must pass through the core. Galactic core inflow is therefore equal to the vortical inflow of Equation (16.13):

$$\Delta_{g_c} = \frac{dM}{dt} = \frac{Q_g L_g}{f_{cn} \varepsilon_U c^2} \tag{16.55}$$

At equilibrium, the rate that material is captured matches the rate it escapes:

$$\Delta_{g_c} = \Delta_{\vee} \tag{16.56}$$

Substitute Equations (16.48) and (16.55) into Equation (16.56), note the temperature dependence of the Coulomb veneer binder, and solve for galactic luminosity:

$$L_g = \frac{\left(\dfrac{f_{cn} \varepsilon_U \mathrm{v}_{\text{F-M}} c^4}{8GQ_g} \right) \left(kT \ln\left(1 + e^{\left(\frac{E_{s\vee} - E_{q\vee(T)}}{kT} \right)} \right) - E_{s\vee} + E_{q\vee(T)} \right)}{m_p c^2 - E_{s\vee} + E_{q\vee(T)}} \tag{16.57}$$

A galaxy's vortical inflow is driven by its luminosity and ability to deliver compound nuclei to its core. Once there, the composition of material entering the core is balanced against the rate of hydrogen leaving it.

Rewrite Equation (16.57), isolating efficiency as a function of temperature:

$$Q_g = \frac{\left(\frac{f_{cn}\varepsilon_U \mathrm{v}_{\text{F-M}}c^4}{8GL_g}\right)\left(kT\ln\left(1+e^{\left(\frac{E_{s\vee}-E_{q\vee(T)}}{kT}\right)}\right)-E_{s\vee}+E_{q\vee(T)}\right)}{m_pc^2-E_{s\vee}+E_{q\vee(T)}} \quad (16.58)$$

Using an estimated luminous output for the Milky Way of $1.4(10)^{37}$ W,[1.14] a Strong veneer binder ($E_{s\vee}$) of 950 EV, a Weak veneer binder ($E_{w\vee}$) of 440 EV, and a Coulomb veneer binder defined by Equation (16.54), galactic efficiency is shown below as a function of veneer temperature for three different inflow compositions:

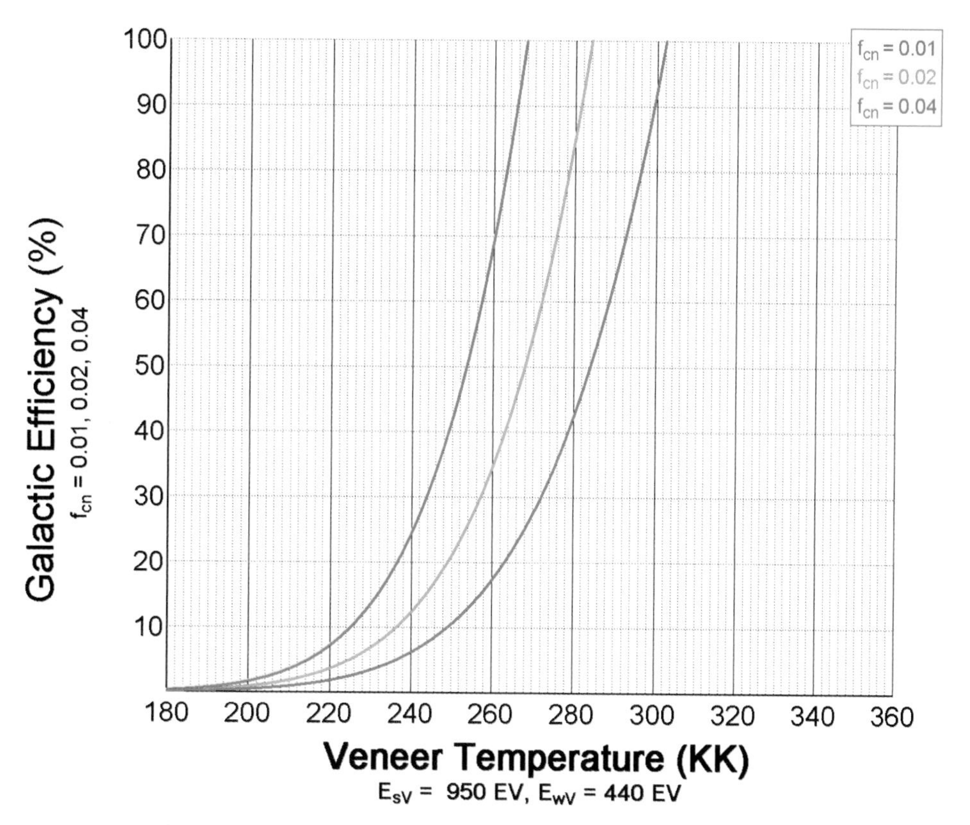

Figure (16.11) Galactic efficiency versus veneer temperature for core inflows of f_{cn} = 0.01, 0.02, 0.04

Compound nuclear fraction decreases from left to right. The green trace is 4% compound nuclei, 96% hydrogen. The red trace is 1% compound nuclei, 99% hydrogen. Both require over 250,000 °K to achieve nuclear disassociation at any appreciable efficiency. Even at an efficiency of only 10%, the temperature for any conceivable inflow composition is in excess of 220,000 °K. *The Milky Way's veneer is hot, but not blazingly hot.* Stellar cores are orders of magnitude hotter, consistent with the idea that heat is the agent of nuclear condensation while gravitational expansion drives nuclear dissociation.

The surface of the Milky Way's core is over 250,000 °K, yet with its enormous redshift, is virtually invisible in its violent ambient environment:

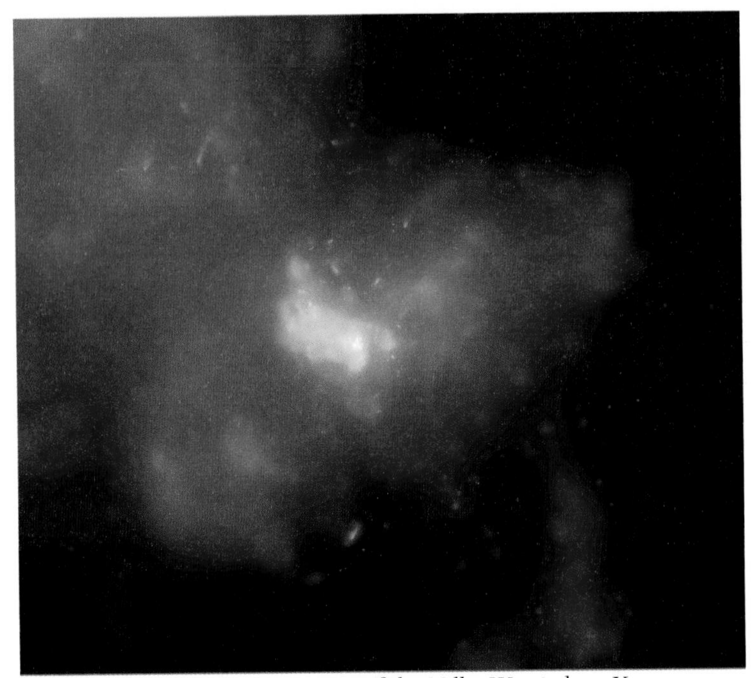

Figure (16.12) Core region of the Milky Way in long X-ray
(Courtesy NASA/CXC/MIT/F.K. Baganoff et al.)

Compound nuclear production is governed purely by a galaxy's luminosity, but the temperature required for nuclear dissolution is controlled by the material composition flowing into a galaxy's core. Since the compound nuclei a galaxy sends to its core are diluted by a high concentration of hydrogen, the only way it can keep pace with its own fusion is to move more material, and the only way to accomplish this is with a higher veneer temperature. That said, a galactic veneer's temperature response to input composition is fairly mild. It takes 268 KK to process a compound fraction of 0.04 at 100% efficiency. Yet a compound fraction twice as small, at 0.02, only raises veneer temperature 7% to 284 KK.

A galactic veneer's temperature is far more sensitive to binder values than compound nuclear fraction. Suppose, as a limiting case, that (a) the Weak binder has no effect on thermal electron migration and (b) protons' intrinsic binding energy is not reduced by the additional bound electrons in a neutral nuclear matrix. In this extreme scenario, veneer binders would be (E_{sv} = 1100 EV) and (E_{wv} = 0 EV), resulting in a core temperature of 0.57

MK for a 2% compound fraction (f_{cn} = 0.02). Hence even with a radically different interpretation of veneer binders, veneer temperature remains well below 1 MK.

Given the close agreement between Equation (16.15)'s radial inflow speed and the radial inflow speeds derived from white dwarf age and galactic disk scale length, Equation (16.15) represents a fairly accurate assessment of our galaxy's vortical flow. It indicates that the Milky Way's core processes a compound nuclear concentration of approximately 2%. This, using the constraints on which Figure (16.11) is based, corresponds to a core temperature of ~280 KK.

A galactic core's luminous power is given by Equation (13.42):

$$L_{\vee} \cong 8(10)^{-40}\left(T^{4}M^{\frac{4}{3}}\right) \qquad \{M >> M_{sun}\}$$

In the case of the Milky Way, a veneer temperature of 280 KK generates a veneer luminosity of $6(10)^{31}$ W, or ~4 ppm of its entire output. Equation (13.43) gives us its peak emission wavelength:

$$\lambda_{\vee} = \left(\frac{z_{\vee}}{4.9651}\right)\left(\frac{hc}{kT}\right)$$

and Equation (13.40) provides its veneer redshift:

$$z_{\vee} \cong 1.9(10)^{-21} M^{\frac{2}{3}} \qquad \{M >> M_{sun}\}$$

Assuming our galaxy is 100% efficient and its inflow is 2% compound nuclei, its core ought to appear as a ~$6(10)^{31}$ W point source of heavily broadened infrared radiation with a peak energy density centered near 0.06 mm. Finding this in the midst of the cataclysmic firestorm near our galaxy's center is a daunting but interesting proposition. For additional analysis of the galactic core environment please refer to Appendix O.

16.8 HYDROGEN EMISSIONS

Measurements of 21 cm radiation from our galaxy's central region indicate the movement of massive volumes of hydrogen gas.[6.23] It flows outward along our galaxy's arms and is ejected from its core region perpendicular to the galactic plane, toward its north and south poles. Under the influence of gravitational and magnetic fields, newly minted hydrogen rains down on the Milky Way's disk from both sides. The *galactic fountain* model supported by a number of astrophysicists maintains that the source of this hydrogen is supernovas.[6.18] This is incorrect. Our galaxy's core generates all of it. *It is the unmistakable and inevitable conclusion of the cosmic proton path.*

A galaxy's hydrogen production isn't always a gradual flow. Violent hydrogen jets have been observed escaping from the cores of a number of galaxies, such as the 5000 ly geyser leaving M87:

Figure (16.13) Hydrogen jet, over 5000 light years in length, emanating from the core of M87
(Courtesy NASA/Hubble Heritage Project)

This could hardly be the result of supernova activity. Furthermore, radio waves from active galactic sources have been linked to bilateral lobes of hydrogen ejected from their cores.[6.13] Galactic cores flood space with hydrogen because that is their principal function in the maintenance of cosmic equilibrium.

The total flow of material through the Milky Way's vortex, as estimated earlier using Equation (16.13):

$$\frac{dM}{dt} = \left(\frac{dn}{dt}\right)\frac{1}{f_{cn}} = \frac{Q_g L_g}{f_{cn}\varepsilon_U c^2}$$

only amounts to about 18 M_{sun}/yr at 100% efficiency and a core inflow of 2% compound nuclei. So while hydrogen emission is clear from both an observational and theoretical standpoint, what is not as clear is where it (and other material) goes once it is released. A fraction of the hydrogen dispersed throughout the Milky Way's center moves at speeds in excess of 700 km/s,⟨6.14⟩ a speed comparable to the galactic escape velocity in this region.⟨1.18⟩ Scientists remain locked in a debate over whether this hydrogen is falling in or blasting out. Hopefully the galactic dynamics presented in this chapter will help resolve this standoff.

Violent and pronounced hydrogen emissions routinely occur in galaxies with Active Galactic Nuclei (AGN).⟨6.13⟩ This is because their hyperactivity exaggerates their core's normal hydrogen production:

Figure (16.14) Active galactic nucleus of NGC 7742
(Courtesy NASA/Hubble Heritage Project)

AGN have spectral signatures surprisingly similar to QSOs (quasars). A discussion of these curious objects and their possible relationship to galactic cores is included in Appendix Q.

16.9 GENERAL CONCLUSIONS

Most of this chapter's conclusions leave virtually no room for alternatives, either theoretical or observational. The form and function of the universe's galactic cogs are inextricably linked and wholly unambiguous.

- Galaxies are vortices. They carry the material of their disk region slowly into their core where it is absorbed into a gravitationally-expanded neutron superfluid that exudes hydrogen through its degenerate surface. This provides a renewable source of fuel for the perpetual cosmic engine.

- All true galaxies have massive, electrically charged black holes in the innermost depths of their central regions. Black holes are a necessary and integral component of galactic function because they are the only objects in the universe with a gravitational potential large enough to provide an environment capable of low-temperature nuclear disassociation with virtually no radiative energy loss. The cosmic fusion cycle is not possible without them.

- The reason why the universe appears 12 – 18 billion years old is because this is the average time required for its material to cycle through its galactic systems. The universe is infinitely older than this, but the compound nuclei of which its luminous material is composed are continuously renewed every 12 – 18 billion years.

- Assuming the Milky Way has (a) a galactic efficiency of 100%, (b) a Strong veneer binder of 950 EV, (c) a Weak veneer binder of 440 EV, and (d) a core inflow of 2% compound nuclei, its core temperature is approximately 280,000 °K. This would give it a luminosity of $6(10)^{31}$ W and a radiative peak wavelength near 0.06 mm in the far infrared.

- The galactic cores of stable galaxies expel newly generated hydrogen at a rate comparable to the pace it is consumed by fusion throughout their disk and central regions.

For the last sixty years the Big Bang has obscured the truth. It violates energy conservation, and in so doing leaves reality quickly and irrevocably behind. In stark contrast, *Null Physics'* unwavering conformance to energy conservation has revealed galactic functionality far beyond the limited reach of the Big Bang or similar theories.

Modern cosmology's options are straightforward:

a) Continue to support the Big Bang, an old, inarticulate theory that doesn't even begin to explain the motion profiles, disk scale lengths, banding, hydrogen jets, and massive cores of galaxies or the ultimate source of the universe and its laws; or

b) Switch to *Null Physics*, where the motion profiles, disk scale lengths, banding, hydrogen jets, and massive cores of galaxies are all integral and necessary components of a quantifiable equilibrium system.

The choice is simple, the Big Bang's incompleteness or the science and rationality of *Null Physics*. Theorists need to embrace the concept of cosmology as a true science, bound by all of the same rules other physical sciences adhere to, in particular energy conservation. Why not start now? An exciting and undiscovered universe awaits us.

Conclusion

Opinion is a flitting thing
But truth, outlasts the Sun-
If then we cannot own them both-
Possess the oldest one-

Emily Dickinson, 1830-1886

Part I derived our universe from nothingness and connected the infinitely small and large, leaving us suspended in the balance. It provided unit hypervolume - the placeholder of finiteness, and the necessity of three dimensional space, all directly from geometry. Part II began by debunking quantum reality, then built space and energy from nothing. Using only Planck's constant, it gave us a value for unit hypervolume directly from a photon's topology. It then revealed unit polarvolume as the quantizing agent for matter. Part III unveiled the particle core, leading inexorably to all four universal forces. Matter's ultimate application, the black hole, emerged as a seamless amalgamation of these forces. Part IV presented an eternal celestial equilibrium, showing how all of the key players, from cosmic microwaves to galactic cores, fit together inside the universal engine.

There is a wealth of support, both empirical and theoretical, for the revolutionary discoveries presented in *Null Physics*. But even as ground-breaking as many of these concepts might be, they, individually or collectively, are not this book's most central message. It is twofold. First, and most importantly, is the fact that our phenomenological universe rests atop a single, perfect, underlying reality. The notion that some new physical theory will always come along to replace an older one is only true until such time as a theory actually connects to reality's invariant bedrock. To say that all theories are passing fads is to say either (a) reality has no immutable foundation or (b) we will never be able to reach it. Both of these statements are grossly overreaching. They taint physical science with an air of ambivalence, leaving everyone convinced that no great, incontrovertible truths remain undiscovered. In light of this supreme arrogance, theories become a mediocre reshuffling of old ideas, not the quest for the deepest perception, and the beauty of *why* is ground to bits by the mediocrity of consensus. *Null Physics* shows us a brighter path.

This book's second message is more subtle, but just as powerful. Einstein once said that imagination is more important than knowledge. This much is true, but the more global reality is that *reason* outshines both. There is no problem, large or small, that will not eventually succumb to a focused analysis. If your theory runs aground *conceptually*, back up, and keep backing up until the wrong turn presents itself. Never settle for an incomplete solution, because an incomplete solution is invariably an incorrect solution. The true test of a theory isn't whether or not it fits the experimental data. Any construct, given enough constants and other artificial machinations, can be made to fit. But does it also make sense? This is the acid test, and is a far greater challenge than what amounts to little more than organized numerology. Empiricism is an indispensable tool, but reasoning can penetrate anything, even a particle's core boundary.

Scientific progress all comes down to conceptual *leverage*, the most eloquent application of Ockham's razor. Nature's existential substrate supports everything we observe. This is why its fundamental truths touch so many seemingly disparate phenomena. The concept of the atom, for instance, explains legions of chemical and physical interactions. Although the details of an atom's internal dynamics still need to be worked out, the idea that our material surroundings are composed of atomic subunits has required no revision since it was firmly established. Isotopes might have different masses, but no atoms are more *atomic* than other atoms, and no theory will ever come along to reveal that a carbon atom is only an approximation of something else. In much the same way, the concept of galaxy as vortex explains virtually all (previously mysterious) galactic properties, and is not subject to future revisionism. Either galaxies are vortices or they are not, and as it turns out, they are. And the particle core accomplishes for particle, nuclear, and atomic physics what the galactic vortex does for cosmology and astrophysics. Both are essential pieces of the universal puzzle, and both, most importantly, tell us *why* certain things are the way they are. The intellectual satisfaction of seeing a concept click into place is a tangible reaffirmation of the survival skill that has been most abused by the march of science - common sense.

Null Physics is the barest beginning, and colossal truths remain hidden throughout nature. From the basic, like a detailed understanding of photon emission and absorption, to the sublime, like a derivation of the proton/electron mass ratio. Even with all of the questions that this book has answered, we still don't know something as simple as how a moving particle stores kinetic energy. *But we can learn.* The universe is real, and so is everything within it. So many discoveries await us, yet without the implicit realization that the tiniest pieces of reality have a genuine, enduring essence, none of these intellectual treasures are accessible, let alone comprehensible. Seeing where the next dot falls on a graph *is not a theory*. We've been measuring things for hundreds of years. *Null Physics* is the first chance we have to put it all together. But until now we didn't even know what energy was, so let's get to work.

APPENDIXES

Predictions, supporting evidence, and works in progress

- *Predictions*
- *Tables of parameters and constants*
- *Universal variability*
- *Lorentz transform*
- *Generalized particle field and core dynamics*
- *The superluminal criterion*
- *Particle field equations - reference*
- *Deep space photons, annihilation and fusion*
- *White dwarf history*
- *Annihilation cycle*
- *Energy and surface brightness loss in a redshifted blackbody spectrum*
- *Neutrinos and dark matter*
- *Galactic core - power loss and thermal currents*
- *Material flux*
- *Quasi-stellar objects*

Reality is that which, when you stop believing in it,
doesn't go away.

Philip Dick, 1928-1982

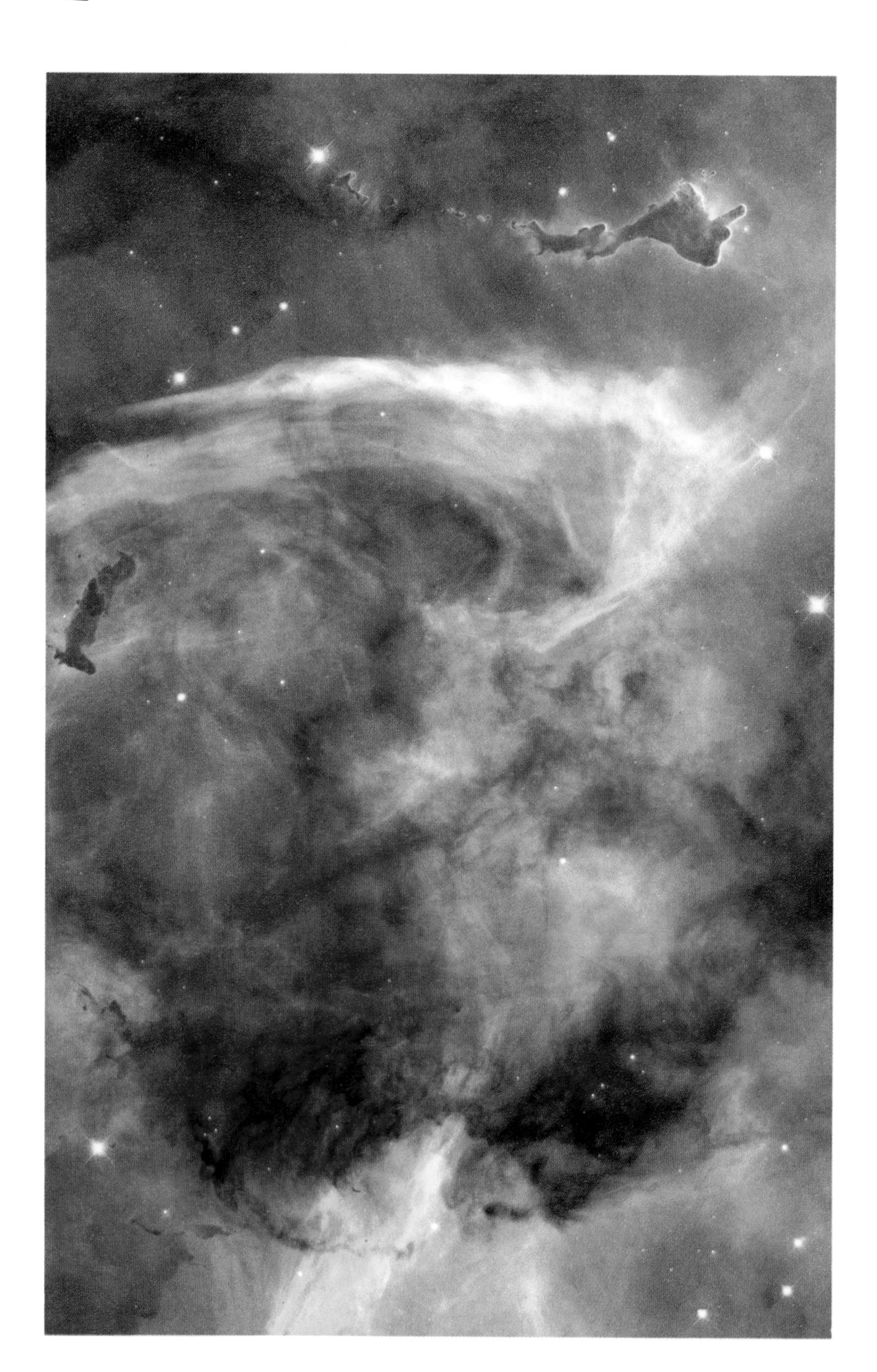

Figure (V) Keyhole nebula
(Courtesy NASA/Hubble Heritage Project)

A. PREDICTIONS

As progressively more powerful telescopes go online, reaching further and further into space, they will find:

- An exponential relationship between energy loss and distance in intergalactic redshift. The JDEM (Joint Dark Energy Mission), as currently proposed for ~2012, should have enough range and resolution to validate lumetic decay.

- The Milky Way's core - a massive black hole with a radiant output of $\sim 6(10)^{31}$ W, peaking in the infrared near ~0.06 mm.

- Galaxies at extreme distances that contain stellar populations similar to local, normal galaxies.

- An increasing number density of local, dim stars and brown dwarfs.

- Extragalactic radiation with a wavelength near 1700 m, corresponding to the lumetic decay of 21 cm radio waves.

If a careful survey of the Milky Way's disk dynamics is performed, it will show that:

- The Milky Way's stars and disk material have an average inward motion of ~1.5 km/s toward its galactic core.

- The total amount of hydrogen flowing outward from the Milky Way's galactic core region is comparable to the amount that all of its stars consume in nuclear fusion.

On the frontier of particle physics:

- Scattering experiments will demonstrate that the inter-nucleon spacing (effective range between nucleons) of the bound state of ^{3}He is 1.639 F.

- Evidence will continue to amass supporting free-space core radii for protons at 0.946 Fermi and electrons at 1740 Fermi.

- The study of refraction in low-density gases will validate the energy density distribution of the particle field.

B. UNIVERSAL AND PHYSICAL CONSTANTS

The following table lists the values of universal and physical constants used in *Null Physics*. They are taken directly from or derived from values found in the CRC Handbook of Chemistry & Physics, 81st edition.[5.2]

Universal and Physical Constants				
	Name	**Value**	**Units**	**Δ(ppm)**
$\diamondsuit_4$	Unit hypervolume	$9.4780168439(10)^{-18}$	$J\text{-}m^2/s$	0.078
$\diamondsuit_q$	Unit polarvolume	$1.5807630564(10)^{-26}$	J-m	0.078
G	Gravitational Constant	$6.673(10)^{-11}$	$m^3/(kg\text{-}s^2)$	1500
q	Elementary Charge	$1.602176462(10)^{-19}$	C	0.039
c	Speed of Light	299,792,458	m/s	Exact
$v_{F\text{-}M}$	Fermi-Maxfield Speed	$0.684853c$	m/s	2
h	Planck's Constant	$6.6260687652(10)^{-34}$	J-s	0.078
k	Boltzmann Constant	$1.380650324(10)^{-23}$	J/K	1.7
σ	Stefan-Boltzmann Constant	$5.67040040(10)^{-8}$	$W/(m^2K^4)$	7
ε_0	Permittivity of vacuum	$8.854187187(10)^{-12}$	F/m	Exact
μ_0	Permeability of vacuum	$4\pi(10)^{-7}$	N/A^2	Exact
m_e	Electron Rest Mass	$9.1093816672(10)^{-31}$	kg	0.08
R_e	Electron Core Radius	$1737.7167307(10)^{-15}$	m	0.08
m_p	Proton Rest Mass	$1.6726215813(10)^{-27}$	kg	0.08
R_p	Proton Core Radius	$0.94639009245(10)^{-15}$	m	0.08
m_n	Neutron Mass	$1.6749271613(10)^{-27}$	kg	0.08
m_d	Deuteron Mass	$3.34358309(10)^{-27}$	kg	0.08

Table (B.1) Universal and physical constants

C. UNIVERSAL PARAMETERS AND ASTROPHYSICAL CONSTANTS

The following tables list the parameters and constants used throughout Null Cosmology.

Universal Parameters				
Name		Value	Units	Error(%)
H_0	Hubble Constant	60	Hz-km/Mpc	50⟨1.2⟩
H_0	Hubble Constant	$1.95(10)^{-18}$	Hz	50⟨1.2⟩
ρ_U	Universal Energy Density	$4.0(10)^{-10}$	J/m^3	40(Chapter 15)
ρ_M	Universal Mass Density	$4.5(10)^{-27}$	kg/m^3	40(Chapter 15)
j_B	Optical Luminosity Density	$1.4(10)^{-33}$	W/m^3	80⟨1.12⟩
j_I	Infrared Luminosity Density	$0.7(10)^{-33}$	W/m^3	80⟨1.13⟩
j_ν	Neutrino Luminosity Density	$<5.6(10)^{-35}$	W/m^3	80(4% of optical) ⟨2.3⟩
j_R	Total Luminosity Density	$2.1(10)^{-33}$	W/m^3	80⟨1.12⟩⟨1.13⟩
$\rho_{\gamma\gamma}$	Gamma/Xray Energy Density	$1.5(10)^{-17}$	J/m^3	80⟨2.1⟩
$\rho_{B\gamma}$	Optical Energy Density	$1.6(10)^{-15}$	J/m^3	80⟨2.1⟩
$\rho_{I\gamma}$	Infrared Energy Density	$1.6(10)^{-14}$	J/m^3	80⟨2.1⟩
ρ_{CMB}	CMB Energy Density	$4.165(10)^{-14}$	J/m^3	0.05⟨2.2⟩
$\rho_{K\gamma}$	Radio Wave Energy Density	$1.6(10)^{-20}$	J/m^3	80⟨2.1⟩
ρ_ν	Neutrino Energy Density	$<6.4(10)^{-17}$	J/m^3	80(4% of optical) ⟨2.3⟩
ρ_R	Total Radiant Density	$5.9(10)^{-14}$	J/m^3	5⟨2.1⟩⟨2.2⟩

Table (C.1) Universal Parameters

Astrophysical Constants				
Name		Value	Units	Error(%)
Ly	Light Year	$9.46(10)^{15}$	m	
pc	Parsec	3.26	Ly	
M_{sun}	Solar Mass	$1.9891(10)^{30}$	kg	
L_{sun}	Solar Luminosity	$3.845(10)^{26}$	W	0.2⟨1.20⟩
R_{sun}	Solar Radius	$6.9599(10)^{8}$	m	
M_g	Galactic Mass ($R \leq 114$ Kly)	$8(10)^{41}$	kg	20⟨1.16⟩
L_g	Galactic Luminosity	$1.4(10)^{37}$	W	20⟨1.14⟩
v_c	Galactic Circular Speed (R_O=28 Kly)	220	km/s	5⟨3.4⟩

Table (C.2) Astrophysical Constants

D. UNIVERSAL VARIABILITY

Any type of object has two kinds of universal variability:

Causal variability, σ_T. Also referred to as *temporal variability*. This arrives from the cause and effect along an object's lifespan - the natural consequence of aging. The difference between Earth in 1957 and Earth in 2957 is an example of causal variability. Its total magnitude is the number of instances in an object's lifespan (instance number):

$$\sigma_T = N_\tau = \tau \infty \qquad \text{(D.1)}$$

where τ is in absolute seconds.

Permutational variability, σ_P. This is the compositional variability within a type of object. Mars and Earth vary permutationally because they are both examples of planets and Mars was never a moment of Earth's history. Since all variability can be delineated into one of two categories, permutational variability can also be called *non-causal variability*. It is the total number of independent lifespans for a given type of object. Two lifespans are defined as *independent* if they do not share a single instance. Permutational variability is given by the ratio of total universal variability and causal variability:

$$\sigma_P = \frac{\rho_N \infty^3}{\tau \infty} = \frac{\rho_N \infty^2}{\tau} \qquad \text{(D.2)}$$

Space and time maintain a symmetric balance on all levels. Galaxies have far fewer independent permutations than planets because they have lower spatial density and longer individual histories. All of the histories within the universe are inextricably connected, but Equations (D.1) and (D.2) show the general relationship among the portions of the universe with a direct causal relationship and those without.

Think of all of the universe's planets as a set of lifespans, built of planetary instances causally linked to each other throughout infinite space. Each lifespan has an average length of ($\tau \infty$) planetary instances. These are strings of planets that are causally connected across the entire universe. Earth history is one of these lifespans. It contains Earth in 1969, Earth in 2845, as well as Earth in 82,335,586 BC. The total number of *separate* planetary lifespans is given by the permutational variability of planets.

The ratio of an object's causal σ_T and permutational σ_P variability will be called its *temporal convergence* and can be expressed in terms of its lifespan and volume density as:

$$\Im_\sigma = \frac{\sigma_T}{\sigma_P} = \frac{\tau\infty}{\left(\frac{\rho_N \infty^2}{\tau}\right)} = \frac{\tau^2}{\rho_N \infty} \tag{D.3}$$

where τ is in absolute seconds and ρ_N is volumetric number density in objects/m_a^3. This is referred to as *convergence* because as an object's size and lifespan increase, the universal fraction of its instances directly related by causality expands markedly. Temporal convergence is the relationship between the large-scale spatial and dynamical properties of time.

Ψ THEOREM D.1 - TEMPORAL CONVERGENCE, $\Im_\sigma$ {Ψ4.8}

THE RATIO OF THE CAUSAL AND PERMUTATIONAL VARIABILITY OF ANY OBJECT IS EQUAL TO $\tau^2/(\rho_N\infty)$

Temporal convergence is valid for omnielements as well as finite objects. Substitute ($\tau=\infty$) and ($\rho_N=1/\infty$) into Equation (D.3):

$$\Im_{\sigma_\Xi} = \infty^2 \tag{D.4}$$

This means omnielements are ∞^2 times more likely to be related by causality than permutation. Since the omnipattern contains a total of ∞^2 omnielements, *their variational distribution is governed entirely by causality*. This is consistent with their direct relationship to ultrastasis.

The *relative temporal convergence* of two objects is given by the finite ratio of their temporal convergences:

$$Q_\Im = \frac{\Im_1}{\Im_2} = \left(\frac{\tau_1}{\tau_2}\right)^2 \frac{\rho_{N_2}}{\rho_{N_1}} \tag{D.5}$$

Compare the Earth and the Milky Way. If the number density of planets is on the order of a trillion times as large as the number density of galaxies, while the average planetary history is perhaps a twentieth of a galaxy's duration, a galaxy's relative temporal convergence is on the order of four hundred trillion times as large as that of a planet. *This is how much more often galaxies are causally related to each other than planets.*

E. LORENTZ TRANSFORM AND ABSOLUTE SPACE

Special Relativity is based on the Lorentz transform between space and time and has been hugely successful in describing high-speed phenomena. A close review of the assumptions required by this transform allows us to identify where it departs from absolute space.

Special Relativity's key premise can be stated: *The speed of light is equal to c in any direction of a uniformly moving coordinate system.*

where *uniform motion* means constant velocity, the absence of acceleration. The relationship between two coordinate systems $\boldsymbol{K}(x, y, z, t)$ and $\boldsymbol{K}'(x', y', z', t')$ moving uniformly relative to each other is called the Lorentz transformation. It is derived using the constancy of the speed of light and a few other reasonable assumptions. The derivation begins with light traveling down the positive and negative directions of the x axis and the $\boldsymbol{K}'$ reference frame moving along the x axis relative to $\boldsymbol{K}$.

Assumption 1: Constancy of light. Light moves at c in both reference frames. The position of the light moving along the positive x axis at any time t is given by $(x - ct = 0)$; the position of light moving along the negative x axis is given by $(x + ct = 0)$. The same is true in the $\boldsymbol{K}'$ reference frame because light moves at c there also, so $(x' - ct' = 0)$ and $(x' + ct' = 0)$.

Assumption 2: Independent direction. Motions down different directions of the same x axis are mathematically independent of each other.

Assumption 3: Linearity. The relationship between the reference frames is linear. The solutions to $(x - ct = 0)$ are related to $(x' - ct' = 0)$ by a constant:

$$a = \frac{(x' - ct')}{(x - ct)} \tag{E.1}$$

as are $(x + ct = 0)$ and $(x' + ct' = 0)$:

$$b = \frac{(x' + ct')}{(x + ct)} \tag{E.2}$$

Assumption 4: Equivalence. This is the fourth and final assumption. A distance of unity in ***K′*** is equal to a distance of unity in ***K***.

To begin the derivation, solve Equations (E.1) and (E.2) for x' and t' in terms of x and t:

$$x' = \left(\frac{a+b}{2}\right)x - c\left(\frac{a-b}{2}\right)t \tag{E.3}$$

$$t' = \left(\frac{a+b}{2}\right)t - \left(\frac{1}{c}\right)\left(\frac{a-b}{2}\right)x \tag{E.4}$$

The constant motion of ***K*** at the origin of ***K′*** is given by Equation (E.3) with ($x' = 0$):

$$\frac{x}{t} = \mathrm{v} = c\left(\frac{a-b}{a+b}\right) \tag{E.5}$$

Since the coordinate systems have uniform motion, this is the motion of ***K*** relative to any fixed point x'. *This is the velocity of **K** with respect to **K′**.* Similarly, the motion of ***K′*** at the origin of ***K*** is given by Equation (E.3) and (E.4) with ($x = 0$):

$$\frac{x'}{t'} = \mathrm{v}' = -c\left(\frac{a-b}{a+b}\right) = -\mathrm{v} \tag{E.6}$$

Note it is opposite the motion of ***K*** with respect to ***K′***. At any time (t, t') or location (x, x'), the relative motion of the reference frames is constant. While this does not prove the linearity assumption, the relative motion it provides is certainly consistent.

Next, instead of looking at the motion at fixed points, time will be fixed so distances can be compared. At $t = 0$ Equation (E.3) scales x in terms of x' as:

$$x = \left(\frac{2}{a+b}\right)x' \tag{E.7}$$

At $t' = 0$, Equations (E.3) and (E.4) scale x' in terms of x [solve for t in Equation (E.4) and substitute into Equation (E.3)]:

$$x' = \left(\frac{a+b}{2}\right)\left(1 - \left(\frac{a-b}{a+b}\right)^2\right)x \tag{E.8}$$

Equivalence requires that a distance of unity in each coordinate system appear the same size in the other, so set Equations (E.7) and (E.8) equal and solve:

$$1=\left(\frac{a+b}{2}\right)^2\left(1-\left(\frac{a-b}{a+b}\right)^2\right) \tag{E.9}$$

This gives us one equation in two unknowns. Note the same expression results from the use of equivalent times instead of equivalent distances.

Velocity is the relationship between the $(a + b)$ and $(a - b)$ terms, so Equation (E.9) can be written in terms of relative motion with the help of Equation (E.5), yielding two identities of a familiar form:

$$\left(\frac{a+b}{2}\right)=\frac{1}{\sqrt{1-\frac{\mathrm{v}^2}{c^2}}} \tag{E.10}$$

$$\left(\frac{a-b}{2}\right)=\frac{\left(\frac{\mathrm{v}}{c}\right)}{\sqrt{1-\frac{\mathrm{v}^2}{c^2}}} \tag{E.11}$$

Substituting these into Equations (E.3) and (E.4) yields the Lorentz transformations for distance and time:

$$x'=(x-\mathrm{v}t)\frac{1}{\sqrt{1-\frac{\mathrm{v}^2}{c^2}}} \tag{E.12}$$

$$t'=\left(t-\frac{\mathrm{v}x}{c^2}\right)\frac{1}{\sqrt{1-\frac{\mathrm{v}^2}{c^2}}} \tag{E.13}$$

The assumption where the Lorentz derivation left absolute space behind is #4, Equivalence. It is a necessary and justifiable assumption given that spatial measurements are limited to the third dimension. *The space we are able to observe is not a vector quantity*. However, neither this assumption nor the successful theory that results in any way detracts from the reality of absolute space. Relativity is successful because it approximates a moving reference frame as a scalar field subject to the limitations of the speed of light. It makes no statements about the physical essence of this reference frame. And it fails to reveal why the speed of light is constant. Why do photons move at this maximal speed and why does it govern the passage of time? Relativity provides computational capability but does not give us access to the underlying reality, much like the quantum statistical interpretation of reality that Einstein fought so strenuously to overturn. There is a reason the speed of light is the universe's maximum speed, just as there is a reason the internal change of a system approaching this speed slows to zero. They can be described in terms of the relative motion of the observer but this has nothing to do with why the universe is the way it is. In short, the Lorentz transformation accurately describes how space and time are *measured*. It tells us nothing, however, about their fundamental nature.

ABSOLUTE SPACE

The speed of light is not a universal constant; it is the linear, dimensional relationship between space and time. There is only one such relationship, and as such, there is only one speed for the propagation of energy through space. Objects moving at speeds less than c are amalgamations of static (particle cores) and dynamic (photon/kinetic) energy. The observed "time dilation" in objects moving through absolute space is solely a product of the universe's photonic speed limit.

A moving object is composed of a combination of static (rest) and dynamic (propagating) energy. Its net motion through absolute space constitutes the *external* manifestation of its dynamic energy, and will be symbolized $\mathbf{v}$. Any motion of its dynamic constituents normal to $\mathbf{v}$ has no measurable effect on the net motion of the entire object, and will be called the *internal* component of its dynamic energy, $\mathbf{v}_i$. Since c is the metric of *energy's* propagation through absolute space, and since the external and internal components of an object's dynamic energy are perpendicular to each other, their relationship can be depicted as:

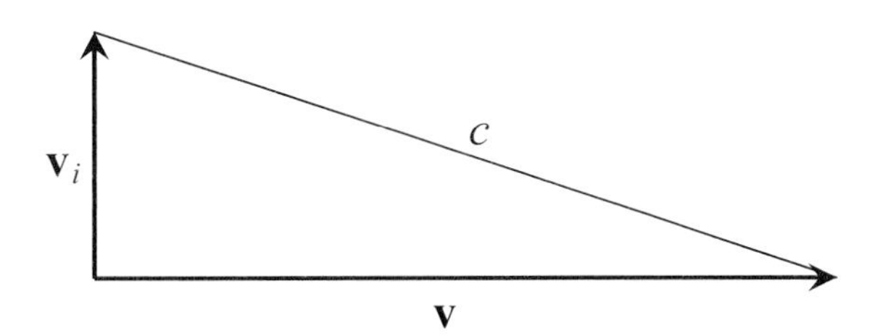

Figure (E.1) Relationship between the internal and external components of an object's motion through absolute space

The speed of an object's internal motion defines the rate of internal change it experiences, such as the decay period for an unstable relativistic particle. An object's internal motion is related to its external motion by:

$$\mathrm{v}_i^2 = c^2 - \mathrm{v}^2 \qquad \text{(E.14)}$$

The speed of light is the metric of all change. A motion of zero represents no change; a motion of c is maximum change. The normalized rate of an object's *internal change* is the ratio of the speed of its internal components to the speed of light:

$$\delta = \frac{v_i}{c} \tag{E.15}$$

when ($\mathbf{v}_i = 0$) internal change goes to zero. The relative length of a unit time interval in this system is the magnitude of its *time dilation*, Θ. This is the inverse of the rate of internal change:

$$\Theta = \frac{t}{t_0} = \frac{1}{\delta} = \frac{c}{v_i} \tag{E.16}$$

where t_0 is the length of a time interval at rest ($\mathbf{v} = 0$). If internal change is reduced to half of normal, for instance, an internal event takes twice as long to occur. Substitute Equation (E.16) into Equation (E.14) and solve:

$$\frac{t}{t_0} = \Theta = \frac{1}{\sqrt{1 - \frac{v^2}{c^2}}} \tag{E.17}$$

As external motion goes to c, the rate of internal motion goes to zero. This is the same time dilation the Lorentz transformation produces, and is based solely on the real, physical limitations of a moving system. No reconfiguration of space and time is necessary.

VELOCITY ADDITION

Another way to express the universe's speed limit is through *velocity addition.* The dimensional relationship between time and space means that *vector addition cannot be used except as an approximation for adding velocities.* Vector addition works fine for distance because space is infinite and the addition of any two distances will always reside in space. Not so with velocity. Its addition can be derived by performing two successive Lorentz transforms:

$$v = \frac{v_1 + v_2}{\left(1 + \frac{v_1 v_2}{c^2}\right)} \tag{E.18}$$

Velocity addition is *the addition of ratios, not magnitudes*. If one of the motions of Equation (E.18) is that of light:

$$v = c\left(\frac{c + v_2}{c + v_2}\right) = c \tag{E.19}$$

Thus the speed of light appears the same in any direction in any moving reference frame. *The only case where velocities would add exactly like distances is if there were no upper limit to their magnitudes* ($c = \infty$). It is safe to say that relativistic effects would not exist in a universe with no speed restriction.

RELATIVISTIC MASS

Although relativistic effects are often considered counterintuitive, there is really no mystery associated with the mass gained by high-speed particles or their inability to exceed the speed of light. Both are the direct result of momentum and energy conservation. A motionless proton has a momentum of zero. Let's call it p_0^+. A high-speed proton, p_1^+, can be viewed as a composite of p_0^+ and a Compton scattering of the form:

$$p_0^+ + \gamma_0 \leftrightarrow p_1^+ + \gamma_1 \tag{E.20}$$

or:

$$p_1^+ \leftrightarrow p_0^+ + (\gamma_0 - \gamma_1) \tag{E.21}$$

Regardless of how much energy γ_0 loses, the momentum of p_1^+ is the sum of its initial (zero) momentum and the difference in the scattered photon's momentum.

A photon's momentum is proportional to its energy:

$$p_\gamma = \frac{E}{c} \tag{E.22}$$

The only way a particle can sustain this momentum as (mv) *for arbitrarily large energy* is to either increase its mass or move faster than light. It cannot exceed c because the composite particle of Equation (E.21) contains no component velocities greater than c. The sole avenue for momentum conservation is a mass increase commensurate with intrinsic photon momentum. A fast-moving particle *has* to incur a mass increase. This is also why free particles tend to scatter photons instead of absorbing them. The relationship between kinetic energy and momentum differs in particles and photons, precluding the possibility of direct absorption.

F. THE GENERALIZED PARTICLE FIELD AND CORE DYNAMICS

F.1 PARTICLE FIELD DISTRIBUTION

A particle field is defined by:

$$t = \frac{\Diamond_q}{V} \tag{F.1}$$

where $\Diamond_q$ is unit polarvolume, the constant product of spatial volume and the magnitude of its fourth-dimensional deflection:

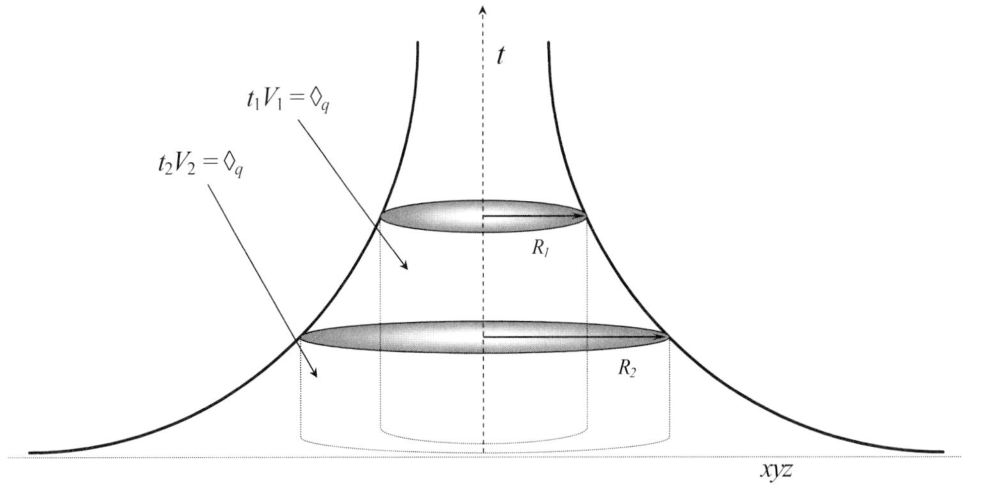

Figure (F.1) Isolated particle as family of concentric spherical volumes with differing fourth-dimensional elevations

Unit polarvolume's spatial component V has a spherical topology in an isolated particle, so its profile is given by:

$$t = \frac{3\Diamond_q}{4\pi r^3} \tag{F.2}$$

This is the direct relationship between external deflection and radial distance from a particle's center. External deflection attenuates *by virtue of its dimensional dilution into increasing volume*. As such, there ought to be a more general expression that describes the way external deflection radiates from any area deflected into the fourth dimension, spherical or otherwise. In order to solve the general case of a source of arbitrary shape, it must first be decomposed into an infinite number of infinitesimal hypervolumetric sources. This approach will be called the *differential source*.

DIFFERENTIAL SOURCE

Unit polarvolume attenuates through volume, decreasing as the cube of distance, not the square. It is a finite value that, like any other finite value, is composed of an infinite number of differential elements. This can be described by treating a particle's core as a hypersphere of radius R_c infused with a uniform *hypervolumetric density*:

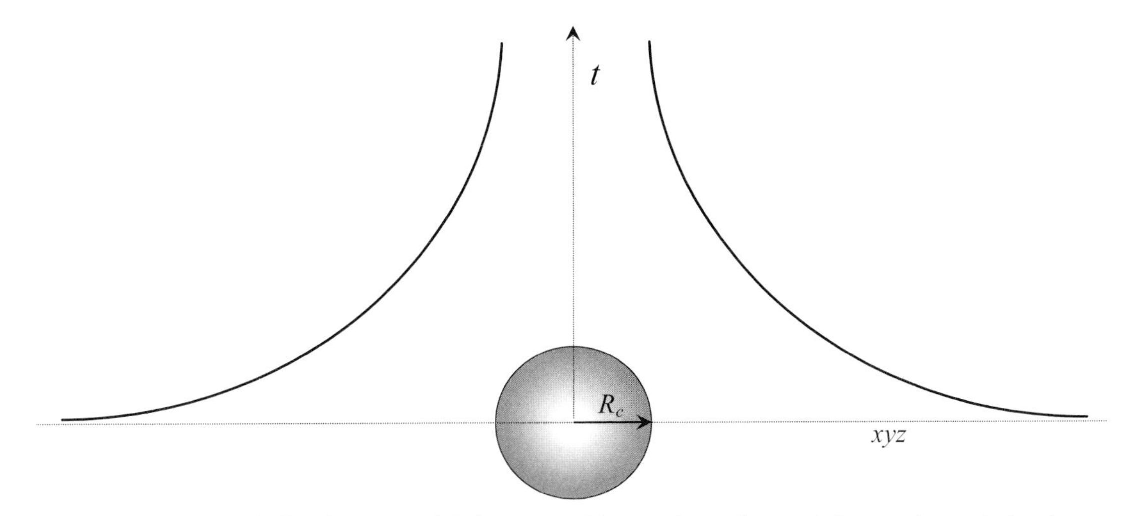

Figure (F.2) Particle distribution modeled as a central hypersphere of a certain hypervolumetric density

Hypervolumetric density is the unitless ratio of unit polarvolume to the hypervolume of a central hypersphere, $\Diamond_c$:

$$\rho_\Diamond = \frac{\Diamond_q}{\Diamond_c} \tag{F.3}$$

The product of a core's hypervolume and hypervolumetric density is unit polarvolume and its net effect is a field that varies as the cube of distance. *A central hypersphere is the only case where the sum of differential elements of hypervolume produces a uniform cubic deflection distribution from infinity to arbitrarily close to the hypersphere's surface.*

The generalized particle field is given by:

$$t(x,y,z) = \left| \int_\Diamond \frac{3\vec{u}_d \rho_\Diamond d\Diamond}{4\pi d^3} \right| \tag{F.4}$$

where $\rho_\Diamond$ is hypervolumetric density, $(\rho_\Diamond d\Diamond)$ is a hypervolumetric differential, d is the distance from the point (x, y, z) to this differential, and $\mathbf{u}_d$ is a unit vector in d's direction. Equation (F.4) applies both to spherical and nonspherical source regions, and holds as long as the point (x, y, z) is exterior to the hypercore's surface. A particle field's central hypersphere will be called its *hypercore*:

Ψ THEOREM F.1 - HYPERCORE {Ψ9.1}

A PARTICLE FIELD CAN BE DESCRIBED AS THE CUMULATIVE EFFECT OF A CENTRAL HYPERSPHERE OF HYPERVOLUME $\Diamond_c$ WITH DISTRIBUTED POSITIVE OR NEGATIVE HYPERVOLUMETRIC DENSITY $\Diamond_q/\Diamond_c$

F.2 THE HYPERCORE

A spherical hypercore's hypervolume is given by integrating four spatial coordinates over its interior region:

$$\Diamond_c = \int_{x^2+y^2+z^2+w^2 \le R_c^2} dxdydzdw = \frac{\pi^2 R_c^4}{2} \qquad \text{(F.5)}$$

Its hypervolumetric density is given by applying Equation (F.3):

$$\rho_\Diamond = \frac{\Diamond_q}{\Diamond_c} = \frac{2\Diamond_q}{\pi^2 R_c^4} \qquad \text{(F.6)}$$

Like energy density, hypervolumetric density varies as the fourth power of radius.

A hypercore produces external deflection at any point $(d, 0, 0)$ exterior to its surface, $(d > R_c)$, so the deflection at d is the sum of its differential sources:

$$\int_{x^2+y^2+z^2+w^2 \le R_c^2} \left(\frac{3(x-d)}{4\pi\left((x-d)^2 + y^2 + z^2 + w^2\right)^{\frac{5}{2}}} \right) \rho_\Diamond dxdydzdw = \frac{3\Diamond_q}{4\pi d^3} \qquad \text{(F.7)}$$

where $\rho_\Diamond$ is given by Equation (F.6). The integrand is separated into two terms for the sake of clarity. The first is the cubic attenuation of the deflection field in the x direction. The second is a differential hypervolume source as the product of hypervolumetric density and differential hypervolume. Equation (F.7) successfully passed numerical integration tests for d over a range of three orders of magnitude $[1.01R_c < d < 1000R_c]$.

Since any point in a spherically symmetric particle field can be expressed as (d, 0, 0) with the proper choice of coordinates:

- The hypercore defines the entire particle field.
- The integrated hypervolumetric density of the hypercore of an isolated particle is equal to unit polarvolume.

The hypercore, as a direct reflection of half of space's totality, is the ultimate description of a particle field:

Ψ THEOREM F.2 - HYPERCORE HYPERVOLUME {ΨF.1}

THE PRODUCT OF THE HYPERVOLUME AND UNIT POLARVOLUMETRIC DENSITY OF A PARTICLE'S HYPERCORE IS EQUAL TO POSITIVE OR NEGATIVE UNIT POLARVOLUME

This is the basis of charge conservation, and is irrespective of the environment in which a particle finds itself. Whether it floats alone in deep space or lies deep in the heart of a star, a hypercore's hypervolume is fixed. Also, since exterior energy can't pass through a particle's core boundary, a hypercore's hypervolumetric density is uniform:

Ψ THEOREM F.3 - HYPERCORE HOMOGENEITY {Ψ9.2}

THE HYPERVOLUMETRIC DENSITY OF A PARTICLE'S HYPERCORE IN ANY ENVIRONMENT IS UNIFORM THROUGHOUT ITS HYPERVOLUME AND IS EQUAL TO THE RATIO OF UNIT POLARVOLUME TO CORE HYPERVOLUME

As before, this is exclusive of annihilation.

F.3 COULOMB DISTORTION

When a particle is exposed to other particles' external deflection, there are two profound effects:

- The product of (deflection)(volume) at its core boundary varies.
- Its isoexternal boundary shifts with the gradient it experiences, giving its core a nonspherical deformation in addition to that caused by applied deflectional force.

Theorems (ΨF.2) and (ΨF.3) tell us that regardless of a core's shape, the following is true:

$$\lozenge_q = \rho_\lozenge \int_{\lozenge_c} dxdydzdw \tag{F.8}$$

where the region the hypercore subtends, $\lozenge_c$, is not necessarily a hypersphere.

When a hypercore is distorted, it produces a slightly different field. When this is combined with an applied field, it produces a slightly different isoexternal surface, which changes the shape of the hypercore, which in turn produces a slightly different field, and so on. The solution to a superimposition-induced core deformation is a hypercore topology that satisfies the following criteria:

- Has an integrated hypervolumetric density equal to unit polarvolume.
- Produces a boundary deflection, which when combined with the applied field, is isoexternal.

All of the universe's dimensions are inherently symmetric, so two hypercores separated along one axis have perfect symmetry along all other axes. This property will be referred to as *hypersymmetry*. So if a particle core's isoexternal boundary is defined in three-space, its corresponding hypercore is defined in four-space:

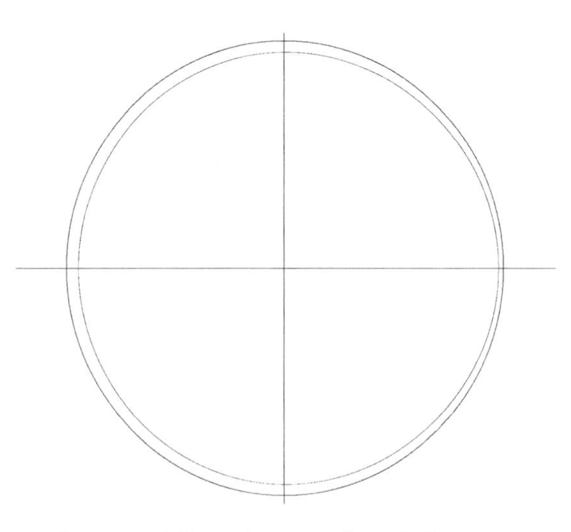

Figure (F.3) An isoexternal contour defines a hypercore's equivalent *x-y*, *x-z*, and *x-w* cross-sections

Hypersymmetry can be exploited to calculate the external field a distorted hypercore generates in space, as is necessary for a full quantification of the Coulomb force. The computational requirements are daunting, however.

F.4 A MOVING HYPERCORE

A small percentage of cosmic rays reaching Earth's upper atmosphere have an unthinkable amount of energy - single elementary particles with more momentum than a flying bird. If a proton's field energy can increase by a factor of a trillion, for instance, there has to be a significant structural change in its four-dimensional topology.

According to the Special Theory of Relativity, the electric field of a moving charge is distorted by its motion as:

$$r_r = r\frac{\left(1-\frac{v^2}{c^2}\right)}{\left(1-\left(\frac{v\sin\theta}{c}\right)^2\right)^{\frac{3}{2}}} \qquad \text{(F.9)}$$

where r_r is the radial distance with the same electric field strength (isopotential) as r in a stationary field and θ is the angle between the charge's trajectory and r. The shape of a charged particle's isopotential contour is spherical when it is at rest. When it has a significant speed, it is flattened in the direction of its motion:

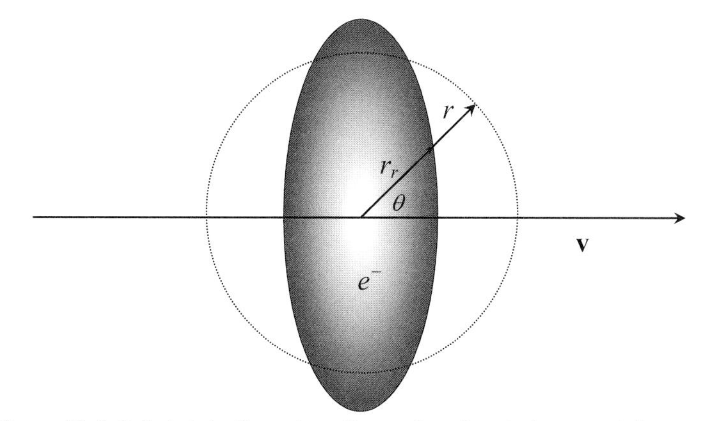

Figure (F.4) Relativistic distortion of a moving electric isopotential contour

Whether or not this is a reflection of the physical distortion of the underlying particle field remains to be seen. Equation (F.9) might only represent the effect of the field and not the field itself. In any event, the field extends normal to its motion ($\theta = 90$) by:

$$r_r = \frac{r}{\sqrt{1-\frac{v^2}{c^2}}} \qquad \text{(F.10)}$$

This is the same as the expression for the ratio of relativistic to rest mass. The contraction of the field parallel to its motion ($\theta = 0$) is even more pronounced than its extension, and is given by:

$$r_r = r\left(1 - \frac{v^2}{c^2}\right) \quad \text{(F.11)}$$

A free electron moving at 0.92c, for instance, suffers a 2.5x lateral extension and 6.5x parallel contraction of its field strength. Does this field asymmetry extend all the way to its core? Certainly the fact that a particle can carry so much kinetic energy indicates a major change occurs there.

CORE STORAGE

The kinetic energy in a particle field is subject to two rigorous constraints:

- Charge density. There is a dimensional relationship between deflection and a particle's core volume. Gradational distance alters this by redefining the metric of space, but an increase in peak deflection requires an attendant increase in hypervolumetric density.

- Field integrity. The energy a particle carries is a finite, bounded quantity, and the size or shape of a particle's core boundary *defines* its energy.

Whether or not particle cores change shape as a result of motion remains open for discussion, but the kinetic energy they carry is stored as some form of core recession. The only way a particle can carry the kind of energy exhibited by cosmic rays is by a radical increase in energy density. This in turn requires a comparable increase in curvature and force density. Core recession with peak deflection increasing according to the unit polarvolume distribution is the only agent available to fulfill this requirement.

Ψ THEOREM F.4 - KINETIC RECESSION {Ψ9.2}

MOVING PARTICLES' KINETIC ENERGY IS STORED AS A FORM OF CORE RECESSION

Energy cannot exist without energy density, energy density cannot exist without external slope, and external slope cannot exist without a terminating boundary and immense source deflection. The particle core of an energetic proton carrying a trillion times its rest energy bears little resemblance to that of its free-space stationary state.

KINETIC CORE DILATION

A particle's field, *or at least the effect of its field*, dilates with motion. Does the core dilate as it recedes to store kinetic energy? As shown in Equation (F.10), the lateral expansion of a particle's field is precisely equal to its increase in mass; they are governed by the same velocity term:

$$r_r = \frac{r}{\sqrt{1 - \frac{v^2}{c^2}}}$$

Suppose as a particle gained kinetic energy its core suffered a relativistic dilation, *but also decreased in volume in accordance with the unit polarvolume distribution*. Thus a proton whose relativistic mass was a hundred times its rest mass would have a core size of 0.0095 F but would also have a hundredfold lateral expansion, giving its cross-section a radius of 0.95 F, the same as its resting state. Since its volume decreases through this process, its overall energy density grows, consistent with the properties of fast-moving particles. This premise will be called the *kinetic hypercore hypothesis*:

Ω HYPOTHESIS F.1 - KINETIC HYPERCORE {Ψ9.2}

THE ELEMENTARY CORE OF A MOVING PARTICLE IS COMPRESSED ALONG ITS DIRECTION OF MOTION

If this is true, a proton's core volume would be compressed in the direction of its motion as it recesses to carry its kinetic energy:

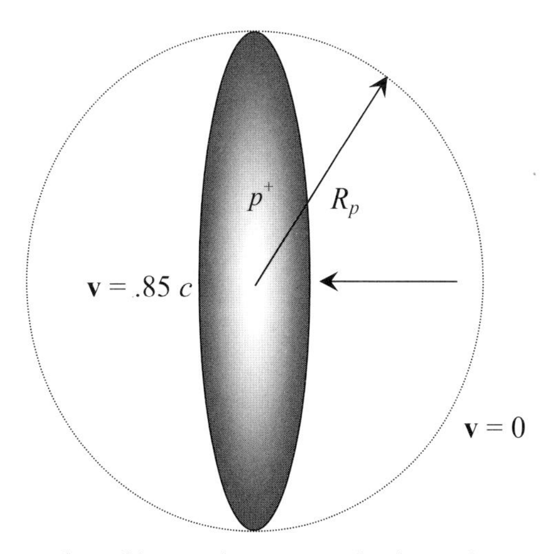

Figure (F.5) Hypothetical kinetic distortion in the shape of a proton's core

The net force in a uniformly moving particle field is zero, as is the net force in a symmetrically compressed core.

Other arguments for core dilation are:

- The directional nature of a particle's momentum. If a moving particle represented a radially symmetric core reduction with no dilation, what is the difference between it and a stationary particle of a similar mass? The momentum of a moving particle is *a difference of energy along space*. It has magnitude, direction, and is just as tangible as the rest energy that carries it. As such, it corresponds to a physical (and directional) change in the underlying core.

- The fact that only two stable particles of matter exist. If a moving proton is simply a stationary proton with a smaller core radius, what makes the proton radius so unique? A moving proton, like a stationary one, is stable. If its structure is independent of its motion then an infinite variety of stable particles should exist, not just the proton and electron.

The premise of coincident core dilation/recession provides directional asymmetry, constant collision cross-section, and consistent energy storage capability. It is not sufficiently conclusive to document as a theorem but it is certainly compelling. Kinetic core dilation could potentially be tested by comparing collision cross-sections at various angles of incidence. Two beams of protons with enough energy to achieve core proximity should have a slightly smaller core cross-section when passing through each other at 90° than when they collide head on.

F.5 MAGNETISM

Moving charges exert magnetic forces on each other. Since differential external deflection is the only force that can be exerted on a particle's core, it follows that magnetism is due, like the Coulomb force, to an asymmetrical distribution of this force about a particle core. Motion produces a physical distortion of a particle's field, which in turn changes the way it superimposes on a neighboring particle. *This means magnetism provides clues about the nature of motion-induced changes in the core boundary region.*

Let's briefly look at some of the aspects of the magnetism of elementary particles.

Our discussion will be simplified by restricting the motion of interacting particles to a single axis and limiting their locations to a plane normal to this axis:

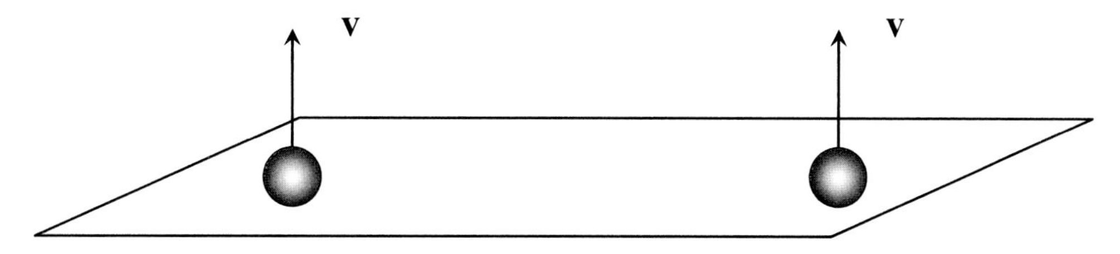

Figure (F.6) A single motion axis and position plane allow removal of magnetism's vector context

This eliminates the need for vector notation.

A particle moving with velocity **v** produces a magnetic field in the plane normal to its motion of the form:

$$B = \frac{\mu_0 q \mathrm{v}}{4\pi d^2} \tag{F.12}$$

where μ_0 is known as the *permeability of free space* and d is the distance from the particle.

The force another particle experiences in this plane, when moving along an axis parallel to the first, is given by:

$$F_B = q\mathrm{v}B \tag{F.13}$$

The force the two particles exert on each other as a result of their respective motion is given by combining Equations (F.12) and (F.13):

$$F_B = \frac{\mu_0 q^2 \mathrm{v}^2}{4\pi d^2} \tag{F.14}$$

When the particles have the same charge and are moving in the same direction this force is attractive; when they move in opposite directions it is repulsive. The situation is reversed when the two particles have opposite charge. Equation (F.14) will be referred to as the *elementary magnetic force*.

Two positive charges moving parallel to each other would normally have a repulsive Coulomb force between them. If they are moving at a speed near c, this force is *nullified* by magnetic attraction. Similarly, two positive charges moving anti-parallel have the same repulsive Coulomb potential, but when they move at a speed close to c their total repulsion doubles, reinforced by magnetic force.

Magnetic force has two significant aspects:

- Magnetic effects cannot *reverse* the polarity of the electric force; they can only nullify or reinforce it. Two electrons repel each other at rest. When they move, the net force between them can be twice their normal repulsion or nothing, depending on their relative velocity.

- Magnetic force is charge dependent, so it cannot be due to a coordinate transformation (such as gradational distance) that changes the distance between particles. Two electrons moving in parallel have reduced repulsion, a positron and electron moving in parallel experience increased attraction. Magnetic force is therefore *not* a motion-induced transformation of the effective distance between charges, $d^* = T^*(d)$.

The fact that the direction of the magnetic force varies with charge suggests it is a direct by-product of the Coulomb force. The ratio between the elementary magnetic force and Coulomb force is:

$$f = \frac{\left(\dfrac{\mu_0 q^2 \mathrm{v}^2}{4\pi d^2}\right)}{\left(\dfrac{q^2}{4\pi\varepsilon_0 d^2}\right)} = \mu_0 \varepsilon_0 \mathrm{v}^2 \qquad \text{(F.15)}$$

When the particles are moving at a speed close to that of light, this reduces to the relationship between the permittivity and permeability of free space:

$$f = \mu_0 \varepsilon_0 c^2 = 1 \qquad \text{(F.16)}$$

This is not fortuitous; these constants were defined this way to incorporate SI units into electromagnetic expressions. The governing phenomenon is the interesting part. When particles are moving slower than *c*, the magnetic force they exert on each other is less than the Coulomb force; when they are moving close to *c, the two approach equality*. This suggests that magnetic force is caused by a direct interaction between the core distortion induced by the Coulomb force and the core distortion caused by kinetic energy storage.

G. THE SUPERLUMINAL CRITERION

A particle's energy field is unbounded whereas a photon's energy is not. This means that when a particle and antiparticle annihilate and release gamma photons, energy transfer occurs between the creation/destruction site and the distributed fields of matter:

$$p^{+} + p^{-} \rightarrow 2\gamma \quad \text{(G.1)}$$

How quickly does this transfer occur?

Particles have unbounded energy fields whose spatial attenuation is purely a function of dimensionality. When a particle and antiparticle are produced by pair production, their fields materialize in the form of a *creation front* that emanates away from the system's center of gravity. Now suppose one of these recently minted particles finds an antiparticle and immediately annihilates, generating an *annihilation front* as its field dissipates.

This particle's creation front echoes outward through space, followed in lock step by the front of its recent destruction:

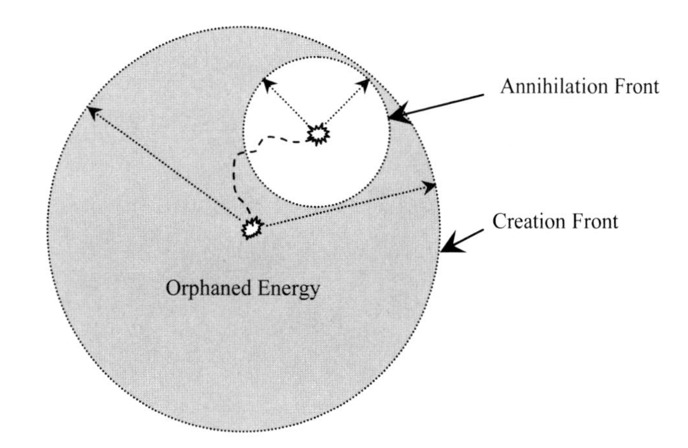

Figure (G.1) Asymmetry of creation/annihilation at finite propagation speed

The field energy difference between creation and annihilation fronts is *orphaned energy*. If the fronts move at infinite speed, orphaned energy is zero. However, if they move at finite speed, regardless of how large this speed might be, orphaned energy would exist. Matter/energy conversion at finite propagation speed is *asymmetric* because annihilation can never release the same amount of energy originally committed to a particle's construction. There would always be energy scattered between the loci of the creation and annihilation fronts.

Another problem with orphaned energy is causality. A radial energy distribution cannot exist in space without a source - a discrete boundary condition. The expanding fronts in Figure (G.1) are certainly not low-energy particle cores devoid of space. Field energy is derived from and dependent on a core boundary and cannot exist without one. *Symmetry can only be achieved if particle fields emerge and dissolve at infinite speed.*

Ψ THEOREM G.1 - SUPERLUMINAL FIELD DISTRIBUTION {Ψ9.1}

THE TRANSFER OF ENERGY BETWEEN A PARTICLE'S CORE AND FIELD OCCURS AT INFINITE SPEED

The speed of light is the universe's limit for energy propagation, but this is only for its *net* movement. When a photon moves from one location to the next, its net energy transfer occurs at c. The same is true for moving particles. At the subquantum level, however, differential energy can redistribute at speeds far in excess of c as long as its *net displacement* does not exceed it. The creation of a stationary particle, for instance, consists of energy radiating isotropically outward at infinite speed. Infinite displacement in one direction is balanced by a comparable transfer in the opposite direction, while the center of mass of the entire system remains unchanged. Since energy density falls off inversely as the fourth power of radius, the quantity of energy transported to the far reaches of the universe gets progressively smaller with distance.

H. PARTICLE FIELD REFERENCE

The equations of the particle field are listed below. Terms that are a function of r are appended with (r) and represent an amount from a radius r to infinity. All other terms are fixed values. The rest energy of a particle, $E_{\wedge}$, for example, could also be written in radial form as $E_{\wedge}(R_c)$. Unit polarvolume, $\Diamond_q$, scales every equation and is a function of Planck's constant via Equation (8.16):

$$\Diamond_q = \frac{hc}{4\pi} \qquad \text{(Cp.1)}$$

DEFLECTION/EXTENTS

External deflection:

$$t(r) = \frac{3\Diamond_q}{4\pi r^3} \qquad \text{(Dp.1)}$$

Hypervolume between radii:

$$\Diamond(r_1 \rightarrow r_2) = 3\Diamond_q\left(\ln(r_2) - \ln(r_1)\right) \qquad \text{(Dp.2)}$$

Particle core radius:

$$R_c = \frac{9\Diamond_q}{E_{\wedge}} \qquad \text{(Dp.3)}$$

Particle core surface area:

$$A_c = \frac{324\pi\Diamond_q^2}{E_{\wedge}^2} \qquad \text{(Dp.4)}$$

Particle core volume:

$$V_c = \frac{972\pi\Diamond_q^3}{E_{\wedge}^3} \qquad \text{(Dp.5)}$$

ENERGY

Energy density by radius:

$$\rho_E(r) = |t'(r)| = \frac{9\Diamond_q}{4\pi r^4} \quad \text{(Ep.1)}$$

Energy between radii:

$$E_\wedge(r_1 \to r_2) = 9\Diamond_q\left(\frac{1}{r_1} - \frac{1}{r_2}\right) \quad \text{(Ep.2)}$$

Rest energy:

$$E_\wedge = \frac{9\Diamond_q}{R_c} \quad \text{(Ep.3)}$$

Intrinsic energy density (core pressure):

$$\rho_{\langle c \rangle} = \frac{27\Diamond_q}{4\pi R_c^4} = \frac{E_\wedge^4}{972\pi \Diamond_q^3} \quad \text{(Ep.4)}$$

Peak energy density:

$$\rho_{E_c} = \frac{9\Diamond_q}{4\pi R_c^4} = \frac{E_\wedge^4}{2916\pi \Diamond_q^3} \quad \text{(Ep.5)}$$

Average energy density:

$$\overline{\rho_E}(R) = \left(\frac{1}{E_\wedge(R)}\right)\int_{r=R}^{\infty} \rho_E(r)\,dE = \frac{9\Diamond_q}{20\pi R^4} \quad \text{(Ep.6)}$$

FORCE

Deflectional force from an excised interval:

$$F_\Delta(r_1 \to r_2) = -\frac{3\Diamond_q}{8\pi}\left(\frac{1}{r_1^2} - \frac{1}{r_2^2}\right) \quad \text{(Fp.1)}$$

Total deflectional force:

$$F_{\Delta_T} = -\frac{3\Diamond_q}{4\pi R_c^2} = -\frac{E_\wedge^2}{104\pi \Diamond_q} \quad \text{(Fp.2)}$$

I. DEEP SPACE PHOTONS - ANNIHILATION AND FUSION

I.1 ANNIHILATION FLUX

It is difficult to quantify the level of activity related to matter-antimatter annihilation in deep space. There is a background gamma radiation flux with a clearly extragalactic origin, but it contains no pronounced peak near proton rest energy as would be expected if its principal source were antimatter. It can, however, be used to establish an upper limit of dark annihilation by assuming that the total energy density of high-energy photons up to the rest energy of a proton is the product of DZ emission. The following calculation is based on the near-Earth gamma background measured by the Energetic Gamma-Ray Experiment Telescope (EGRET), as described by Sreekumar et al. in the Astrophysical Journal (1998).[39]

The extragalactic gamma energy distribution follows a power law of the form:

$$f(E) = b\left(\frac{E}{E_0}\right)^{-a} = 7.32(10)^{-9}\left(\frac{E}{451}\right)^{-2.1} \tag{I.1}$$

in photons/(MEV-cm^2-sr-s), where sr represents a *steradian*. Eliminate Equation (I.1)'s angular dependence by multiplying by the number of steradians in a spherical area (4π), convert area from square centimeters to square meters (10000), convert radiance to energy density with a factor of ($4/c$), and simplify, yielding photon number density:

$$f(E) = 4.6(10)^{-6} E^{-2.1} \tag{I.2}$$

in photons/(MEV-m^3).

The total extragalactic gamma energy density within a range from E_1 to E_2 is given by the weighted integral:

$$\rho_{\gamma\gamma} = \int_{E_1}^{E_2} Ef(E)dE = 4.6(10)^{-5}\left(E_1^{-0.1} - E_2^{-0.1}\right) \tag{I.3}$$

in units of MEV.

The dark annihilation of protons produces gamma rays with an initial energy of 940 MEV. Depending on how far this radiation has to travel prior to reaching Earth, its energy can be anywhere from EGRET's lower limit of 30 MEV up to a maximum of close to 940 MEV. Equation (I.3) gives the energy density in this interval as $1.5(10)^{-18}$ J/m^3, which is about 10% of the total gamma background estimated earlier by Silk.[2.1] To put the intensity of this background into perspective, the energy density of the CMB, at $4.2(10)^{-14}$ J/m^3, is over four orders of magnitude greater.

The rarity of gamma photons increases rapidly with respect to their individual energy, but they persist nonetheless. EGRET's upper energy limit is 120 GEV, over 125 times the rest energy of a proton, and the extragalactic gamma background follows Equation (I.2) smoothly up to (and therefore past) this limit. This means that the number density of 100 GEV photons, for example, is $1.45(10)^{-16}$ per cubic meter. There are, on average throughout the universe, 150,000 photons whose energy exceeds 100 GEV in any spatial volume comparable to the size of the Earth. And this is just the high-energy gamma environment. Space is a very dangerous place, to an extent not often appreciated by the people living under the blanket of our planet's atmosphere.

I.2 FUSION FLUX

The universe's average fusion output is typically called the *integrated starlight*. It can be approximated as an attenuated 10,000 °K blackbody. In general, a blackbody spectrum's average photon energy is the ratio of its energy density to number density:

$$\overline{E}_\gamma = \frac{\rho_E}{n_\gamma} = \frac{h^3 c^2 \sigma T}{2\pi(2.404)k^3} = \frac{\pi^4 kT}{15(2.404)} \cong 3.73(10)^{-23} T \tag{I.4}$$

The average photon energy in integrated starlight, at 10,000 °K, is therefore $3.73(10)^{-19}$ J. This corresponds to a wavelength of 536 nm. So although the peak energy density of this spectrum lays in the ultraviolet, its average photon is red, much like the stellar population of a typical spiral galaxy.

The average photon number density of integrated starlight follows as the ratio of its energy density to average photon energy:

$$n_{B\gamma} = \frac{\rho_{B\gamma}}{\overline{E}_{B\gamma}} \tag{I.5}$$

This amounts to ~4300 photons/m^3 using the optical energy density listed in Appendix C.

J. WHITE DWARF HISTORY

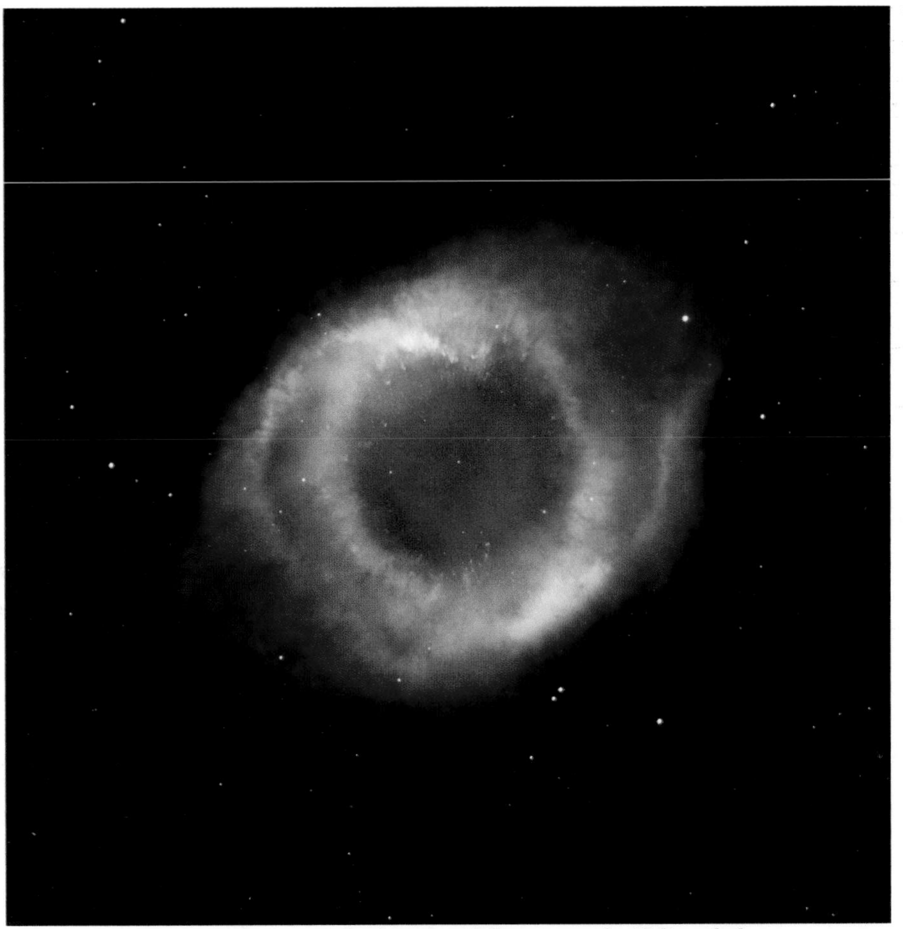

Figure (J.1) A central white dwarf illuminates the Helix nebula
(Courtesy NASA/Hubble Heritage Project)

The cooling of white dwarf stars has been described by a number of comprehensive theories, most of which rely on the equilibrium between a dwarf's hot core and degenerate electron atmosphere. A simplified model of this process will be presented here, resulting in an interesting discovery.

A white dwarf's age, τ_w, is defined as the time since it was first created by a supernova. This varies with its luminosity as:[6.16]

$$L = L_0\left(1 + \alpha\tau_w\right)^{-7/5} \tag{J.1}$$

where L_0 is its initial luminosity and α is a time constant.

A white dwarf's luminosity is related to the temperature of its electronically degenerate surface according to:[6.16]

$$L \propto T_e^{7/2} A \qquad \text{(J.2)}$$

where A is its surface area and T_e is its *effective surface temperature*. This is the temperature originally presented in Figure (16.3), the population of local white dwarf stars compiled by Sion and McCook:

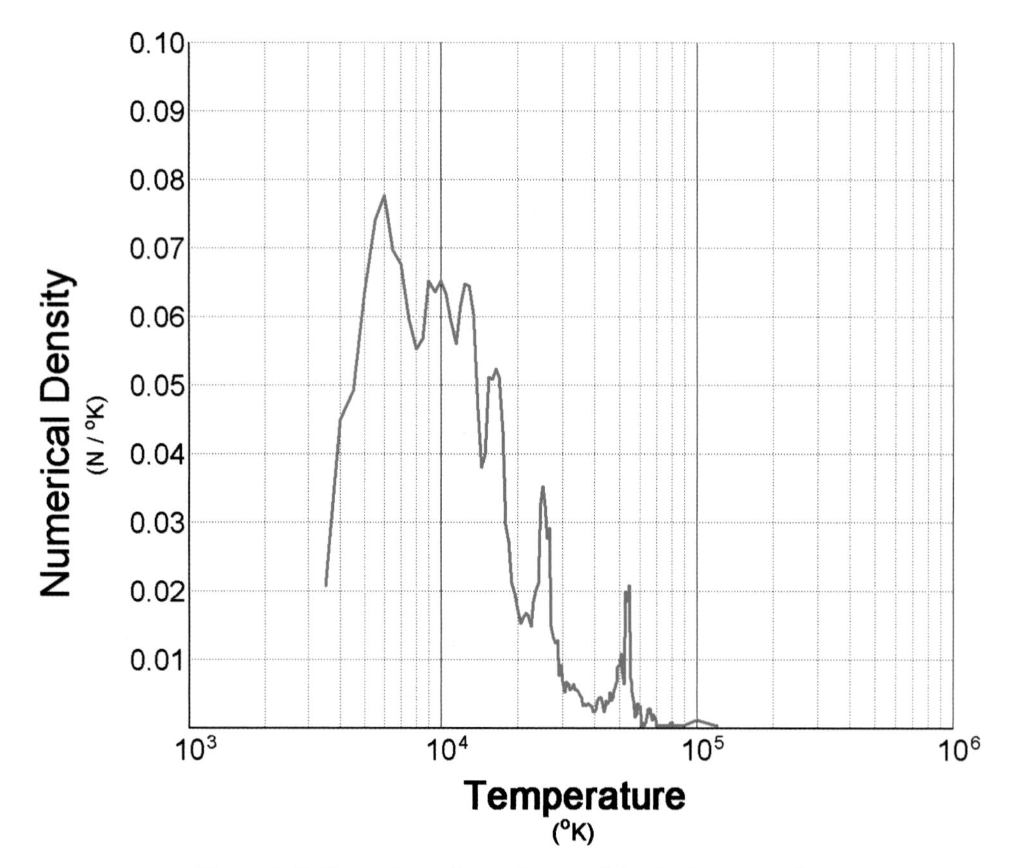

Figure (J.2) Thermal population density of the 1262 star sample set

A white dwarf's size is controlled by its mass, not its temperature, so its luminous area remains essentially constant as it cools. Replace the luminosity terms of Equation (J.1) with Equation (J.2) and solve for effective temperature:

$$T_e = T_{\max}\left(1 + \alpha\tau_w\right)^{-2/5} \qquad \text{(J.3)}$$

where $T_{\max}$ is a maximum temperature of about 120,000 °K, corresponding to a white dwarf's initial luminosity, L_0. The older the dwarf, the lower its surface temperature.

White dwarf age is given by solving Equation (J.3) for τ_w:

$$\tau_w = \left(\frac{1}{\alpha}\right)\left(\left(\frac{T_{max}}{T_e}\right)^{\frac{5}{2}} - 1\right) \tag{J.4}$$

Rewrite Equation (J.4) in terms of α, minimum temperature, and maximum age:

$$\alpha = \left(\frac{1}{\tau_{w_{max}}}\right)\left(\left(\frac{T_{max}}{T_{min}}\right)^{\frac{5}{2}} - 1\right) \tag{J.5}$$

The oldest dwarfs are the coolest ones in the dataset, with temperatures of 3600 °K. If the oldest local stars are 11 Gyr old,⟨1.15⟩ α has a value of 580 Gyr^{-1}. Applying Equation (J.4) to the white dwarf dataset of Figure (J.2) produces an interesting result:

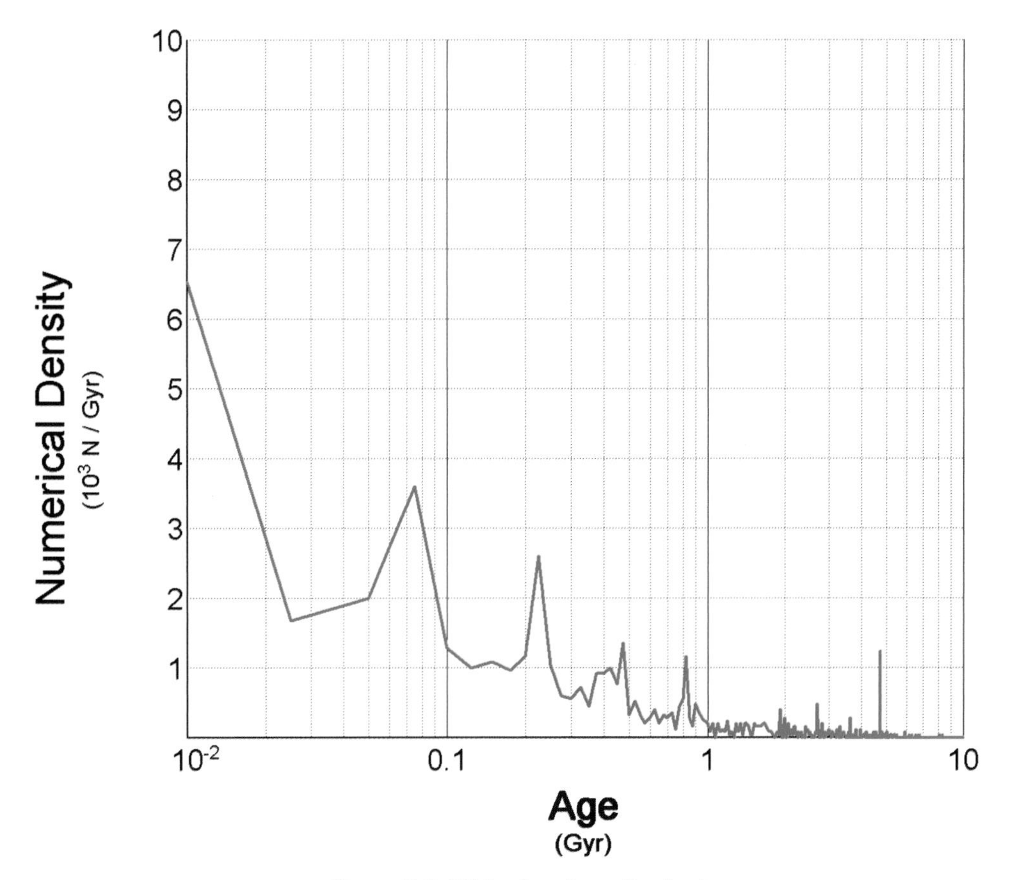

Figure (J.3) White dwarf age distribution

There is an unmistakable spike (30 times background) of white dwarf ages at 4.7 Gyr, *the age of our solar system.*[6.17] Moreover, the coordinates of the stars in this population spike span about 60% of the sky. So not only do the oldest white dwarfs reveal the maximum age of our local neighborhood, the conspicuous peak shown in Figure (J.3) provides compelling evidence of an intense period of stellar genesis on or about the time our solar system was born.

The Milky Way's galactic vortex defines minimum white dwarf temperature as a function of distance from its galactic rim:

Figure (J.4) The closer to the galactic core, the cooler the coldest white dwarf.

Since the cooling time for white dwarfs exceeds the galactic transit time, the coldest white dwarf defines the age of fusion initiation for any stellar location. No dwarfs in the galactic disk will be older than the galactic transit time, about 16 billion years. At this age, Equation (J.3) puts our galaxy's lowest possible white dwarf temperature at 3100 °K, ~500 °K cooler than the coldest objects in the dataset. Dwarfs formed on a galaxy's rim will be referred to as *original dwarfs*.

The cooling rate at some temperature T_e is given by Equation (J.3)'s derivative:

$$\frac{dT_e}{d\tau_w} = -\frac{2\alpha}{5} T_{\max} \left(1+\alpha\tau_w\right)^{-7/5} = -\left(\frac{2\alpha}{5} T_{\max}^{-5/2}\right) T_e^{7/2} \tag{J.6}$$

with ($\alpha = 580$ Gyr^{-1}) and ($T_{\max} = 120{,}000$ K) this is:

$$\frac{dT_e}{d\tau_w} \cong -4.7(10)^{-11}\, T_e^{7/2} \tag{J.7}$$

In °K/Gyr. At this rate, it takes a white dwarf at 3000 °K about *15 million years to cool a single degree.* It goes without saying that the galactic vortex's existence can't be demonstrated by measuring temperature gradients in the original dwarf distribution scattered throughout our galaxy. Stars this cold are currently only visible at ranges of less than a few dozen light years. *Two original dwarfs separated by a thousand light years will only differ in temperature by about 14 °K due to their vortical age displacement.* Even if this could be measured, it is so small that it would be completely obscured by white dwarfs' natural temperature variability due to differences in mass and composition.

K. ANNIHILATION CYCLE

As introduced at the beginning of Part IV, the universe has two energy cycles, the fusion cycle and the annihilation cycle. The first has to do with the reshuffling of protons; the second deals with their creation and destruction. Of the two, the fusion cycle is far more manifest in space. Annihilation, in sharp contrast, is dark and enigmatic.

K.1 ANTIMATTER, THE GENERAL CASE

The Null Axiom requires a universe of 50% antimatter. It could be characterized as missing at present, although there is really no way to know for certain because an antigalaxy would look the same as a galaxy. Our astronomy books might be full of photos of antimatter. Antihydrogen, antihelium, and anticarbon all have the same spectral characteristics, melting points, and chemical properties as hydrogen, helium, and carbon. Because of its extremely volatile nature, antimatter cannot exist in close proximity to matter. The only thing keeping the universe's matter and antimatter stable is a large spatial separation as well as some form of gravitational and/or electromagnetic containment.

The search for direct evidence of cosmic antimatter has been ongoing since the late 1970s, culminating in the 2006 launch of the PAMELA satellite (Payload for Antimatter Matter Exploration and Light-nuclei Astrophysics). This and similar space-borne instruments are designed to detect the presence of antihelium and other antinuclei in cosmic rays. In 1997, Pascal Chardonnet calculated a virtually nonexistent probability for the random creation of antihelium by cosmic ray collisions. This means that *antihelium originates only when antistars burn antihydrogen*. Antistars lead naturally to antigalaxies, so the discovery of even a single compound antinucleus would constitute irrefutable evidence of an abundance of cosmic antimatter. Unfortunately, we currently have no way of knowing whether or not it is even possible for antihelium to reach Earth from deep space.

Unlike PAMELA and similar efforts, the focus of this appendix is *indirect* evidence of cosmic antimatter. It will investigate the luminosity generated from the interaction between celestial packages of antimatter and matter in deep space. This will be referred to as *dark annihilation*. The boundary between the mirror images of matter is the *disintegration zone*, or DZ. The luminosity in question is *primary annihilation* luminosity - radiation caused by the annihilation of otherwise stable, relatively low temperature matter and antimatter. This is certainly not the only source of annihilation radiation, however. Trace amounts of antimatter are continuously generated by cosmic rays throughout space and in interactions near the galactic core. Once these antiparticles find matter they annihilate, causing a background gamma noise that might partially mask the primary annihilation process.

ANTIMATTER DISTRIBUTION

The universe maintains a stable level of matter-antimatter annihilation reactions, at least as far as can be observed from Earth. How large do cosmic containment units need to be in order for gravitation or electromagnetism to maintain this stability? There are two ways matter and antimatter might be distributed across the heavens en masse:

- Finite scale. Diffuse finite regions scattered across space. If this is true, there are vast cosmic membranes where antimatter comes into contact with matter and annihilates in the most frenetic interaction possible. This annihilation will produce intermittent gamma ray energy coming from virtually all directions of space. As it turns out, gamma bursts have been observed throughout deep space, and their possible relationship to dark annihilation will be investigated shortly.

- Infinite scale. If this is true no trace of the disintegration zone will be found because matter's containment units are unbounded within a larger unbounded context. Omnielements' spatial symmetry is satisfied by infinite size and distribution. Their temporal symmetry might require their material distributions of matter and antimatter to be indistinguishable, which in turn may only be possible with infinite largeness.

Since the infinite case is untestable, let's search for a telltale signature of the finite case. Lumetic decay, however, creates an unbreachable range limit for this information.

K.2 ELECTROMAGNETIC RANGE

The mean free path of the 940 MEV photons released by proton annihilation is given by:

$$l = \left(\frac{m_H}{\rho_M \sigma_{pp}} \right) \tag{K.1}$$

where σ_{pp} is the pair production cross-section of a gamma ray, its principle mode of interaction at energies above 1 MEV. The value of this cross-section is $\sim 2(10)^{-30}$ m^2, so at an average universal density of $4.5(10)^{-27}$ kg/m^3, a 940 MEV gamma's mean free path is a staggering $1.85(10)^{29}$ m, or about 20,000 Gly.

In reality, the cross-section of Equation (K.1) only applies until the photon has decayed to ~1 MEV. This occurs in a distance given by Equation (15.10):

$$x = \left(\frac{c}{H_0}\right)\ln\left(\frac{E_0}{E}\right) = \left(\frac{c}{H_0}\right)\ln(z+1)$$

or about 110 Gly. Once the annihilation photon drops below ~1 MEV it can no longer provide enough energy for pair production. In another 320 Gly it will be a 1 mm microwave (~0.001 EV). Once an annihilation photon is redshifted down to the CMB the prevailing thermalization will quickly destroy any information it originally carried.

The universe's *observational horizon* is the distance required for a gamma ray emitted from proton annihilation to decay into a 1 mm microwave:

$$R_\lambda = \left(\frac{c}{H_0}\right)\ln\left(\frac{m_p c^2}{\frac{hc}{0.001}}\right) = \left(\frac{c}{H_0}\right)\ln\left(\frac{m_p c}{1000h}\right) \tag{K.2}$$

This is equal to ~430 Gly, depending on the correct value of the Hubble constant.

The only way information can penetrate this observational horizon is if it begins with more energy than a proton:

Ψ THEOREM K.1 - OBSERVATIONAL HORIZON {Ψ15.2}

THE MAXIMUM RANGE OF ANNIHILATION INFORMATION IS THE DISTANCE REQUIRED FOR A PHOTON OF PROTON ENERGY TO DECAY TO CMB ENERGY

Any massive collections of antimatter located more than ~400 Gly from our solar system will be virtually impossible to detect.

K.3 SEARCHING FOR A FINITE GAMMASTRUCTURE

The universe's celestial quanta of matter and antimatter will be referred to as *gammastructures* since their boundaries might, at least in the finite case, be delineated by the gamma radiation produced in annihilation reactions. So again:

If gammastructures are finite, what is their size?

The largest visible structure of the universe, as revealed by redshift surveys, is a foam-like lattice that surrounds empty voids about 300 Mly in diameter.[8.1] Is this structure due to the presence of antimatter or strictly a by-product of fusion and gravity? The annihilation of antimatter produces about 140 times the power of fusion. Does this mean gammastructures are 140 times larger than galaxies? Or are they even more immense? The luminosity density of space, in power/distance^3, is related to luminous density by Equation (15.33):

$$j_B = H_0 \rho_{B\gamma}$$

This shows a relationship between the rate of a process (fusion) and the density of its product (light). The same is true of dark annihilation reactions. The uncertainty is whether or not the product of dark annihilation has even been observed.

Since gamma and X-ray radiation have a low energy density in space, and since annihilation releases so much more energy than fusion, Equation (15.33) tells us the universal rate of annihilation is much lower than fusion. The rate of dark annihilation, in turn, is related to a gammastructure's surface area to volume ratio. It is therefore possible to estimate a lower limit of gammastructure size based on residual energy density.

The problem is what is, specifically, the residual energy density due to dark annihilation? Space contains a diffuse and isotropic flux of X-ray and gamma radiation. Galactic processes are responsible for some of this energy, perhaps even the lion's share of it. None of it shows the clear signature of dark annihilation reactions - spikes near electron and proton rest energies. A proton energy signature would be particularly telling since fewer cosmic processes produce antiprotons than positrons. Furthermore, if the boundary of our local gammastructure is sufficiently distant from us, lumetic decay would redshift its annihilation energy into the X-ray or even ultraviolet band. The only way to identify a distant edge of a gammastructure is by dual electromagnetic flux peaks with an energy ratio equal to the electron/proton mass ratio. Unfortunately, none of the currently deployed gamma ray observatories are designed to look for this composite signature.

GAMMA FLUX

The area associated with a gammastructure's DZ will be denoted A_{DZ}. The material flux moving across its surface is:

$$M_{flux} = \left(\frac{\overline{\mathrm{v}}_{DZ}}{4} \right) A_{DZ} \rho_{DZ} \tag{K.3}$$

where ρ_{DZ} is the material density at the gammastructure boundary and v_{DZ} is its mean velocity. The factor of four in the first term is the standard relationship between density and flux. Most of the universe's material is composed of hydrogen. Calculations done in Appendix P indicate that its average speed, as a gas in deep space, is 240 m/s.

A gammastructure's effective area is subject to some interpretation. In a random arrangement, adjacent gammastructures will be the same type of matter half of the time. This means half of a gammastructure's surface area is subject to annihilation as shown in the rectangular cell in Figure (K.1):

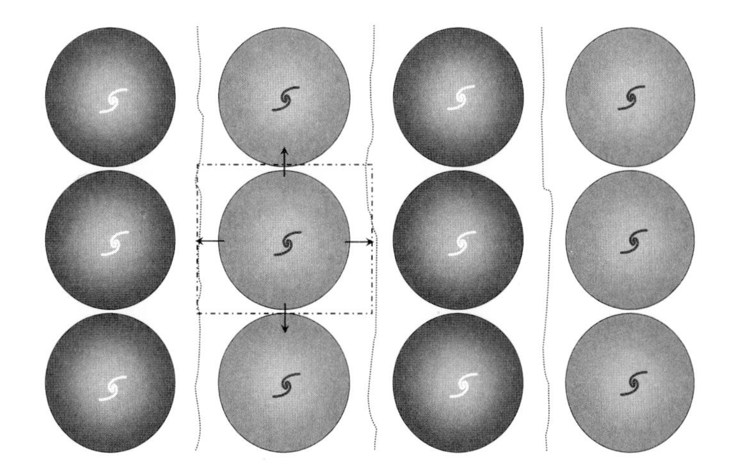

Figure (K.1) Schematic representation of material flux into the DZ of a finite gammastructure

However, flux moves in both directions across half of the area, so the effective surface is the DZ's entire area. Since flux in both cases is moving in the direction of a predominantly opposite type of matter, the annihilation rate will be taken as 100%. The *dark gamma luminosity*, L_{DZ}, arising from the material moving across the DZ is given by:

$$L_{DZ} = \frac{\overline{v}_{DZ} c^2 \rho_{DZ} A_{DZ}}{4} \qquad \text{(K.4)}$$

The subatomic cross-section or rate of interaction is not important because at equilibrium particles are annihilated at the same rate they cross the DZ. Otherwise the material density near the DZ boundary would continually increase.

Convert dark gamma luminosity into a dark gamma luminosity density by dividing by the gammastructure volume:

$$\frac{L_{DZ}}{V_{DZ}} = \frac{\overline{v}_{DZ} c^2 \rho_{DZ}}{4} \left(\frac{A_{DZ}}{V_{DZ}} \right) \qquad \text{(K.5)}$$

GAMMASTRUCTURE GEOMETRY

Dark gamma luminosity is sensitive to the ratio between a DZ's area and volume, and therefore linked to its topography. Surface/volume ratios for various three-dimensional structures are shown in the following table. The surface area of filaments is exclusive of their ends; sheets are exclusive of their edges.

Surface to Volume Ratios for Various Geometries				
Topography	**Metric**	**Volume**	**Area**	**Area/Volume**
Filament	Radius=R	$\pi R^2 L$	$2\pi RL$	$2/R$
Sheet	Thickness=R	LWR	$2LW$	$2/R$
Sphere	Radius=R	$4/3\pi R^3$	$4\pi R^2$	$3/R$

Table (K.1) Surface area to volume ratios by topography

Since matter and antimatter's distribution is random, not spherically symmetrical, the area to volume ratio of the finite gammastructure case will be defined as $(2/R)$. Substituting this term into Equation (K.5) yields:

$$\frac{L_{DZ}}{V_{DZ}} = \frac{\overline{\mathrm{v}}_{DZ} c^2 \rho_{DZ}}{2R_{DZ}} \tag{K.6}$$

As a finite gammastructure's radius increases, its light to volume ratio decreases because a DZ's area increases less rapidly than its volume, and area determines the total flux.

All photons whose energy is appreciably greater than those in the CMB have the same relationship between luminosity density and luminous energy density, as described by Equation (15.33). Apply this provision to the DZ's gamma flux:

$$\frac{L_{DZ}}{V_{DZ}} = j_{DZ} = H_0 \rho_{DZ\gamma} \tag{K.7}$$

Substitute this into Equation (K.6) and solve for R_{DZ}:

$$R_{DZ} = \frac{\overline{\mathrm{v}}_{DZ} \rho_{DZ} c^2}{2H_0 \rho_{DZ\gamma}} \tag{K.8}$$

As described in Appendix I, the upper limit of the dark annihilation photon energy near Earth (from Sreekumar, 1998)[39] amounts to an energy density of $1.5(10)^{-18}$ J/m^3. This is about 10% of the gamma background published earlier by Silk, included in Appendix C, of

$1.5(10)^{-17}$ J/m^3.[2.1] The greater the dark gamma density, the smaller the radius of the finite gammastructure. Its minimum value can be established by using Silk's estimate and assuming dark annihilation is its sole source. But what of intergalactic material density?

Fully 98% of the universe's dark matter is hydrogen. The calculations in Chapter 16 suggest that most of it passes through the galactic vortex, and as such it is an integral part of galactic structure. If so, galaxies contain as much as ~98% of the universe's matter. The Milky Way's halo reaches out to at least 750 Kly[1.19] from its center but the universally average intergalactic spacing is ~14 Mly,[3.2] ~19 times as great. This means ~98% of the universe's material is concentrated into only 0.015% of its space, or conversely, 2% of its material is distributed into 99.985% of its space. This in turn means deep space's material density is ~2% of the universal average, putting ρ_{DZ} near $9(10)^{-29}$ kg/m^3 when ρ_M=$4.5(10)^{-27}$ kg/m^3. Using this density, Silk's gamma estimates, and an average DZ material speed of 240 m/s, Equation (K.8) yields a *minimum* gammastructure size of 3.4 billion light years.

If gammastructures are finite, and the Milky Way is relatively close to one of their boundaries, Earth would receive gamma radiation from annihilation events:

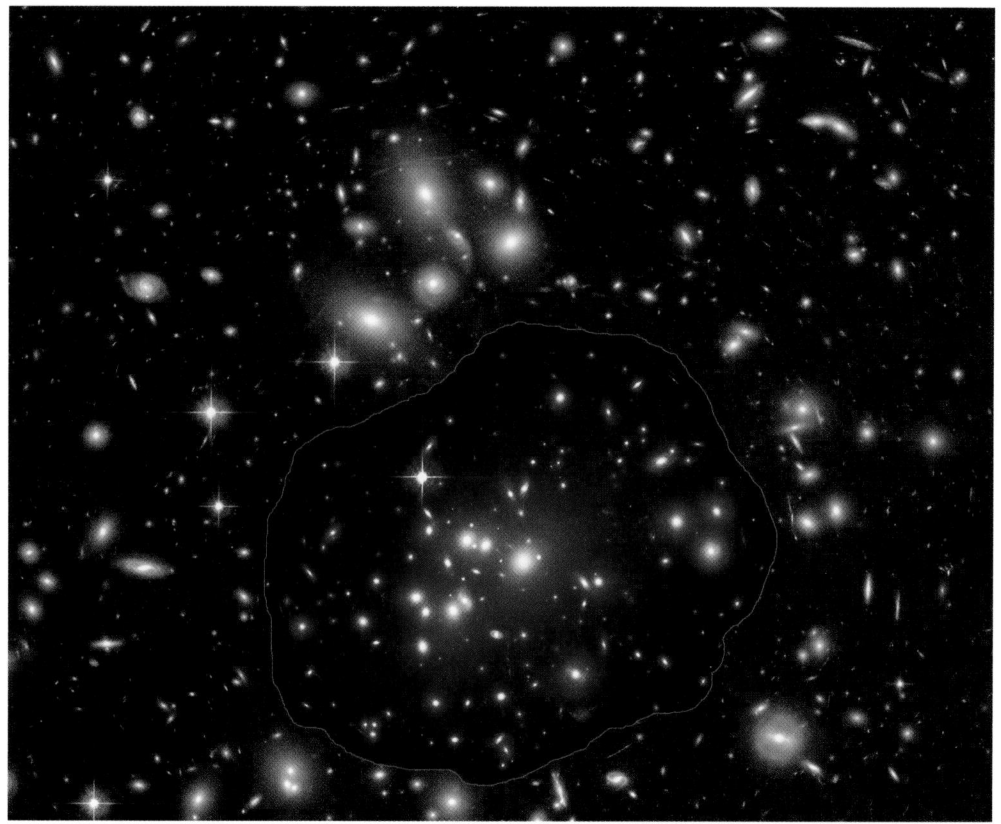

Figure (K.2) Are gammastructures small enough to produce a measurable interaction?

Let's look at some additional data.

K.4 GAMMA RAY BURSTERS

On an average of about once a day, Earth is bathed in a short burst of energetic gamma rays.[6.2] These bursts have a typical duration of 0.01 to 1000 seconds, rise time of 0.0001 seconds followed by an exponential decay, and photon energy between 1 KEV and 100 MEV. Although it has been difficult to pinpoint the objects responsible for this radiation, the Compton Gamma Ray Observatory (CGRO) and similar instruments have gathered a wealth of data about their general nature. Gamma bursts have an isotropic distribution throughout the sky and less than 0.1% repeat. The sources of this energy are referred to as *gamma ray bursters*.

The spectral composition of nonrepeating bursters varies widely and is usually complex and multi-peaked. Repeating bursters, in contrast, tend to have simpler, singly peaked outputs. This has led most astronomers to believe that the two types originate from different cosmic processes. Some repeating bursts have been linked to the known location of neutron stars, whereas the nonrepeating majority has no evident connection to any visible objects. Are nonrepeating bursters direct evidence of dark annihilation? Perhaps a rare few are, but the abrupt events studied by CGRO do not correspond well with the diffuse, low-density annihilation characterized earlier for the disintegration zone. Moreover, the mapping of these sources is difficult to associate with dark annihilation because:

- Lumetic decay induces significant signal distortion.
- The large-scale distribution of matter/antimatter might be complex.

The energy of the gamma photons in the bursts tends to be much lower than the annihilation energy of protons, so either it is an extraordinarily distant form of dark annihilation or originates from an entirely different process.

Since no method is available to measure a burster's distance, their intensity is described in terms of energy per area, or *fluence*. Burster fluence spans a fairly broad intensity range, from a low of 10^{-12} J/m^2 all the way to a high of 10^{-6} J/m^2. In general, the weaker a burster, the longer its duration. This suggests that they originate from cosmological distances, since space's average curvature causes photons to disperse as they expand.

Bursters have irregular spectral structures, blazing luminosity (again assuming great distance), and precipitous rise times, so if any are caused by annihilation, they are probably borne of the chance encounters of macroscopic objects in the disintegration zone, such as a meteor and anti-meteor. They certainly don't have the appearance of the reaction between ultra-low density hydrogen and antihydrogen. Indeed, any electromagnetic process that minimizes hydrogen's density in the disintegration zone is ill-suited for trapping a meteor or anti-meteor. Just the fact bursters are almost exclusively nonrepeating means they are either isolated annihilation collisions or rare irreversible transitions of distant celestial objects.

Over a mission spanning six years, CGRO collected about four years of data in four gamma energy ranges: 20-50, 50-100, 100-300, and >300 KEV. This information was published as the BATSE Burster Catalog 4B, which lists a total of 1292 different burst events. Each burst typically generates fluence in at least two out of the four detector channels with a maximum fluence in a particular channel. Figure (K.3) shows the numerical distribution of peak fluence energy as well as average fluence for each of the four channels:

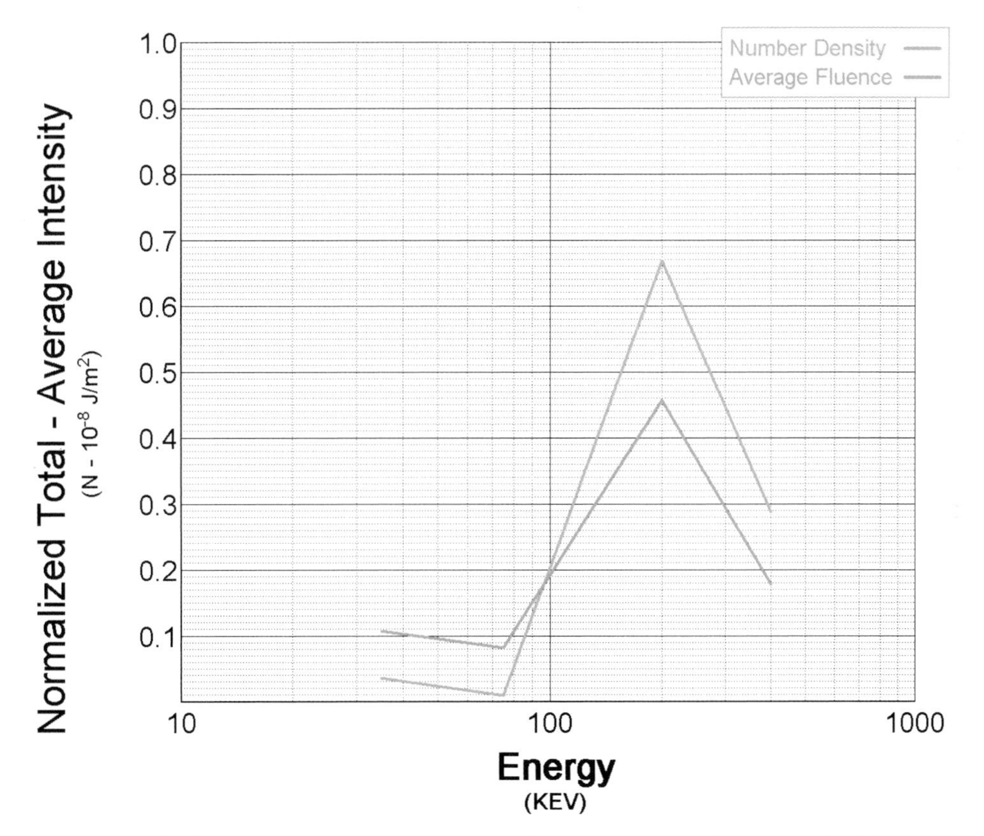

Figure (K.3) Photon energy population density of peak and average fluence for 1292 bursters

The green trace is the normalized total number of bursters; the orange trace is average burster fluence. The energy axis denotes the center energy of a channel, such as (20–50)/2=35 KEV for channel 1. Curiously, the maximum average fluence and peak burster population density both occurred on channel 3, 100–300 KEV. About 67% of the 1292 bursters had maximum fluence in this range.

In general, bursters exhibit characteristics that can be attributed specifically to either a neutron star environment or great cosmological distance:

- Photon energy of ~300 KEV (neutron star).
- Rise time of 0.0001 (neutron star).
- Association between repeaters and neutron stars (neutron star).
- Rough proportionality between intensity and duration (cosmological distance).
- Isotropic distribution (cosmological distance).
- No correlation with visible objects (cosmological distance).

These characteristics suggest that most bursters probably originate from rare state changes in neutron stars located cosmological distances from Earth. Perhaps a burster is the energy a binary neutron star system releases when it coalesces.

The general location of bursters is currently a hotly debated topic in astrophysics. While many scientists believe they originate from cosmological distances, others are convinced they are distributed in the Milky Way's galactic halo. The halo idea is suspect for two reasons. First, our galaxy's material has a continuous distribution; the delineation between its halo and outer rim is largely a matter of convention. As such it is difficult to believe the halo's burster distribution would be as uniform as it is. It should at least demonstrate a slight correlation to the galactic plane, and it does not. Secondly, if there are bursters in the Milky Way's halo there should also be bursters in Andromeda's halo. These would appear as a diffuse source of low-energy signals originating from the general direction of our massive spiral neighbor. The distribution of known bursters shows no such asymmetry.

Bursters represent a great deal of energy density, something not typically available in deep space at 2.7 °K. Their rapid rise time requires emission from a spatially discrete region (at least by astronomical standards). Both of these requirements can be satisfied by either neutron stars or bulk annihilation in the DZ, but the observed photon energy places severe limitations on the latter. The only way bursters could be bulk annihilation events is if they occurred at distances sufficient for lumetic decay to shift either electron or proton rest energy to between 100 and 300 KEV. For positron-electron annihilation, gamma photons originating at 511 KEV decay to 300 KEV after traversing 8.5 Gly. For proton-antiproton annihilation, it takes 940 MEV photons an unfathomable journey of 130 Gly to decay down to 300 KEV. In general, the fluence I at some distance R is defined by intrinsic burster energy, E_B, as follows:

$$I = \frac{E_B}{4\pi R^2} e^{-\frac{H_0 R}{c}} \quad \text{(K.9)}$$

Equation (K.9) is shown below for intrinsic burster energies of 10^{44} J (green), 10^{45} J (orange), and 10^{46} J (red):

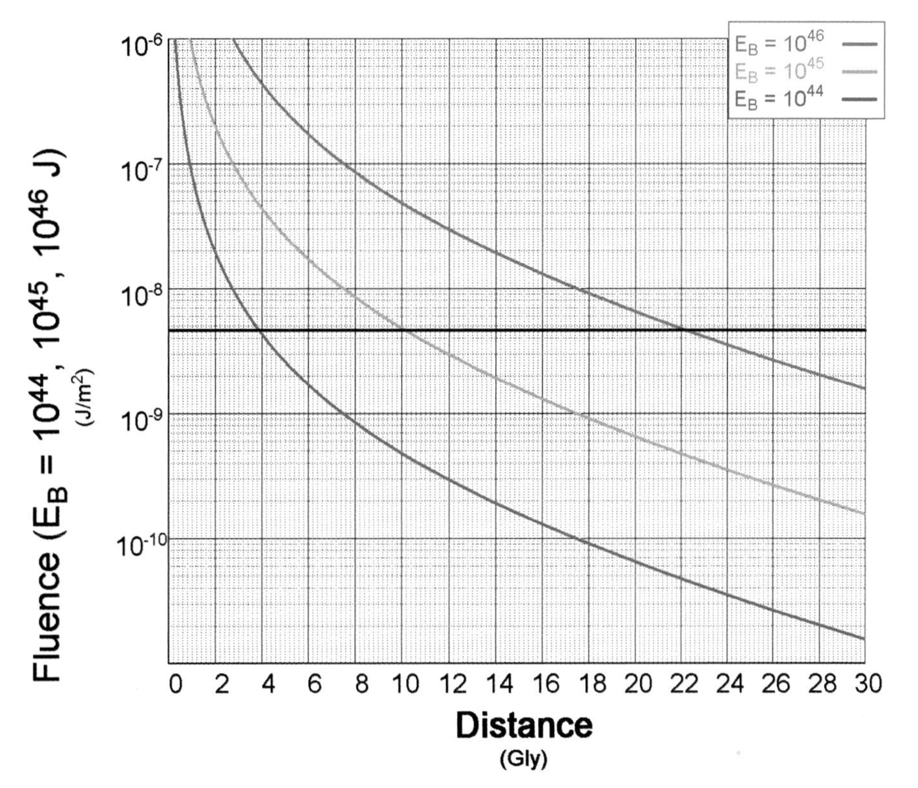

Figure (K.4) Burster fluence as a function of distance, $E_B = \{10^{44}\ \mathrm{J},\ 10^{45}\ \mathrm{J},\ 10^{46}\ \mathrm{J}\}$

Bursters' isotropic and intensity/duration characteristics suggest a range of distribution across the observable universe. If they are a cosmologically rare transition for neutron stars, their intrinsic energy and spatial distribution ought to have a great deal of uniformity. The horizontal line in the above graph represents the peak average fluence $4.6(10)^{-9}$ J/m^2 of the orange trace of Figure (K.3). If the broadening of burster duration is due to their cosmological redshift, which is also consistent with their isotropic distribution, then their distance is often several Gly. This puts their intrinsic energy at least on the order of $\sim 10^{45}$ J. The energy in our sun's rest mass amounts to $2(10)^{47}$ J. It is not inconceivable for a change of state of a 1 M_{sun} neutron star to release ~0.5% of its total energy in a fraction of a second, given the other short periodicities they have, such as the rotation rate of some pulsars.

The original purpose of our burster investigation was to determine whether or not they are related to the cold dark annihilation of the DZ. Judging from our analysis, the bursters measured by CGRO are entirely unrelated to this. They remain an intriguing celestial phenomenon, probably more closely related to the life cycle of a compact object than the bulk disintegration of matter.

K.5 GAMMASTRUCTURE SIZE

The available data does not provide compelling evidence for finite gammastructures, but the galactic scenario can be effectively ruled out by the energy and energy density of gamma background photons. Three scenarios remain:

- Gammastructures are astronomically large. If galactic clusters or superclusters are gammastructures, then there should be a positive correlation for increased gamma luminosity in a fraction of the spaces between them.

- Gammastructures are immense beyond current observational range. If our solar system lies within a detectable distance of a DZ boundary, it should be possible to identify a structured anisotropy in gamma, X-ray, or ultraviolet bands. No such effect has as yet been reported. There is a general tendency, as the energy scale rises, to have grainier background content, but the available data look reasonably random.

- Gammastructures are infinite. This is a real possibility. The gamma background of the universe is not inconsistent with the known density of high-energy interactions. *None of it need be attributed to a local DZ.* Even a billionth of an omnielement's volume is still infinite. How large do gammastructures need to be in order to provide omnielement ultrasymmetry? Or rather how large can they be without violating this symmetry?

In the absence of observational confirmation, perhaps computer modeling of the universe's energy flow will eventually isolate the balance between matter and its enigmatic twin. If annihilation is essential to the cosmic energy flow at a local level, gammastructures are indeed finite.

Determination: Space's gamma background is just too weak and too broad spectrum to support small gammastructures. Although the evidence is largely circumstantial, it leans in the direction of a vast level of structure significantly beyond the galactic scale:

Ω HYPOTHESIS K.1 - MINIMUM GAMMASTRUCTURE SIZE {Ψ15.2}

THE MINIMUM SIZE OF GAMMASTRUCTURES IS AT LEAST AN ORDER OF MAGNITUDE LARGER THAN THE GALACTIC SCALE

Determining the size of gammastructures may ultimately play a key role in understanding the universe's largest finite structure.

L. ENERGY LOSS IN A REDSHIFTED BLACKBODY SPECTRUM

L.1 BLACKBODY BASICS

A blackbody spectrum's energy density is related to its temperature by:

$$\rho_E = \frac{4\sigma T^4}{c} \tag{L.1}$$

where σ is the Stefan-Boltzmann constant.

The equation describing this energy density as a function of wavelength is:

$$\rho_E(\lambda)d\lambda = \left(\frac{hc}{\lambda}\right)\left(\frac{8\pi}{\lambda^4\left(e^{\frac{hc}{\lambda kT}}-1\right)}\right)d\lambda \tag{L.2}$$

Energy density is the product of individual photon energy (first term) and numerical density (second term). Since a frequency interval corresponds to an energy interval in accordance with the Planck relation ($E = hv$), it is often appropriate to describe thermal distributions in terms of frequency:

$$\rho_E(v)dv = (hv)\left(\frac{8\pi v^2}{c^3\left(e^{\frac{hv}{kT}}-1\right)}\right)dv \tag{L.3}$$

Again, the first term is individual photon energy (hv) and the second is their numerical density as a function of frequency.

A photon's energy is inversely proportional to its wavelength, so thermal spectrums have different peaks depending on whether energy density is expressed as a function of wavelength or frequency. The wavelength peak (maximum energy per wavelength interval) of a blackbody is:

$$\lambda_\lambda = \left(\frac{1}{4.9651}\right)\left(\frac{hc}{kT}\right) \tag{L.4}$$

Alternatively, the frequency peak (maximum energy per frequency interval) is:

$$\lambda_\nu = \left(\frac{1}{2.8216}\right)\left(\frac{hc}{kT}\right) \tag{L.5}$$

The wavelength dividing a thermal spectrum into two equal amounts of energy is:

$$\lambda_h = \left(\frac{1}{3.5021}\right)\left(\frac{hc}{kT}\right) \tag{L.6}$$

Blackbody photon number density is:

$$n_\gamma = (2.404)8\pi\left(\frac{kT}{hc}\right)^3 \tag{L.7}$$

Average photon energy is the ratio of energy density to number density:

$$\overline{E}_\gamma = \frac{\rho_E}{n_\gamma} = \left(\frac{\pi^4}{36.06}\right)kT = (2.701)kT \tag{L.8}$$

Average photonic volume is the inverse of number density, written here in terms of the peak energy density wavelength, Equation (L.4):

$$\overline{V}_\gamma = \frac{1}{n_\gamma} = \frac{1}{(2.404)8\pi}\left(\frac{hc}{kT}\right)^3 = (2.026)\lambda_\lambda^3 \tag{L.9}$$

Regardless of which characteristic wavelength is chosen, from Equation (L.4) to Equation (L.6), *the average volume of a photon in a thermal spectrum is proportional to the cube of its wavelength*.

L.2 TOTAL RADIANCY OF A REDSHIFTED BLACKBODY SPECTRUM

A red-shifted thermal spectrum can be expressed as a function of wavelength as:

$$R_T(\lambda)d\lambda = \left(\frac{hc(z+1)}{\lambda}\right)\left(\frac{2\pi c(z+1)^2}{\lambda^4\left(e^{\frac{hc(z+1)}{\lambda kT}}-1\right)}\right)d\lambda \qquad \text{(L.10)}$$

Even a slight redshift of (z = 0.005) distorts the CMB's Planck spectrum into a nonthermal distribution:

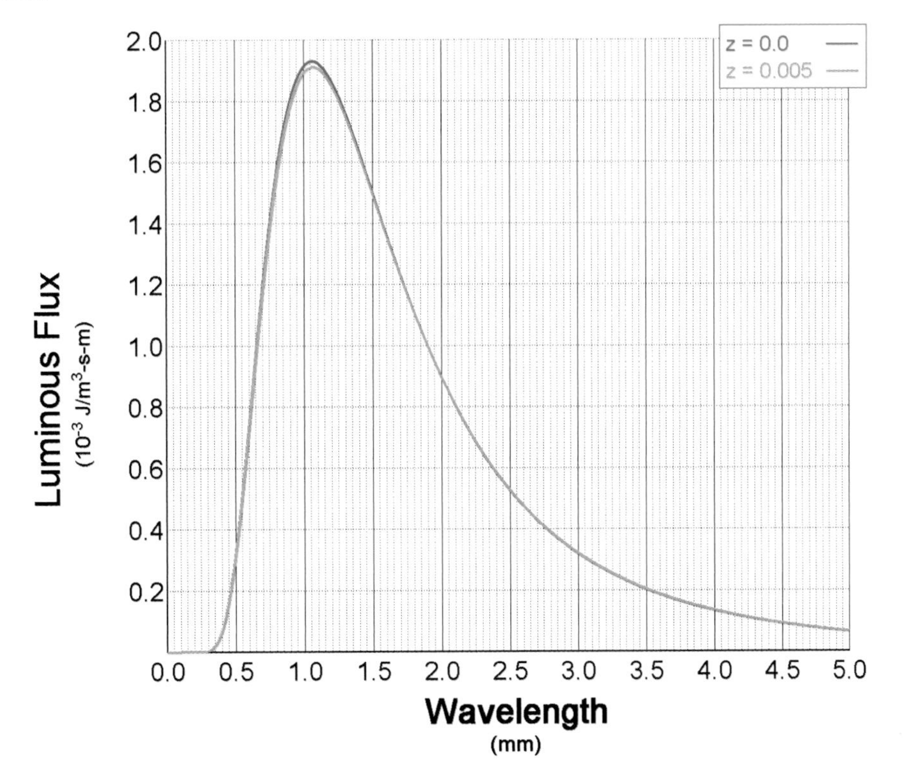

Figure (L.1) CMB spectrum (green) compared to slightly redshifted (z = 0.005) CMB spectrum (orange)

Using Equation (14.2):

$$H_0 \cong \frac{cz}{d} \qquad \{z < 0.2\}$$

a redshift of 0.005 amounts to a distance of ~80 Mly (at H_0 = 60 Hz-km/Mpc). *The CMB contains no redshift components representing distances in excess of ~80 Mly.*

In the frequency domain, a redshifted blackbody presents a similar form, although more directly tied to a photon's energy:

$$R_{\mathrm{T}}(v)dv = (hv)(z+1)\left(\frac{2\pi v^2(z+1)^2}{c^2\left(e^{\frac{hv(z+1)}{kT}}-1\right)}\right)dv \quad \text{(L.11)}$$

The total radiancy in this spectrum is the integral from zero to infinity. Factor in the necessary terms:

$$R_{\mathrm{T}} = \int_0^{\infty} R_{\mathrm{T}}(v)dv = \left(\frac{2\pi k^3T^3}{c^2h^2(z+1)}\right)\int_0^{\infty}\left(\frac{\left(\frac{hv(z+1)}{kT}\right)^3}{e^{\frac{hv(z+1)}{kT}}-1}\right)dv = \frac{2\pi^5k^4T^4}{15c^2h^3(z+1)} \quad \text{(L.12)}$$

This was integrated using the definite integral:

$$R_{\mathrm{T}} = \int_0^{\infty}\left(\frac{x^3}{e^x-1}\right)dx = \frac{\pi^4}{15} \quad \text{(L.13)}$$

where:

$$x = \frac{hv(z+1)}{kT} \qquad dx = \frac{h(z+1)}{kT}dv \quad \text{(L.14)}$$

The total radiancy of Equation (L.12):

$$R_{\mathrm{T}} = \frac{2\pi^5k^4T^4}{15c^2h^3(z+1)}$$

can be written in a much more compact form as:

$$R_T = \frac{\sigma T^4}{(z+1)} \quad \text{(L.15)}$$

where σ is the Stefan-Boltzmann constant. The total energy present in the spectrum decreases inversely as $(z + 1)$. This is as expected since it is the individual energy loss of all of its photons.

M. SURFACE BRIGHTNESS LOSS IN IMAGES OF DISTANT OBJECTS

The surface brightness reduction in images from astronomically distant objects is severe, and is thought to vary with redshift as $\sim 1/(z+1)^4$.⟨8.7⟩ Lumetic decay is responsible for the lion's share of this loss, as a combination of two effects:

- It lowers the energy of individual photons and therefore the energy in the entire spectrum by a factor of $1/(z+1)$, as derived in Appendix L.

- It shifts photon populations into new bandwidths, resulting in energy loss in certain bands far in excess of $1/(z+1)$.

Let's derive the magnitude of the optical spectrum's energy loss as a function of z.

There are a number of different photometric systems used to quantify celestial images. One enjoying widespread use is called the *UBV* system. It covers the spectral range from Ultraviolet [365 ± 34 nm] to Blue [440 ± 49 nm] and finally Visual [550 ± 45 nm]. The ratio of energy in the various bands defines a celestial object's *color*.

When the light from stars and galaxies is averaged, its radiancy distribution is similar to an attenuated 10,000 °K blackbody spectrum:

$$R_T(\lambda)d\lambda \cong \alpha_B\left(\frac{hc}{\lambda}\right)\left(\frac{2\pi c}{\lambda^4\left(e^{\frac{hc}{\lambda kT}}-1\right)}\right)d\lambda \qquad \text{(M.1)}$$

where the attenuation factor α_B is the ratio of the observed energy density of integrated starlight and the energy density of a 10,000 °K blackbody field:

$$\alpha_B = \frac{\rho_{B\gamma}}{\rho_E} = \frac{c\rho_{B\gamma}}{4\sigma T^4} \qquad \text{(M.2)}$$

where $\rho_{B\gamma}$ is the density of luminous radiation in space, from Appendix C.

The radiancy of redshifted integrated starlight is approximated by substituting Equation (M.2) into Equation (L.10):

$$R_{\mathrm{T}}(\lambda)d\lambda \cong \left(\frac{c\rho_{B\gamma}}{4\sigma T^4}\right)\left(\frac{hc(z+1)}{\lambda}\right)\left(\frac{2\pi c(z+1)^2}{\lambda^4\left(e^{\frac{hc(z+1)}{\lambda kT}}-1\right)}\right)d\lambda \tag{M.3}$$

Radiancy (also called luminous flux) is measured as power/distance^2. The following graph shows the attenuated 10,000 °K blackbody spectrum of Equation (M.3) at redshifts of $z = 0$, 1, 2, 3, and 4 marked with the full width of the *UBV* photometric system:

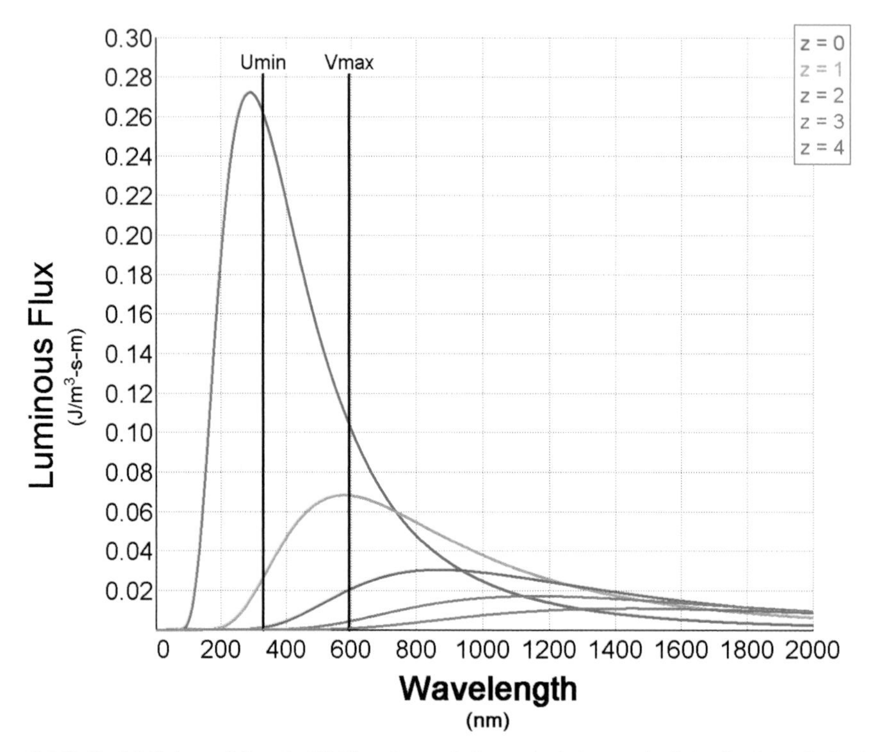

Figure (M.1) Redshift loss of flux in *UBV* band, $z = 0$ (green), 1 (orange), 2 (red), 3 (violet), 4 (blue)

At ($z = 1$), the flux in the fairly wide *UBV* band (331-595 nm) has fallen to 31% of its original magnitude, not 50% as expected by $1/(z + 1)$. At ($z = 2$) the situation is much worse; only 5.4% of the initial energy is present instead of the 33% given by $1/(z + 1)$. This effect will be referred to as *photon migration*.

Ψ THEOREM M.1 - PHOTON MIGRATION {Ψ15.2}

A FIXED WAVELENGTH BAND'S ENERGY DENSITY LOSS DUE TO LUMETIC DECAY IS MARKEDLY GREATER THAN THE ENERGY LOSS OF ITS INDIVIDUAL PHOTONS

The following shows the fraction of energy remaining in the *UBV* band following photon migration. It is depicted as a function of z:

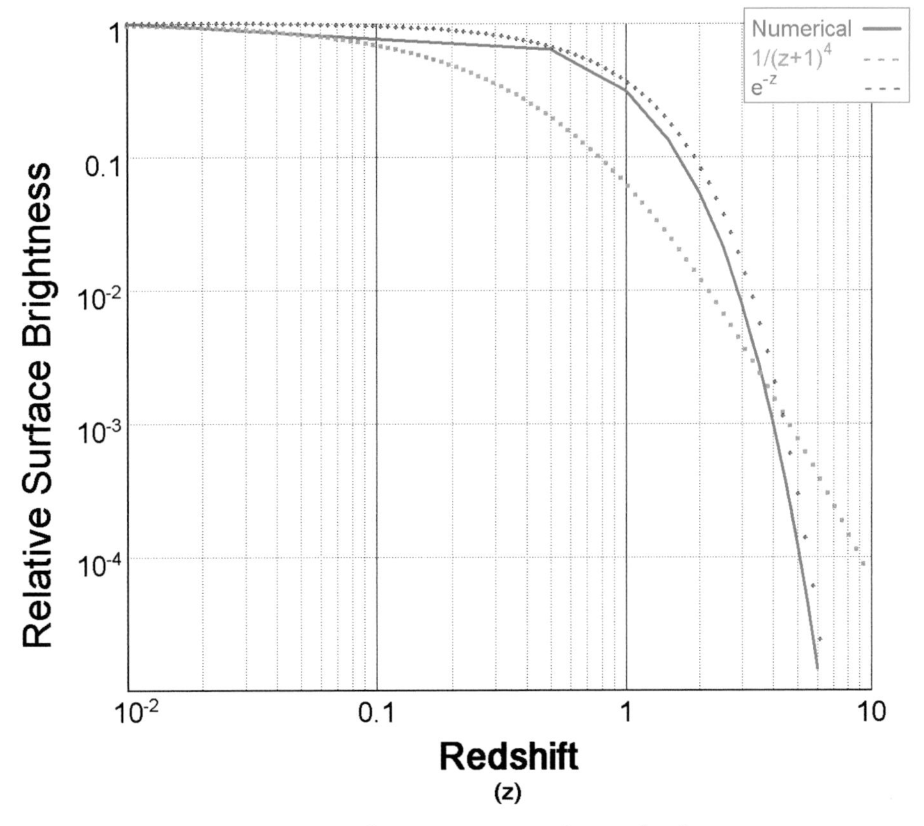

Figure (M.2) Photon migration in the *UBV* band

The solid trace is a numerical calculation of a 10,000 °K thermal spectrum's migration through the *UBV*. The orange trace is $1/(z+1)^4$, which is thought to approximate the loss of surface brightness in distant objects. As shown in Figure (M.2), the $1/(z+1)^4$ profile actually attenuates more slowly with high values of z than the photon migration effect. Migration is more accurately characterized by the exponential form:

$$\frac{I}{I_0} = e^{-z^{1.3}} \qquad \text{(M.4)}$$

This is the violet trace in Figure (M.2) that fits the numerical solution (green).

Intergalactic redshift has a devastating impact on the effective luminosity of distant objects. The fact that our instruments can image galaxies with redshifts in excess of ($z = 5$) is nothing short of astonishing.

N. NEUTRINOS AND DARK MATTER

According to galactic rotation profiles, universal composition calculations, and estimates of the Hubble constant, at least 95% of the universe's material has gone missing. Where is it?

There are two possible candidates:

- Cosmic neutrino background.
- Hydrogen.

N.1 NEUTRINOS

The existence of neutrinos isn't typically contested, but virtually all of their basic properties are. Their reality was originally demonstrated using energy conservation. *When a block of beta-emitting radioactive material is placed in a sealed chamber designed to be opaque to gamma radiation, the amount of mass the block loses over time is greater than the amount that the chamber is heated.* Some entity other than gamma radiation carries energy away from beta decay, and it was designated the *neutrino* in the early 1930s, although they weren't actually detected until 1956.

Neutrinos are electrically neutral bundles of energy, so their quantization is governed by unit hypervolume, not unit polarvolume. This also means that they need to propagate at c in order to maintain stability. So if a neutrino is a neutral energy packet, quantized by unit hypervolume, moving at the velocity of light, what differentiates it from a photon?

Another curious thing about neutrinos is the spectral characteristics of beta decay. When a neutron decays in free space, it liberates a proton, electron, and neutrino. With the exception of a very small proton component, the binding energy associated with this decay is spectrally distributed between the electron and neutrino. In some cases the electron receives virtually no energy and the neutrino carries 0.78 MEV. In others, the electron streaks away from the decay site with 0.78 MEV of kinetic energy and the neutrino has negligible energy. These are the two limits, and the neutron's decay energy can fall on them or anywhere between.

So the question is this:

> *If a neutron can, if only rarely, decay into a proton and fast-moving electron, and a fast-moving electron is equivalent to the Compton scattering of a motionless electron by a high-energy photon, what is the essential difference between a photon and a neutrino?*

This leads to the realization that a neutrino is a photon *in a different state*:

Ω HYPOTHESIS N.1 - NEUTRINO {Ψ8.1}

> *A NEUTRINO IS A PHOTON'S BOUND STATE. IT CONSISTS OF TWIN PHOTONS IN ULTRA-CLOSE PROXIMITY, PROPAGATING ALONG THE SAME TRAJECTORY WITH VIRTUALLY THE SAME ENERGY AND MOMENTUM*

There are only two possible formulations of universal finiteness, open and closed. Just as unstable particles are bound combinations of cores and anticores, neutrinos consist of two bound photons in exceptionally close proximity. The difference between neutrinos and neutral elementary particles, however, is that bound photons move at the speed of light and are therefore stable by default. It is no coincidence that bound particles are the only source of neutrinos.

Just as bound electrons exist in an entirely different state than free electrons, bound photons are uniquely different from their free-space manifestation.

When a particle and antiparticle annihilate, they release twin photons traveling in opposite directions to conserve momentum. When a bound particle's core expands, it releases twin photons moving in the *same* direction because it is an asymmetric interaction. *A neutrino is an inverted form of pair-polarization.* The reason why it has such a small cross-section in space is because the proximity of its component photons cancels virtually all of their field deflection. Even though photons are electrically neutral particles, neutrinos take this neutrality to an entirely new level by collapsing the photons' fields to far smaller than their usual spatial footprint. In much the same way that a proton reduces an electron's fields to nuclear dimensions in a neutron, the complementary photons of a neutrino's substructure reduce its fields to sub-nuclear extent. The only energy form that can bind a photon is another photon, because it is the only expression of unit hypervolume that:

- Has temporal symmetry.
- Moves at the speed of light.

NEUTRINO DENSITY

The reason why our eternal universe isn't completely inundated with neutrinos is because they decay just like free photons. Lumetic decay requires that an energy form have three characteristics:

- Spatially distributed quantized energy.
- Propagation through space.
- Nonthermal relationship to the IGM.

Most photons, with the sole exception being the microwaves of the CMB, have all three of these properties, and by extension, so do neutrinos, regardless of how spatially discrete they might be.

As neutrinos decay, their photon-photon binding deteriorates accordingly. It is possible, however, that this binding falls apart faster than the rate of intergalactic redshift. Circumstantial evidence suggests that the neutrinos our sun emits actually change form on their brief journey from its surface to the detectors we have buried deep at the bottoms of several mine shafts. Bound photons experience no Coulomb attraction; their bond is limited to their momentum and spatial proximity. Any velocity dispersion in a neutrino's composition will cause it to break down into its constituent photons in a timeframe consistent with the magnitude of said dispersion.

Discrete radiant energy bands exist at a luminous limit defined by the Hubble constant, as shown in the optical band of Equation (15.32):

$$\rho_{B\gamma} = \frac{j_B}{H_0}$$

Like starlight, the universal energy density of neutrinos is governed entirely by luminosity density and the Hubble constant. If universal luminosity contains 4% neutrinos, based on our sun's fractional output and the idea that all main-sequence stars exhibit similar nuclear processes, the universal density of neutrino energy is about 4% of its optical energy density. This is consistent with matter-energy correspondence (Ψ15.3), which also tells us that the universe's cosmic neutrino energy density is negligible. This in turn means that dark matter is predominantly hydrogen - *dark hydrogen*. Let's see what our galaxy's visible material can tell us about where all of this dark hydrogen might be hiding.

N.2 DARK HYDROGEN

The number density/mass distribution of the Milky Way's stars is given by the following graph:⟨1.8⟩⟨1.10⟩

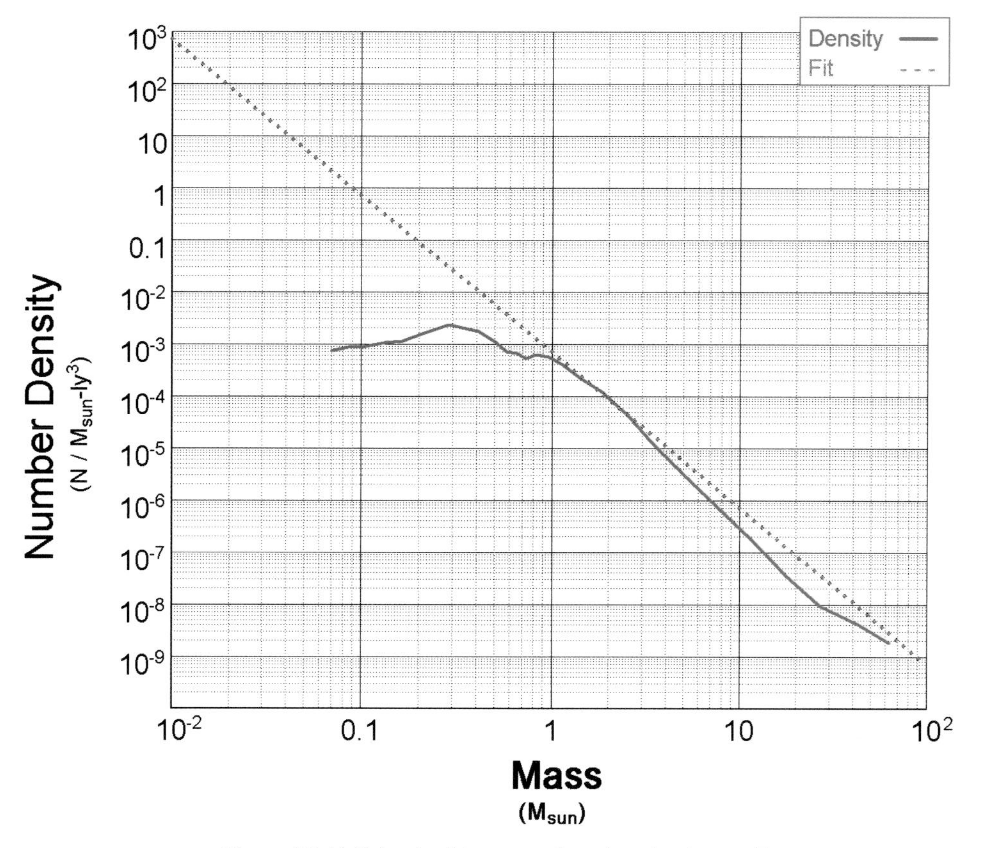

Figure (N.1) Galactic object mass/number density profile

Beginning at far right with supergiant stars and moving down the mass scale, the solid trace shows number density increasing smoothly until it approaches the mass of our sun, 1 M_{sun}. In the interval including and below our sun's mass (1.0 M_{sun}–0.07 M_{sun}) number density is fairly constant at ~0.001/ly^3, eventually vanishing at the theoretical lower limit for fusion, a mass of ~0.07 M_{sun}.

The dotted fitting function shown in Figure (N.1) is given by:

$$\frac{d\rho_{s_m}}{dm} = \frac{S_M}{m^3} \tag{N.1}$$

where m is mass and S_M is a *stellar mass distribution* constant of (S_M = 0.0007 $M_{sun}^{\;3}/ly^3$). The mass distribution function for luminous material of mass greater than 1 M_{sun} is *inversely proportional to the cube of mass*. Total mass density is given by the integral of Equation (N.1) over a mass range:

$$\rho_{s_m} = \int_{m_1}^{m_2} \left(\frac{S_M}{m^3} \right) dm = \frac{S_M}{2} \left(\frac{1}{m_1^2} - \frac{1}{m_2^2} \right) \quad \text{(N.2)}$$

where $m_1 < m_2$. For a mass range of ($1\ M_{sun} - 100\ M_{sun}$) and $S_M = 0.0007\ M_{sun}{}^3/\text{ly}^3$, this yields a total stellar mass density of $0.00035\ M_{sun}/\text{ly}^3$, consistent with current estimates.⟨1.11⟩ However, the observed number density for stars with masses less than M_{sun} is markedly lower than Equation (N.1) predicts. These stars either don't exist or are for some reason difficult to detect. Two factors could contribute to these "missing" low-luminosity stars:

- Instrumentation limitations. This can certainly account for a fair number of faint stars, perhaps even close to a factor of a thousand for stars of mass $0.08\ M_{sun}$ as indicated by the extrapolation in Figure (N.1).

- Fusion limit variability. This is already known to be true, but not to what extent. An object composed of pure hydrogen will not evolve into a star even with a mass of $100\ M_{sun}$. Stellar ignition requires a small fraction of heavy elements. Although the distribution of naturally occurring elements is generally uniform throughout space, massive clouds of hydrogen and helium are often incapable of producing stars. Perhaps what Figure (N.1) is telling us is, due to compositional variation, the fraction of non-stellar and therefore nonluminous objects increases markedly as fusion's lower mass limit is approached.

Dark mass is everywhere luminous mass is, it's just not bright enough to see.

N.3 DARK MINIMUM

Equation (N.2) represents stellar population density in our galaxy for ($m \geq 1\ M_{sun}$). If its divergence from the observed number densities of low-mass stars is an artifact of either instrumentation or stellar ignition criteria, stellar mass distribution correlates with Equation (N.2) down to a fraction of a solar mass. It can't track it indefinitely, however, as to do so would result in infinite universal mass density. There is a lower mass limit that corresponds to the universe's mass density. It will be referred to as the *dark minimum*.

The first step in calculating the dark minimum is to extrapolate the Milky Way's stellar mass distribution to the universe at large. This is possible because *the size of stars within galaxies is unrelated to the size of the galaxies themselves*. Figure (N.1) is therefore similar to the universe's star distribution. The universal distribution constant will differ to some extent from S_M, but not the basic inverse cubic mass relationship.

Suppose a fraction f_{sm} of the universe's mass is accumulated into objects whose mass/numerical densities conform to a distribution similar to Equation (N.2), and that the majority of these objects exist in galaxies. Suppose also that the fraction of space occupied by galaxies is f_{gv}. This factor *converts the average material density of galactic space to the average material density of the universe.*

This means that the universe's mass density is related to the numerical density of galactic objects by:

$$\rho_M = \left(\frac{S_{M_U} f_{gv}}{2f_{sm}}\right)\left(\frac{1}{M_1^2} - \frac{1}{M_2^2}\right) \tag{N.3}$$

where the mass m of a galactic star in Equation (N.2) is replaced by the more general mass of a universal object M and the Milky Way's mass distribution constant S_M is replaced by the universal average S_{M_U}.

Stars at the high end of the mass scale (M_2) are about three orders of magnitude more massive then the lower stellar limit of 0.08 M_{sun}. However, since they are ~9 orders of magnitude less common, their total mass contribution is negligible and the second term of Equation (N.3) can be discarded to yield:

$$\rho_M = \left(\frac{f_{gv}}{2f_{sm}}\right)\left(\frac{S_{M_U}}{M_{dm}^2}\right) \tag{N.4}$$

where $M_1 = M_{dm}$, the dark minimum of the universal mass distribution. Solving for M_{dm}:

$$M_{dm} = \sqrt{\frac{f_{gm} S_{M_U}}{2f_{sm}\rho_M}} \tag{N.5}$$

The universe's average energy density, as estimated earlier using a Hubble constant of 60 Hz-km/Mpc, is $4.5(10)^{-27}$ kg/m^3. This amounts to a universally average material density of ($\rho_M = 2(10)^{-9}$ M_{sun}/ly^3). The current estimate of the mean spacing between galaxies is on the order of a hundred times their diameter, so if they were perfectly spherical they would occupy roughly a millionth of the universe's volume. Since most have a high degree of flattening, their actual spatial extent is about 10% of this, or ($f_{gm} \sim 10^{-7}$). Assuming that (a) most of the universe's mass exists in stellar and sub-stellar objects and (b) the universal distribution constant of these objects is equal to that of our galaxy's objects yields ($f_{sm} \sim 1$) and ($S_{M_U} = S_M = 0.0007$). Substituting all of these values into Equation (N.5) results in a dark minimum of ~0.1 M_{sun}. This is about 100 times the mass of Jupiter and slightly greater than the lower limit for fusion. Although this is certainly a rough estimate, it suggests that *the majority of the universe's mass is stored in small red dwarfs or heavy brown dwarfs.*

The dark minimum is not a precipitous cutoff where less massive objects are simply nowhere to be found. Matter distribution is just that, *distribution*, but the dark minimum at least shows us where to begin looking for the missing mass.

N.4 DARK MATTER IN OUR SOLAR SYSTEM

Speed measurements of the deep space probes Pioneer 10 and 11 indicate that they are slowing down more than expected, based on the estimated mass distribution of our solar system and the Oort cloud.[34] Some scientists have attributed this to a perturbation in the gravitational field generated by matter. Their idea, which has no theoretical basis, is that gravity becomes progressively stronger at a certain range. A better explanation, which is more compelling because of its brute simplicity, is that our solar system is surrounded by a greater mass density than current estimates would indicate. The Oort cloud, hypothetical home to so many far flung comets, is not visible from Earth. Nor is it possible (except with deep-space probes) to accurately measure its mass. The deceleration of these far-flung instruments should be used to upgrade models of the Oort cloud, not challenge Newtonian physics. Indeed, perhaps some of the universe's dark hydrogen is distributed in the Oort clouds of its innumerable stars.

Visible matter, such as that so gloriously illuminated by the many nebulae scattered across the heavens, has high concentrations of hydrogen, but this is only a hint of its true cosmic abundance:

Figure (N.2) Only a tiny fraction of the universe's hydrogen is visible
(Crab nebula, M1, courtesy NASA/Hubble Heritage Project)

O. GALACTIC CORE - POWER LOSSES AND THERMAL CURRENTS

O.1 FLUX POWER LOSS

When a galactic core's energy distribution extends too far beyond the escape velocity of its veneer, an intense flux of ultra-high-energy protons carries energy away, cooling it. Even in a steady-state situation, however, there is a small core power loss associated with this high-energy tail.

VENEER POWER LOSS

A veneer flux's total energy loss is given by the product of proton flux and the average energy lost per proton:

$$\Gamma_\vee = \Delta_p\left(\overline{E}_p - m_p c^2\right) \tag{O.1}$$

where $\Gamma_\vee$ is the *veneer power loss*.

Total proton flux is given by converting the veneer flux of Equation (16.48) to proton flux by dividing by the proton rest mass:

$$\Delta_p = \frac{\Delta_\vee}{m_p} = \frac{\left(\dfrac{\mathrm{v}_{\text{F-M}}c^2}{8Gm_p}\right)\left(kT\ln\left(1+e^{\left(\frac{E_{s\vee}-E_{q\vee(T)}}{kT}\right)}\right)-E_{s\vee}+E_{q\vee(T)}\right)}{m_p c^2 - E_{s\vee} + E_{q\vee(T)}} \tag{O.2}$$

This converts mass/time to protons/time. Note that the Coulomb binder's temperature dependence has also been incorporated in this expression.

Substitute Equation (O.2) into Equation (O.1), yielding the veneer power loss:

$$\Gamma_{\vee} = \left(\frac{\left(\frac{\mathrm{v}_{\text{F-M}} c^2}{8Gm_p} \right) \left(kT \ln\left(1 + e^{\left(\frac{E_{s\vee} - E_{q\vee(T)}}{kT} \right)} \right) - E_{s\vee} + E_{q\vee(T)} \right)}{m_p c^2 - E_{s\vee} + E_{q\vee(T)}} \right) \left(\overline{E}_p - m_p c^2 \right) \tag{O.3}$$

An ejected proton's average energy is the integration of the Fermi-Maxfield distribution weighted by energy and normalized by area:

$$\overline{E}_p = \frac{\left(\int_{m_p c^2}^{\infty} \frac{E}{\left(e^{\left(\frac{E - m_p c^2 + E_{s\vee} - E_{q\vee(T)}}{kT} \right)} + 1 \right)} dE \right)}{\left(kT \ln\left(1 + e^{\left(\frac{E_{s\vee} - E_{q\vee(T)}}{kT} \right)} \right) - E_{s\vee} + E_{q\vee(T)} \right)} \tag{O.4}$$

No known analytic solution exists for the energy-weighted term in the numerator; the denominator is the integrated Fermi-Maxfield distribution above a particle's rest mass.

GALACTIC POWER LOSS

A galaxy's power capture is related to its core temperature and veneer binders by a slight modification of Equation (16.57):

$$P_g = L_g Q_g = \frac{\left(\frac{f_{cn} \varepsilon_U \mathrm{v}_{\text{F-M}} c^4}{8G} \right) \left(kT \ln\left(1 + e^{\left(\frac{E_{s\vee} - E_{q\vee(T)}}{kT} \right)} \right) - E_{s\vee} + E_{q\vee(T)} \right)}{m_p c^2 - E_{s\vee} + E_{q\vee(T)}} \tag{O.5}$$

Galactic power loss is the ratio between veneer loss and power input:

$$\Gamma_g = \frac{\Gamma_\vee}{L_g Q_g} = \frac{\overline{E}_p - m_p c^2}{f_{cn} \varepsilon_U m_p c^2} \tag{O.6}$$

The higher the concentration of compound nuclei for a given heat loss, the lower the galactic power loss. Substitute for the average proton energy using Equation (O.4), bringing core temperature and the veneer binders explicitly into the expression:

$$\Gamma_g = \left(\frac{1}{f_{cn}\varepsilon_U}\right)\left(\frac{\left(\int_{m_p c^2}^{\infty} \frac{E}{\left(e^{\left(\frac{E - m_p c^2 + E_{s\vee} - E_{q\vee(T)}}{kT}\right)} + 1\right)} dE\right)}{m_p c^2 \left(kT \ln\left(1 + e^{\left(\frac{E_{s\vee} - E_{q\vee(T)}}{kT}\right)}\right) - E_{s\vee} + E_{q\vee(T)}\right)} - 1\right) \tag{O.7}$$

The Milky Way's galactic power loss can be determined using estimated inflow composition at a galactic efficiency of 100% ($Q_g = 1$).

Numerical evaluation of Equation (O.7) for a compound fraction of ($f_{cn} = 0.02$) yields:

- At its most likely values - a core temperature of 284,000 °K from binders of ($E_{s\vee}$=950 EV) and ($E_{w\vee}$=440 EV), the Milky Way's galactic power loss is 186 ppm at a thermal proton current of $1.072(10)^{32}$ A.

- At its most extreme values - a core temperature of 570,000 °K from binders of ($E_{s\vee}$=1100 EV) and ($E_{w\vee}$=0 EV), the Milky Way's galactic power loss is 373 ppm at a thermal proton current of $1.066(10)^{32}$ A.

Both are many orders of magnitude less than our galaxy's total bolometric output.

A galactic core's thermal proton power loss is greater than its radiative loss of ~4 ppm, as given by Equation (13.42):

$$L_{\vee} \cong 8(10)^{-40}\left(T^{4}M^{\frac{4}{3}}\right) \qquad \{M >> M_{sun}\}$$

Galactic cores are the de facto unification of nature's fundamental forces, and are extraordinarily efficient in their quiescent state.

O.2 ELECTRICAL POWER LOSS

Thus far our main focus has been the energy loss caused by material flow. The power loss caused by thermal electrons can be calculated by retrofitting Equation (O.1):

$$\Gamma_{\vee_e} = \Delta_e\left(\overline{E}_e - m_e c^2\right) \tag{O.8}$$

Substitute for flux with a modified Equation (O.3):

$$\Gamma_{\vee_e} = \frac{\left(\frac{v_{\text{F-M}}c^2}{8Gm_p}\right)\left(kT\ln\left(1+e^{\left(\frac{E_{q\vee(T)}-E_{w\vee}}{kT}\right)}\right)-E_{q\vee(T)}+E_{w\vee}\right)}{m_e c^2 - E_{q\vee(T)} + E_{w\vee}}\left(\overline{E}_e - m_e c^2\right) \tag{O.9}$$

where:

$$\overline{E}_e = \frac{\left(\int_{m_e c^2}^{\infty}\frac{E}{\left(e^{\left(\frac{E-m_e c^2+E_{q\vee(T)}-E_{w\vee}}{kT}\right)}+1\right)}dE\right)}{\left(kT\ln\left(1+e^{\left(\frac{E_{q\vee(T)}-E_{w\vee}}{kT}\right)}\right)-E_{q\vee(T)}+E_{w\vee}\right)} \tag{O.10}$$

Numerical solution of Equation (O.9) for a 284,000 °K veneer temperature, ($E_{s\vee}$=950 EV) and ($E_{w\vee}$=440 EV) yields a loss of $2.8(10)^{33}$ W at $1.1(10)^{32}$ A, or ~200 ppm of the Milky Way's bolometric power output. Increasing temperature to 570,000 °K with the limiting case for binders ($E_{s\vee}$=1100 EV and $E_{w\vee}$=0 EV) causes a slight reduction in thermal current, to $9.8(10)^{31}$ A, but increases loss to $4.8(10)^{33}$ W, ~340 ppm of the Milky Way's output.

Our galaxy's thermal core current is far smaller than the galactic disk current it processes, which from Equation (15.60):

$$i_g = \frac{2Q_g L_g q}{m_e \mathrm{v}_c^2 (1-\eta_g)^2 \tan^2(\theta_g) - E_{ex}}$$

is on the order of 10^{40} A. This is because its thermal current is a by-product of its ambient heat, whereas its galactic current carries the core's nuclear disassociation energy.

O.3 CORE CURRENT LIMIT

A gravitational veneer's material sourcing limit is given by its veneer capacity, Equation (13.45):

$$C_\vee = \frac{\overline{\mathrm{v}}c^2}{8G}$$

This has units of mass/time with the understanding that a black hole's size is governed by the mass of its protons, not electrons. To convert to protons/time from mass/time, divide by proton rest mass:

$$C_{\vee p} = \frac{\overline{\mathrm{v}}c^2}{8Gm_p} \tag{O.11}$$

Since bound electrons and protons have comparable number densities, Equation (O.11) represents the capacity of either. Rewriting it in terms of electrons:

$$C_{\vee e} = \frac{\overline{\mathrm{v}}c^2}{8Gm_p} \tag{O.12}$$

All electrons are bound and degenerate in a Maxfield-limited environment, so their average energy is, as is the case with protons, close to their rest energy. Their Fermi energy distribution is flat, so average speed is also the same as protons, 0.6849*c* as given by Equation (16.46). Substitute this for the average velocity and convert to current with unit elementary charge q:

$$C_{\vee i} = \frac{q\mathrm{v}_{\text{F-M}}c^2}{8Gm_p} \tag{O.13}$$

this is a veneer's *current sourcing capacity*, $3.2(10)^{42}$ amps.

The maximum current a large black hole can source is given by the veneer current capacity of Equation (O.13). This is the upper limit of its electron migration, which in turn sets a bound on the maximum luminosity of a single-core galaxy:

Ψ THEOREM O.1 - GALACTIC CORE> CURRENT LIMIT {Ψ5.6}

THE MAXIMUM ELECTRON MIGRATION OF A SINGLE BLACK HOLE IS $3.2(10)^{42}$ AMPS

Equation (O.13) represents the *maximum current* a galactic core can source. Note that it is *irrespective of temperature or mass.* Ultra-luminous galaxies such as giant ellipticals (cD) may routinely support their prodigious output using cores with multiple black holes, as have recently been observed.[32] Even our closest galactic neighbor Andromeda has a dual core, as do many galaxies with active galactic nuclei (AGN). *The core current limit imposes a restriction on the size and/or configuration of galactic systems.*

Mark 315 is one of many AGN with a prominent multiple core:

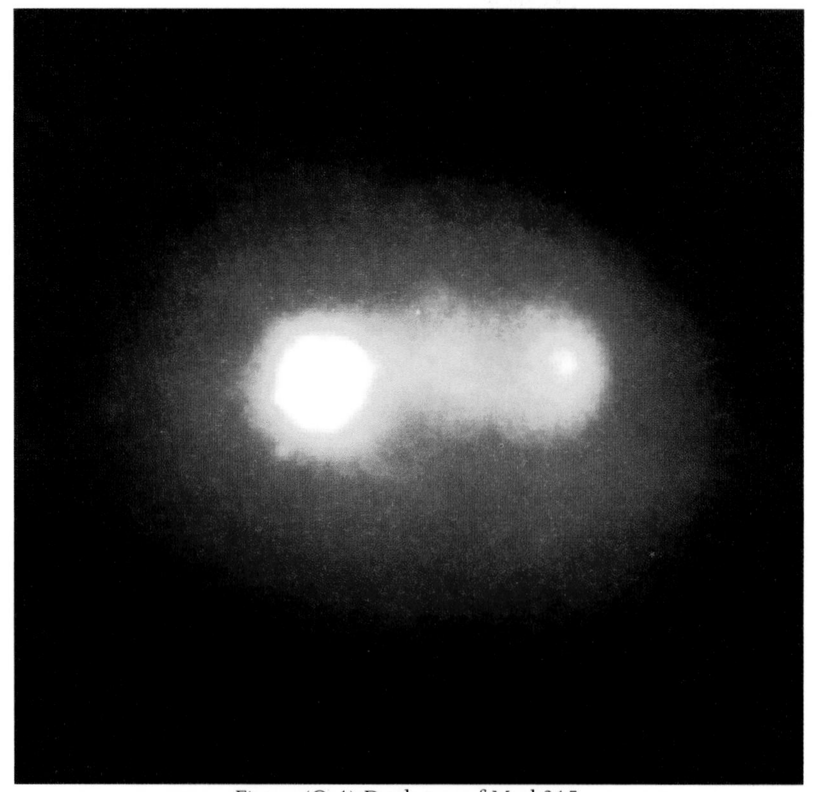

Figure (O.1) Dual core of Mark315
(Courtesy NASA/Hubble Heritage Project)

Andromeda's binary core is less pronounced, but still evident:

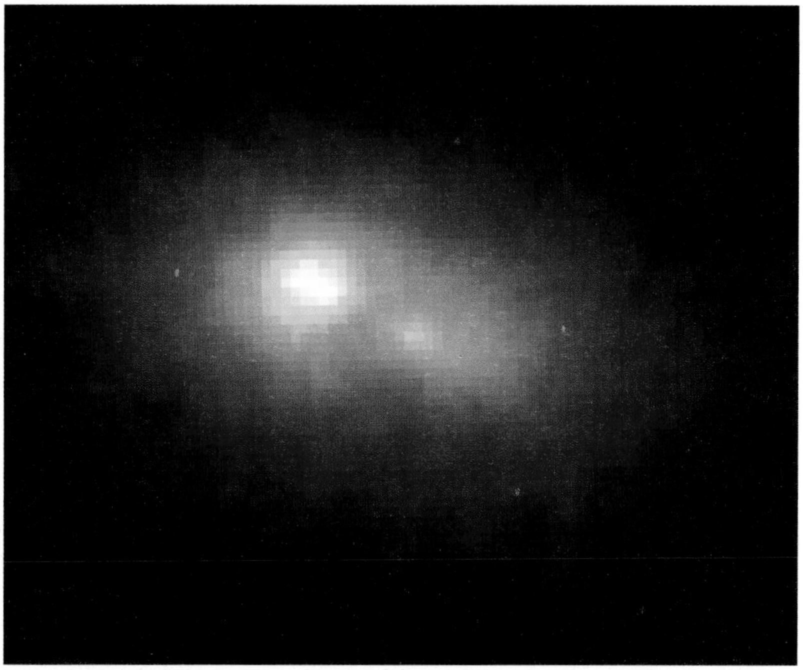

Figure (O.2) Dual core of the Andromeda galaxy, M31
(Courtesy NASA/Hubble Heritage Project)

P. MATERIAL FLUX AND VORTICAL FLOW

P.1 FLUX

An important relationship used in physical cosmology is that between the volume density of moving particles and their flux through a given area. This applies equally to photons, elementary particles, and any other entities moving with a random velocity in thermal equilibrium. Flux is defined as the number of particles moving through a given area in a given time:

$$\Phi = \frac{N}{At} \tag{P.1}$$

This will be calculated using the concept of a *differential flux*. First, determine the flux of particles between velocity $\mathbf{v}$ and $\mathbf{v}+d\mathbf{v}$ moving through some area per unit time, then integrate throughout the entire velocity range to find the total flux. Consider some small area in space with a portion of flux between $\mathbf{v}$ and $\mathbf{v}+d\mathbf{v}$ impinging from some angle θ in relation to the z axis:

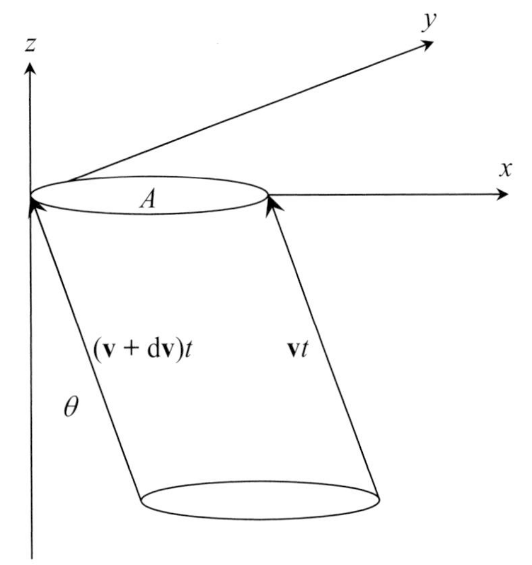

Figure (P.1) Differential material flux through an area at angle θ.

The flux shown is a differential flux, since the velocity range in question is only a small component of the total range of velocities of material moving through this area.

The volume of the region shown in Figure (P.1) is given by:

$$V = Avt\cos(\theta) \quad \text{(P.2)}$$

The number density of particles in the differential velocity range between **v** and **v**+d**v** is defined to be:

$$\left(\frac{N}{V}\right)dv_x dv_y dv_z = f(v)dv_x dv_y dv_z \quad \text{(P.3)}$$

where $f(\mathbf{v})$ is their velocity distribution.

The number of particles in the volume in the differential velocity range is the product of the volume of Equation (P.2) and the density of Equation (P.3):

$$Ndv_x dv_y dv_z = f(v)dv_x dv_y dv_z Avt\cos(\theta) \quad \text{(P.4)}$$

The differential flux, which will be called Φ(v), is therefore:

$$\frac{Ndv_x dv_y dv_z}{At} = \Phi(v) = f(v)v\cos(\theta)dv_x dv_y dv_z \quad \text{(P.5)}$$

Next change the velocity differentials from Cartesian coordinates to velocity space by the transformation:

$$dv_x dv_y dv_z = v^2 dv\sin(\theta)d\theta d\phi \quad \text{(P.6)}$$

to obtain the spherical form of the differential flux:

$$\Phi(v) = f(v)v^3\cos(\theta)\sin(\theta)dvd\theta d\phi \quad \text{(P.7)}$$

Now integrate Equation (P.7) for θ from 0 to $\pi/2$ (the entire negative z axis) and ϕ from 0 to 2π, resulting in a form dependent on velocity alone:

$$\Phi(v) = \pi f(v)v^3 dv \quad \text{(P.8)}$$

Integrating over all possible velocities gives the total material flux:

$$\Phi = \pi\int_0^\infty f(v)v^3 dv \quad \text{(P.9)}$$

This is the total flux due to particles of all velocities moving through the area per unit time.

By definition, the mean speed of any velocity distribution is given by:

$$\bar{v} = \frac{4\pi}{\rho} \int_0^\infty f(v) v^3 dv \tag{P.10}$$

or:

$$\int_0^\infty f(v) v^3 dv = \frac{\rho \bar{v}}{4\pi} \tag{P.11}$$

Substituting Equation (P.11) into Equation (P.9) yields the general solution for the flux:

$$\Phi = \frac{\rho \bar{v}}{4} \tag{P.12}$$

Particles in deep space are in thermal equilibrium and have a Maxwellian velocity distribution of the form:

$$f(v) = \rho \left(\frac{m}{2\pi kT} \right)^{\frac{3}{2}} e^{-\frac{mv^2}{2kT}} dv \tag{P.13}$$

Substituting this into the average velocity of Equation (P.10) yields:

$$\bar{v} = 4\pi \left(\frac{m}{2\pi kT} \right)^{\frac{3}{2}} \left(\int_0^\infty e^{-\frac{mv^2}{2kT}} v^3 dv \right) = \sqrt{\frac{8kT}{\pi m}} \tag{P.14}$$

Protons in thermal equilibrium at 2.7 °K have an average speed of 240 m/s; electrons are moving quite a bit faster at 10.2 km/s.

Lastly, substitute Equation (P.14) into Equation (P.12) to obtain the relationship between material density and flux in the IGM of deep space:

$$\Phi = \frac{\rho}{4} \sqrt{\frac{8kT}{\pi m}} \tag{P.15}$$

At a density of $\sim 5(10)^{-27}$ kg/m^3 and temperature of 2.7 °K, the available material flux amounts to $\sim 3(10)^{-25}$ kg/m^2/s. This certainly seems small, but astronomical distances are vast. The amount of material moving through a square light year is 27 million kg per second, and a light year is tiny by galactic standards.

P.2 AVERAGE KINETIC ENERGY

Energy content is another essential characteristic of deep space material. This is derived using the three-dimensional Maxwellian velocity distribution:

$$f(\mathrm{v})d\mathrm{v}_x d\mathrm{v}_y d\mathrm{v}_z = n\left(\frac{m}{2\pi kT}\right)e^{-\frac{m}{2kT}\left(\mathrm{v}_x^2+\mathrm{v}_y^2+\mathrm{v}_z^2\right)}d\mathrm{v}_x d\mathrm{v}_y d\mathrm{v}_z \quad \text{(P.16)}$$

This is the number density of particles in the velocity space ($\mathbf{v}_x$, $\mathbf{v}_y$, $\mathbf{v}_z$). A particle's average energy in this distribution is given by a kinetic-energy-weighted integration of velocity over three dimensions:

$$\overline{E}_K = \frac{\int_{-\infty}^{\infty}\int_{-\infty}^{\infty}\int_{-\infty}^{\infty}\left(\frac{1}{2}m\left(\mathrm{v}_x^2+\mathrm{v}_y^2+\mathrm{v}_z^2\right)\left(\frac{m}{2\pi kT}\right)^{\frac{3}{2}}e^{-\frac{m}{2kT}\left(\mathrm{v}_x^2+\mathrm{v}_y^2+\mathrm{v}_z^2\right)}\right)d\mathrm{v}_x d\mathrm{v}_y d\mathrm{v}_z}{\int_{-\infty}^{\infty}\int_{-\infty}^{\infty}\int_{-\infty}^{\infty}\left(\left(\frac{m}{2\pi kT}\right)^{\frac{3}{2}}e^{-\frac{m}{2kT}\left(\mathrm{v}_x^2+\mathrm{v}_y^2+\mathrm{v}_z^2\right)}\right)d\mathrm{v}_x d\mathrm{v}_y d\mathrm{v}_z} \quad \text{(P.17)}$$

which simplifies to:

$$\overline{E}_K = \frac{\int_{-\infty}^{\infty}\int_{-\infty}^{\infty}\int_{-\infty}^{\infty}\left(\frac{1}{2}m\left(\mathrm{v}_x^2+\mathrm{v}_y^2+\mathrm{v}_z^2\right)e^{-\frac{m}{2kT}\left(\mathrm{v}_x^2+\mathrm{v}_y^2+\mathrm{v}_z^2\right)}\right)d\mathrm{v}_x d\mathrm{v}_y d\mathrm{v}_z}{\int_{-\infty}^{\infty}\int_{-\infty}^{\infty}\int_{-\infty}^{\infty}\left(e^{-\frac{m}{2kT}\left(\mathrm{v}_x^2+\mathrm{v}_y^2+\mathrm{v}_z^2\right)}\right)d\mathrm{v}_x d\mathrm{v}_y d\mathrm{v}_z} \quad \text{(P.18)}$$

This equation is separable, yielding three distinct integrals of the form:

$$\begin{aligned}\overline{E} = {} & \frac{1}{A}\int_{-\infty}^{\infty}\left(\frac{1}{2}m\mathrm{v}_x^2 e^{-\frac{m\mathrm{v}_x^2}{2kT}}\right)d\mathrm{v}_x \int_{-\infty}^{\infty}e^{-\frac{m\mathrm{v}_y^2}{2kT}}d\mathrm{v}_y \int_{-\infty}^{\infty}e^{-\frac{m\mathrm{v}_z^2}{2kT}}d\mathrm{v}_z + \\ & \frac{1}{A}\int_{-\infty}^{\infty}e^{-\frac{m\mathrm{v}_x^2}{2kT}}d\mathrm{v}_x \int_{-\infty}^{\infty}\left(\frac{1}{2}m\mathrm{v}_y^2 e^{-\frac{m\mathrm{v}_y^2}{2kT}}\right)d\mathrm{v}_y \int_{-\infty}^{\infty}e^{-\frac{m\mathrm{v}_z^2}{2kT}}d\mathrm{v}_z + \\ & \frac{1}{A}\int_{-\infty}^{\infty}e^{-\frac{m\mathrm{v}_x^2}{2kT}}d\mathrm{v}_x \int_{-\infty}^{\infty}e^{-\frac{m\mathrm{v}_y^2}{2kT}}d\mathrm{v}_y \int_{-\infty}^{\infty}\left(\frac{1}{2}m\mathrm{v}_z^2 e^{-\frac{m\mathrm{v}_z^2}{2kT}}\right)d\mathrm{v}_z\end{aligned} \quad \text{(P.19)}$$

Where:

$$A = \int_{-\infty}^{\infty}e^{-\frac{m\mathrm{v}_x^2}{2kT}}d\mathrm{v}_x \int_{-\infty}^{\infty}e^{-\frac{m\mathrm{v}_y^2}{2kT}}d\mathrm{v}_y \int_{-\infty}^{\infty}e^{-\frac{m\mathrm{v}_z^2}{2kT}}d\mathrm{v}_z \quad \text{(P.20)}$$

Simplify:

$$\overline{E}=\left(\frac{\int_{-\infty}^{\infty}\left(\frac{1}{2}m\mathrm{v}_x^2e^{-\frac{m\mathrm{v}_x^2}{2kT}}\right)d\mathrm{v}_x}{\int_{-\infty}^{\infty}e^{-\frac{m\mathrm{v}_x^2}{2kT}}d\mathrm{v}_x}\right)+\left(\frac{\int_{-\infty}^{\infty}\left(\frac{1}{2}m\mathrm{v}_y^2e^{-\frac{m\mathrm{v}_y^2}{2kT}}\right)d\mathrm{v}_y}{\int_{-\infty}^{\infty}e^{-\frac{m\mathrm{v}_y^2}{2kT}}d\mathrm{v}_y}\right)+\left(\frac{\int_{-\infty}^{\infty}\left(\frac{1}{2}m\mathrm{v}_z^2e^{-\frac{m\mathrm{v}_z^2}{2kT}}\right)d\mathrm{v}_z}{\int_{-\infty}^{\infty}e^{-\frac{m\mathrm{v}_z^2}{2kT}}d\mathrm{v}_z}\right) \tag{P.21}$$

Let:

$$\mathrm{v}_n=x_n\sqrt{\frac{2kT}{m}} \qquad d\mathrm{v}_n=\sqrt{\frac{2kT}{m}}dx_n \tag{P.22}$$

Substitute into Equation (P.21):

$$\overline{E}_K=kT\left(\frac{\int_{-\infty}^{\infty}x_1^2e^{-x_1^2}dx_1}{\int_{-\infty}^{\infty}e^{-x_1^2}dx_1}\right)+kT\left(\frac{\int_{-\infty}^{\infty}x_2^2e^{-x_2^2}dx_2}{\int_{-\infty}^{\infty}e^{-x_2^2}dx_2}\right)+kT\left(\frac{\int_{-\infty}^{\infty}x_3^2e^{-x_3^2}dx_3}{\int_{-\infty}^{\infty}e^{-x_3^2}dx_3}\right) \tag{P.23}$$

All of the functions are symmetric about zero, so:

$$\int_{-\infty}^{\infty}x^2e^{-x^2}dx=2\int_0^{\infty}x^2e^{-x^2}dx \qquad \int_{-\infty}^{\infty}e^{-x^2}dx=2\int_0^{\infty}e^{-x^2}dx \tag{P.24}$$

Substituting Equation (P.24) into Equation (P.23):

$$\overline{E}_K=kT\left(\frac{\int_0^{\infty}x_1^2e^{-x_1^2}dx_1}{\int_0^{\infty}e^{-x_1^2}dx_1}\right)+kT\left(\frac{\int_0^{\infty}x_2^2e^{-x_2^2}dx_2}{\int_0^{\infty}e^{-x_2^2}dx_2}\right)+kT\left(\frac{\int_0^{\infty}x_3^2e^{-x_3^2}dx_3}{\int_0^{\infty}e^{-x_3^2}dx_3}\right) \tag{P.25}$$

Since:

$$\int_0^{\infty}x^2e^{-x^2}dx=\frac{\sqrt{\pi}}{4} \qquad \int_0^{\infty}e^{-x^2}dx=\frac{\sqrt{\pi}}{2} \tag{P.26}$$

Equation (P.25) reduces to:

$$\overline{E}_K=\frac{kT}{2}+\frac{kT}{2}+\frac{kT}{2}=\frac{3kT}{2} \tag{P.27}$$

P.3 VORTICAL MASS DRIVER

A galaxy's vortex works as a balance between gravitation and electromagnetism. Gravity pulls its neutral disk material inward while electromagnetic fields remove the charged hydrogen that the galaxy's core generates. This process requires a slight instability in the orbital motion of the entire galactic disk, where the mass that is interior to any given orbit *is slightly greater than what is required to maintain a constant radius for circulating material.* This relationship is derived below in terms of a galaxy's constant circular (v_c) and radial (v_r) velocities.

The difference between a circular orbit and an orbit that spirals inward can be expressed as a difference in acceleration - a difference in velocity over a difference in time. In half of a galactic year, the circular velocity of disk material reverses direction (resulting in a net change in velocity of $2v_c$). During this time, a perfectly circular orbit would remain stable at R_1 while one subjected to radial velocity shrinks from R_1 to R_2. The average *difference in acceleration* between these two cases is half of the total difference of acceleration at the two radii:

$$\Delta a = \left(\frac{1}{2}\right)\left(\frac{\Delta v}{\Delta t_1} - \frac{\Delta v}{\Delta t_2}\right) = \left(\frac{1}{2}\right)\left(\frac{2v_c}{\left(\frac{\pi R_1}{v_c}\right)} - \frac{2v_c}{\left(\frac{\pi R_2}{v_c}\right)}\right) = \left(\frac{v_c^2}{\pi}\right)\left(\frac{1}{R_1} - \frac{1}{R_2}\right) \tag{P.28}$$

The difference between R_1 and R_2 is a function of v_r over half of a radially averaged galactic year:

$$R_2 = R_1 + v_r\left(\frac{\pi(R_1 + R_2)}{v_c 2}\right) \cong R + v_r\left(\frac{\pi R}{v_c}\right) = R\left(1 + \frac{\pi v_r}{v_c}\right) \tag{P.29}$$

where R is the average of R_1 and R_2.

Substitute Equation (P.29) into (P.28), simplify, and normalize by acceleration to identify the size of the mass driver:

$$\frac{\Delta a}{a} = \frac{\Delta M}{M} = \left(\frac{R}{v_c^2}\right)\left(\frac{v_c^2}{\pi}\right)\left(\frac{1}{R} - \frac{1}{R\left(1 + \frac{\pi v_r}{v_c}\right)}\right) = \frac{v_r}{v_c + \pi v_r} \tag{P.30}$$

The Milky Way, at (v_r = ~1.5 km/s) and (v_c = 220 km/s), has a mass driver equal to ~0.7% of the mass density required for its circular disk motion.

Q. QUASI-STELLAR OBJECTS

The heavens have no shortage of mysteries. Some are riddles of absentia, such as antimatter and dark matter. Others are roaring beacons scattered indiscriminately across space, taunting us with their baffling characteristics. Quasi-stellar objects (QSOs) fall into the latter category, with the following curious properties.

QSOs have:

- rapid intensity fluctuations, suggesting small size.
- large intensity fluctuations, suggesting instability.
- high-energy spectra similar to galactic nuclei. Radio-bright QSOs are called *quasars*. They constitute about 10% of all QSOs.⟨6.3⟩
- an enigmatic redshift distribution, with population density increasing to a peak near $z = 2$, then falling off sharply at higher z.
- what might be incredibly intense luminosity. Their true power output depends on a proper interpretation of their redshift. If it were purely a function of distance, for example, HS 1946+7658 would be impossibly bright.⟨6.20⟩ It has a redshift of 3.02 and apparent visual magnitude of $V = 15.85$, corresponding to a power output of $5.7(10)^{41}$ W. This is over ten thousand times the output of the Milky Way, generated in a spatially discrete region of space.

Let's take a closer look at their spectral output and redshift.

Q.1 QSO LUMINOSITY

QSOs emit a broad spectrum of radiation of fairly uniform flux density all the way from radio waves to gamma radiation. Quasars are more powerful in the radio band than other QSOs, but still have a broadband power output. A typical QSO's spectral output⟨6.3⟩ is shown in the following figure by the green trace, in comparison to the relatively narrow spectrum of an ideal 10,000 °K blackbody (orange):

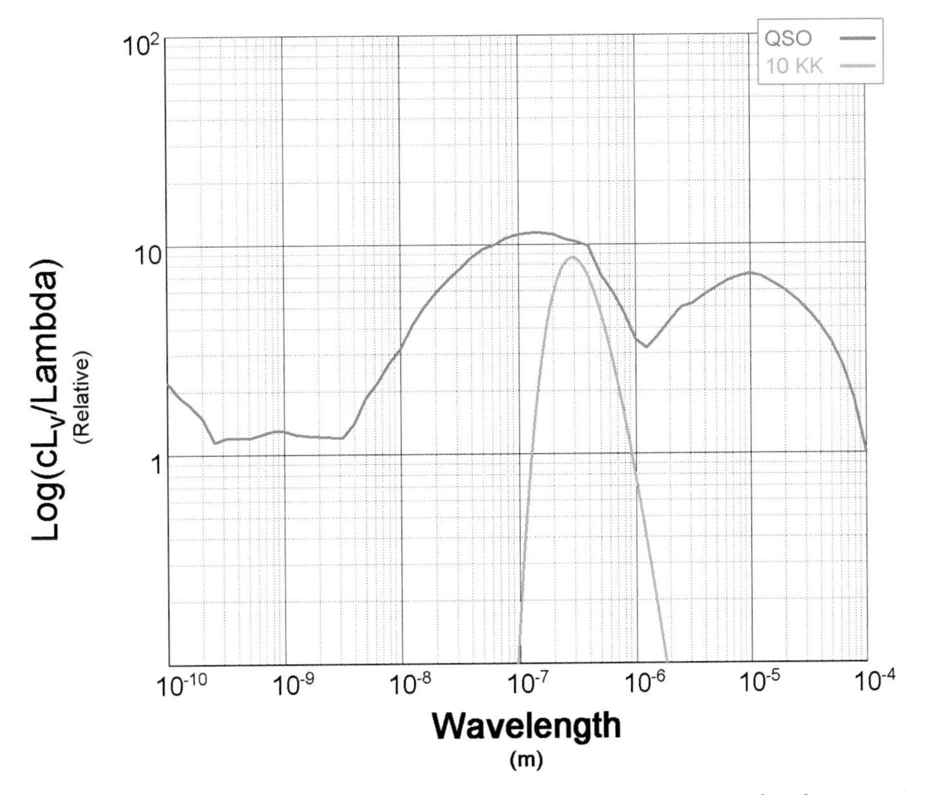

Figure (Q.1) 10,000 °K blackbody compared to typical QSO energy distribution

The 10,000 °K spectrum peaks near a wavelength of $\sim 3(10)^{-7}$ m and is orders of magnitudes narrower than that of a QSO. Although the actual thermal spectrum of a star or galaxy is somewhat broader than the ideal trace shown above, it is still substantially narrower than the frequency distribution that a QSO radiates. QSOs' spectra are for all intents and purposes immune to the surface brightness loss of normal celestial objects. *They are ideal cosmic beacons*. Does their redshift distribution reflect this?

Q.2 QSO REDSHIFT DISTRIBUTION

The relationship between range and redshift is given by Equation (15.10):

$$r = \left(\frac{c}{H_0}\right)\ln(z+1) \qquad \text{(Q.1)}$$

At the uniform spatial density required by the Cosmological Principle, the number of QSOs in some range interval *dr* located a distance *r* from Earth is given by:

$$dN_{qso} = \rho_{qso} 4\pi r^2 dr \qquad (Q.2)$$

where ρ_{qso} is QSO density in units of QSO/distance^3. Use Equation (Q.1) to substitute *z* and *dz* for *r* and *dr*:

$$dN_{qso} = \rho_{qso} 4\pi \left(\frac{c}{H_0}\right)^3 \left(\frac{(\ln(z+1))^2}{z+1}\right) dz \qquad (Q.3)$$

Since the broadness of the QSO spectrum effectively nullifies the effects of photon migration, the number of objects visible to us will be attenuated purely by their energy loss due to lumetic decay, a factor of 1/(*z* + 1). Include this in Equation (Q.3) to yield the population density of QSO objects in a cosmically uniform spatial distribution as a function of redshift:

$$dN_{qso} = \rho_{qso} 4\pi \left(\frac{c}{H_0}\right)^3 \left(\frac{\ln(z+1)}{z+1}\right)^2 dz \qquad (Q.4)$$

The following is the redshift distribution for the more than 16,700 QSOs in the Sloan Digital Sky Survey (SDSS), shown in comparison to the distribution expected of lumetic decay:

Figure (Q.2) SDSS redshift number density distribution versus uniform spatial distribution

The dotted trace is Equation (Q.4), normalized to the Sloan survey. If QSO redshift corresponded to great distance and dazzling luminosity, their population density should decrease slowly past $z = 2$, not fall off precipitously as it does. *The reasonable conclusion is distance is not the dominant component of QSO redshift.*

Only three agents are known to induce redshift in celestial images:

- Lumetic decay.
- Recession velocity.
- Gravitational potential.

The first is ruled out by the observed QSO population/redshift distribution and the second by a nonexpanding universe, leaving gravitational potential as the only remaining option. It is no coincidence QSOs have spectra similar to galactic cores, because they too are sites of crushing gravitational potential.

Ω HYPOTHESIS Q.1 - QSO {Ψ5.6}

A QSO IS AN ULTRA-HOT, DENSE OBJECT WITH A SURFACE GRAVITATIONAL POTENTIAL IN THE RANGE $\sim(-0.3) < \Phi_g/m_0c^2 < \sim(-0.7)$

Since QSOs are closer than their redshifts would otherwise indicate, they are also far less luminous than currently thought. The question of what they are, however, remains unanswered.

Figure (Q.3) Quasar Q3C 273
(Courtesy NASA/Hubble Heritage Project)

Halton Arp has postulated that QSOs are debris expended from active galactic nuclei (AGN).[11] He identified a number of systems consisting of a galaxy flanked symmetrically on two sides with QSOs of strikingly similar redshifts. The redshift of the host galaxies suggested relatively close objects whereas the redshifts of the associated QSO pairs were much larger, corresponding to far greater distance. These configurations are consistent with the idea that galaxies *eject* QSOs from their core regions. Whether or not this is the case remains to be seen.

An alternate, and at least in some cases perhaps more appropriate, explanation is gravitational lensing, where a close foreground object bends light's path enough to produce twin images of a distant background object. This would explain the symmetry of the images as well as the disparity between the galaxies' and associated QSOs' redshifts. It would also resolve the skewed statistical association that Arp found between QSO pairs and the galaxies they flank. Moreover, galaxies have a mass that corresponds to a gravitational focal length far in excess of their own redshift distance.

Although this seems like a reasonable resolution of the redshift mismatch, in most cases QSOs:

- are related to galaxies with hyperactive core regions (AGN).

- have a trajectory coincident with meandering hydrogen jets from galactic cores.

Both of these circumstances could be argued in favor of lensing. The association with active galactic nuclei, for instance, might simply mean these types of galaxies have a thinner halo of dark material, allowing them to act as better lenses. Hydrogen jets might be mere coincidence.

What is far more difficult to explain is the abrupt loss of QSO number density above a redshift of $z = 2$. A non-active galactic nucleus sheds hydrogen gas slowly and evenly, carried away from the core above, within, and below the galactic plane. An active galactic nucleus is less stable, often launching cataclysmic bursts of high-speed hydrogen gas. Perhaps AGN also, on occasion, eject portions of the galactic core itself, and *these are what QSOs are*. The core is a massive black hole at fairly low density. Are QSOs white holes - hot objects with massive gravitational fields, *but insufficient surface potential to subdue visible light at their ~300,000 °K surface temperature?* A QSO might be what happens to a black hole when it loses enough mass to leak torrents of visible light. Perhaps the symmetry Arp found is a consequence of momentum conservation of galactic ejections, not gravitational lensing.

References

BOOKS

[1] *Allen's Astrophysical Quantities*, 4th Edition, Arthur Cox Editor [Springer-Verlag, 2002]
(Astrophysical data)

1. p. 28 (Elemental composition of luminous portion of universe)
2. p. 649 (Hubble constant range)
3. pp. 572, 653 (Globular cluster ages older than galactic disk material)
4. p. 619 (Virgo cluster temperature)
5. p. 632 (Sedouski-Zeldelov [S-Z] effect)
6. p. 602 (Active galactic nuclei electromagnetic emission)
7. p. 573 (Milky Way's rotation profile)
8. p. 485 (Stellar luminosity function, converted to cubic light years)
9. p. 572 (Milky Way core's mass)
10. p. 489 (Stellar mass-luminosity function)
11. p. 487 (Milky Way's stellar mass density distribution)
12. p. 661 (Universal optical luminosity density)
13. p. 570 (Universal IR luminosity density, scaled by Milky Way's component [~33%])
14. p. 570 (Milky Way's luminosity)
15. p. 572 (Age of Milky Way's oldest disk stars, midrange of scale)
16. p. 570 (Milky Way's mass, $R < 114$ Kly)
17. p. 663 (Extragalactic radio background)
18. p. 570 (Milky Way's escape velocities)
19. p. 570 (Milky Way's dark halo radius)
20. p. 340 (Sun's power output)
21. p. 382 (Stellar mass/luminosity relationship)

[2] *Astrophysical Formulae*, Kenneth Lang, Volume I, 3rd Edition [Springer-Verlag, 1999]
(Astrophysical data, radiation and high-energy)

1. p. 457 (Background electromagnetic densities)
2. p. 232 (CMB temperature)
3. pp. 387, 388 (Solar neutrino flux)
4. p. 390 (Neutrino cross-sections)

⟨3⟩ *Astrophysical Formulae*, Kenneth Lang, Volume II, 3rd Edition [Springer-Verlag, 1999]
(Astrophysical data, mass distribution)
1. p. 58 (Universe's luminous mass density)
2. p. 57 (Average intergalactic spacing, Equation [5.210] with $h = 0.6$)
3. p. 87 (Fraction of a star's hydrogen content available to fusion)
4. p. 45 (H I regions in galactic arms, sun's circular speed about Milky Way)
5. p. 46 (Galactic global pattern speed, using 13.5 km/s/Kpc and $R_o = 8.5$ Kpc)

⟨4⟩ *Gravitation*, Charles Misner, Kip Thorne & John Wheeler [Freeman, 1973]
(Relativistic gravitational theory)
1. pp. 935-940 (Environment beneath an event horizon)
2. p. 659 (Relativistic gravitational potential, from z of Equation [25.26] to GM/c^2 units)

⟨5⟩ *CRC Handbook of Chemistry and Physics*, 81st Edition [CRC Press, 1994]
(Physical constants)
1. p. 10-220 (Index of refraction of gases proportional to density, $p \to 0$, t = K)
2. pp. (1-1) - (1-8) (Fundamental physical constants)

⟨6⟩ *Modern Astrophysics*, Bradley Carroll & Dale Ostlie [Addison Wesley Longman, 1996]
(Astrophysics)
1. p. 1243 (CMB dipole)
2. p. 1141 (Gamma-ray burster energy)
3. p. 1170 (QSO spectrum and quasars account for ~10% of all QSOs)
4. p. 528 (Stellar mass/luminosity range)
5. p. 945 (Solar motion)
6. p. 915 (Olbers' paradox, light)
7. p. 1222 (Olbers' paradox, thermodynamic)
8. p. 1016 (Galactic arm density gradient)
9. p. 916 (Milky Way's face-on appearance as rendered from available data)
10. p. 486 (Lifespan of stars that produce white dwarfs
11. p. 917 (Milky Way's radius and distance from the sun to its rim and center)
12. p. 928 (de Vaucouleurs law)
13. pp. 1165, 1166 (Bilateral galactic core hydrogen expulsion)
14. p. 966 (Speed of hydrogen near Milky Way's core)
15. p. 602 (White dwarf density)
16. p. 595 (White dwarf cooling profile)
17. pp. 882, 894 (Solar system's age)
18. p. 925 (Galactic fountain model)
19. p. 1015 (Winding problem)
20. p. 1169 (QSO HS 1946+7658's magnitude)
21. p. 1000 (Pitch angle versus rotation speed)
22. p. 956 (Rigid-body galactic rotation)
23. p. 967 (Movement and presence of central galactic hydrogen content)
24. p. 516 (Supernovae light curves)
25. p. 505 (Stellar mass requirement for supernova)
26. p. 53 (Virial Theorem)

⟨7⟩ *A Different Approach to Cosmology*, Fred Hoyle, Geoffrey Burbidge & Jayant Narlikar [Cambridge, 2000]
(Quasi-Steady State theory)
1. p. 27 (Hubble constant, historical perspective)

⟨8⟩ *Principles of Physical Cosmology*, Phillip Peebles [Princeton, 1993]
(Big Bang theory & supporting evidence)
1. pp. 39, 40 (Large-scale material structure)
2. p. 209 (Evidence that the universe's material is increasingly uniform with increasing scale)
3. p. 158 (CMB photon number density)
4. p. 225 (Tired light)
5. p. 47 (Galactic disk motion contrary to expectation of Newtonian physics)
6. p. 67 (Gravitational instability)
7. p. 92 (Loss of surface brightness with distance)

⟨9⟩ *Quantum Physics*, 2nd Edition, Robert Eisberg & Robert Resnick [Wiley, 1985]
(Nuclear potentials)
1. p. 637 (A neutron's charge distribution)
2. p. 524 (Nuclear binding energy/nucleon)
3. p. 629 (Radius of the Strong Force's repulsive core)
4. p. 570 (Beta decay energy)
5. p. 626 (Deuteron's l=0 bound kinetic energy)

⟨10⟩ *Radiative Processes in Astrophysics*, George Rybicki & Alan Lightman [Wiley, 1979]
(Astrophysical theory)
1. p. 84 (Larmor's formula source, adapted to SI units)
2. p. 36 (Extinction coefficient, assuming Thomson cross-section in deep space)

⟨11⟩ *Seeing Red*, Halton Arp [Apeiron, 1998]
(Quasar redshift anomaly)

⟨12⟩ *Zero, The Biography of a Dangerous Idea*, Charles Seife [Viking, 2000]
(History and application of the concept of zero)
1. p. 144 (Riemann sphere)

⟨13⟩ *The Book of Nothing*, John Barrow [Pantheon, 2000]
(History and application of the concept of zero)
1. p. 158 (Construction from empty sets)

⟨14⟩ *Galaxies in the Universe*, Linda Sparke & John Gallagher [Cambridge, 2000]
(Galactic pitch angle and brightness)
1. p. 206 (Pitch angle definition)
2. p. 207 (HI content of galactic arms)
3. p. 180 (Exponential disk brightness profile)
4. p. 76 (Milky Way's disk scale length)
5. p. 215 (Barred spiral rotation)

[15] *Galactic Astronomy*, James Binney & Michael Merrifield [Princeton, 1998]
(Galactic surface brightness)
1. p. 561 (Milky Way's pitch)
2. p. 201 (Elliptical fine structure)

[16] *Galactic Dynamics*, James Binney & Scott Tremaine [Princeton, 1987]
(Galactic material flow)
1. p. 353 (Pitch angle versus Hubble type)
2. p. 341 (Galactic bands' radio wave emission)

[17] *Subtle is the Lord*, Abraham Pais [Oxford, 1982]
(Historical perspective of Einstein and the development of General Relativity)
1. p. 287 (Relativistic spatial curvature due to mass density, converted to energy density)

[18] *Introductory Nuclear Physics*, Kenneth Krane [Wiley, 1987]
(Nuclear density)
1. pp. 56-57 (Nuclear radii)
2. pp. 56-57 (Nuclear density)
3. p. 95 (Effective range between a deuteron's nucleons)
4. p. 81 (Deuteron binding energy)
5. p. 83 (Deuteron potential well)

[19] *Cosmic Ray Astrophysics*, Reinhard Schlickeiser [Springer, 2002]
(Cosmic rays, general reference)

[20] *Active Galactic Nuclei*, Julian Krolik [Princeton, 1999]
(AGN, general reference)

PAPERS

[21] Cohen & Taylor, *The 1986 adjustment of the fundamental physical constants*,
1987, Review Modern Physics, 59, 1121
(Gravitational constant)

[22] Mohr & Taylor, *CODATA recommended values of the fundamental physical constants: 1998*,
2000, Review Modern Physics, 72, 351
(Gravitational constant)

[23] Michaelis, Haars & Augustin, *A new precise determination of Newton's gravitational constant*,
1995, Metrologia, 32, 267
(Gravitational constant)

[24]Fitzgerald & Armstrong, *The measurement of G using the MSL torsion balance*,
1999, Measurement Science and Technology, 10, 439
(Gravitational constant)

[25]Quinn, Speake, Richman, Davis & Picard, *A new determination of G using two methods*,
2001, Physics Review Letters, 87, 1101
(Gravitational constant)

[26]Schlamminger, Holzschuh & Kundig, *Determination of the gravitational constant with a beam balance*,
2002, Physics Review Letters, 89, 1102
(Gravitational constant)

[27]Nolting, Schurr, Schlamminger & Kundig, *A value for G from beam-balance experiments*,
1999, Measurement Science and Technology, 10, 487
(Gravitational constant, comparison of mercury and water)

[28]Kleinevob, Meyer, Schumacher et al., *Absolute measurement of the Newtonian force and a determination of G*,
1999, Measurement Science and Technology, 10, 492
(Gravitational constant)

[29]Gunlach & Merkowitz, *Measurement of Newton's constant using a torsion balance with angular acceleration feedback*, 2000, Physics Review Letters, 85, 2869
(Gravitational constant)

[30]Bagley & Luther, *Preliminary results of a determination of the Newtonian constant of gravitation: A test of the Kuroda Hypothesis*, 1997, Physics Review Letters, 78, 3047
(Gravitational constant)

[31]Luo, Hu, Fu & Fan, *Determination of the Newtonian gravitational constant G with a nonlinear fitting method*,
1998, Physics Review D, 59, 042001
(Gravitational constant)

[32]Barnard, Kolb & Osborne, *Tracing a Z-track in the M31 X-ray binary RXJ0042.6+4115*
2003, Astronomy & Astrophysics, 411, 553
(Multiple galactic cores)

[33]Harris et al., *New experimental limit on the electric dipole moment of the neutron*,
1999, Physics Review Letters, 82, 904
(Neutron electric dipole moment)

[34]John D. Anderson et al., *Study of the anomalous acceleration of Pioneer 10 and 11*
2002, Physics Review, D65, 082004
(Anomalous deceleration of Pioneer 10 and 11)

[35] Richard Kienberger et al., *Atomic transient recorder*,
February 26, 2004, Nature, V427
(Extreme UV measurement of electron orbital period)

[36] A. Kogut et al., *Dipole anisotropy in the COBE differential microwave radiometers first-year sky maps*,
December 10, 1993, Astrophysical Journal, V419
(Earth's motion through the CMB)

[37] Adam Riess et al., *Observational evidence from supernovae for an accelerating universe and a cosmological constant*,
September, 1998, Astrophysical Journal, V116
(Variation of Hubble constant with distance)

[38] C. Bennett et al., *Four-year COBE CMR cosmic microwave background observations: maps and basic results*,
June 10, 1996, Astrophysical Journal, V464
(Small-scale CMB anisotropies)

[39] P. Sreekumar et al., *EGRET observations of the extragalactic gamma-ray emission*,
February 20, 1998, Astrophysical Journal, V494
(Universal gamma-ray background)

[40] W. G. Tifft & W. J. Cocke, *Quantum Cosmology*,
1996, Astrophysics and Space Science 238: 247-283
(Quantized optical/radio wave redshifts)

[41] W. M. Napier & B. N. G. Guthrie, *Testing for quantized redshifts II. The Local Supercluster*,
1996, Astrophysics and Space Science 244: 111-126
(Quantized radio wave redshifts)

Glossary

∞ - Infinity. The universe's diameter in absolute meters.

∞_0 - Unitless infinity; $\infty/1$.

∞_m - Metric infinity. The universe's diameter in meters.

$\lozenge_4$ - Unit hypervolume. The universe's finite four-dimensional volume. It is the sole boundary condition for light's quantization.

$\lozenge_{||}$ - Internal hypervolume. The product of a volume of internally deflected space and its internal deflection. Internal hypervolume defines the relative strength of a gravitational field.

$\lozenge_q$ - Unit polarvolume. The boundary condition for the quantization and four-dimensional field distribution of elementary particles. Equal to half of unit hypervolume, the finite four-dimensional volume of the universe.

δ_3 - Netherspace. The linear resolution and fourth-dimensional thickness of space, $(1^4/\infty^3)$.

ε - Fusion mass fraction. The fraction of mass converted into energy during hydrogen fusion, 0.0073.

ε_U - Universally averaged fusion mass fraction. The luminosity-weighted fraction of mass converted into energy during all types of fusion, from hydrogen to iron, averaged over the universe. It is approximately equal to ε, 0.0073.

$\kappa_{||}$ - Gravitational index. The constant ratio of internal hypervolume to energy content in all physical objects.

$\rho_{|C|}$ - Hyperdensity. The universe's maximum material density, $1.2(10)^{19}$ kg/m^3.

ρ_{CMB} - The CMB's energy density, $4.16(10)^{-14}$ J/m^3.

ρ_M - The universally average material density throughout space, in units of mass/distance^3.

ρ_U - The universally average energy density throughout space, in units of energy/distance^3.

σ - Stefan-Boltzmann constant. The factor used to scale the relationship between electromagnetic radiation's energy density and temperature.

Absolute meter, m_a - The fourth root of unit hypervolume. Current estimates of universal energy density put the absolute meter's value at approximately 0.1 mm.

Absolute second, s_a - The time it takes light to travel one absolute meter.

Accretion - The slow and continuous capture of matter through gravitational attraction.

Accretion disk - A region of space near a black hole where material is gravitationally captured, emitting energetic electromagnetic radiation as it accelerates toward its gravitational veneer.

AGN - Active galactic nucleus. Hyperactive galactic core region, with higher than normal radiant output, hydrogen production, and radio wave emissions.

Anisotropic - Exhibiting difference in character in different directions from a common center.

Anthropic - Characterizing the universe from a human perspective in terms of priorities, motivations, or requirements.

Antimatter - A material whose elementary particles have the opposite electric charge of matter. In matter electrons are negative and protons are positive. In antimatter electrons are positive, called positrons, and protons are negative, called antiprotons. An isolated galaxy composed exclusively of antimatter would be visually indistinguishable from one of matter, as antimatter has the same physical properties as matter in terms of spectral characteristics, melting point, etc.

Antineutrino - *See* neutrino.

Big Bang theory - An ad hoc, completely erroneous attempt to explain the existence of the universe, intergalactic redshift, the universal concentration of the elements, and the CMB field as the result of a primordial explosion-expansion. Also called the Standard Model.

Binary star - One of a pair of stars bound by gravitational attraction. The majority of stars are thought to have binary companions.

Black hole - A massive object composed of degenerate nuclear matter with a near-unity gravitational potential at its surface. Matter at such a high potential can emit only negligible amounts of electromagnetic radiation. However, particles whose kinetic energy exceeds their rest mass can escape with the resultant loss of nearly one rest mass of energy.

Bound electron - An electron (or positron) compressed from its free core radius of ~1740 F to nuclear size, <1 F, by the powerful, high-energy density fields of protons.

Bound proton - A proton (or antiproton) compressed from its free core radius of 0.946 F to a recessed size of as small as ~0.6 F by the powerful, high-energy density fields of other protons. Core size can be further reduced by the pressure found in the interior of massive compact objects, but can be no smaller than the minimum defined by hyperdensity, 0.23 F.

Brown dwarf - A star-like object with mass greater than a typical planet yet insufficient to ignite nuclear fusion (less than ~8% of a solar mass). Composition also plays a role, as an object composed of pure hydrogen and mass equal to our sun would also fail to achieve fusion.

c - The speed of light. This is the maximum propagation speed of electromagnetic energy and matter through space as well as the linear dimensional relationship between space and time.

Circular velocity - Circular motion around a central mass. The magnitude of an object's circular velocity in a radially symmetric system is determined by the mass within its orbit.

Cluster - A collection of gravitationally bound objects larger than a group.

CMB - Cosmic Microwave Background energy. A universal field of thermal microwave radiation with a temperature of 2.724 °K. It is the ancient equilibrium concentration of electromagnetic energy in space, constantly thermalized by the flow of energy through the universal fusion cycle.

Compound nuclei - Any matter whose nucleus is heavier than hydrogen.

Copenhagenism - The grotesquely erroneous belief that the universe has no deep reality beneath its quantum statistical appearance.

Core - *See* particle or galactic.

Cosmic rays - Charged elementary particles and small atomic nuclei with enormous energy levels, moving close to the speed of light. They are generated in high-energy celestial environments such as supernovae and galactic cores.

Cosmological Principle - The premise the universe has a similar character everywhere in terms of its appearance, composition, and physical laws.

Coulomb force - Interaction caused by the particle core asymmetry induced by proximity with other particle cores.

Coulomb veneer binder - The electrostatic potential energy of a gravitational veneer's net charge (positive for matter, negative for antimatter).

Dark matter - The universe's nonluminous material. It takes the form of red and brown dwarfs and smaller objects distributed throughout galactic and intergalactic space, all of which are composed primarily of hydrogen. Dark matter comprises approximately 97% of the universe's mass.

Degenerate - State of matter whose density is limited by the rest mass of its constituent elementary particles. Material can be either electronically or neutronically degenerate. Electronically degenerate matter is composed of orbital (non-nuclear) electrons compressed against each other and their attendant nuclei. White dwarf stars are electronically degenerate. Neutron stars, by comparison, are neutronically degenerate. They are composed primarily of a neutron superfluid.

DZ - Disintegration zone. The hypothetical site of cold dark matter-antimatter annihilation in a finite gammastructure boundary. It is a natural consequence of a universe composed of half matter and half antimatter, but only exists if their large-scale containment occurs in finite packages.

Effigy - A particle core of the size which, when superimposed into a given energy density distribution, causes no change in total system energy. Effigies are massless mathematical particles, not physical realities.

Electron - The lightest stable elementary form of matter. Unless otherwise specified, the term *electron* refers to either electrons or positrons.

Element - A certain number of bound electrons and protons (or bound positrons and antiprotons) held together at nuclear density by the Strong force. An element's distinctive chemical and physical properties are set by its net nuclear charge. A total of 92 different elements occur naturally in Earth's low-density/low-energy environment.

Energy - Distorted space. A region of space with fourth-dimensional slope. Energy's SI units are joules but its true dimensional units are time-distance^2.

Energy density - The fourth-dimensional slope of a spatial distortion. Energy density is single-valued in space and has infinitely fine resolution.

Equipartition energy - A quantum effect wherein particles in a superfluid at a given temperature all have the same energy level, typically equal to or proportional to kT.

Event horizon - The hypothetical spherical or nearly spherical non-emitting surface of a black hole. Due to the compressibility limits of matter, a gravitational potential of negative unity is not physically possible, so an event horizon is a mathematical limit, not a real boundary.

F - Fermi. A unit used for expressing nuclear distances and the core sizes of elementary particles. One Fermi is equal to $(10)^{-15}$ meters.

G - Universal gravitational constant. The ratio of the product of gravitational potential energy and radius to mass. $G = (\Phi R)/M$.

Galactic center - The inner region of a spiral or lenticular galaxy, typically less than about 15 Kly in radius, where the circular velocity of its disk material varies proportionately with radius.

Galactic cluster - A collection of groups of gravitational-bound galaxies embedded in a hot cloud of hydrogen.

Galactic core - One or more black holes at the center of a galaxy, with masses millions or billions times greater than our sun. Responsible for the reversal, by nuclear disassociation, of the fusion occurring in a galaxy's disk.

Galactic disk - The region of a spiral or lenticular galaxy, typically beyond about 15 Kly from its center, where the circular velocity of its disk material is constant.

Galactic halo - A roughly spherical distribution of material located immediately external to a galaxy's outer luminous edge (rim). It contains globular clusters and dark matter and constitutes a significant fraction of a galaxy's mass.

Galactic rim - The outermost luminous edge of a galaxy. The rim is the region in the galactic vortex that achieves a material and electrical current density sufficient to form and ignite stars.

Galactic vortex - The uniform inward motion of galactic disk material to the galactic core. This allows galaxies to recycle the compound nuclei created by various forms of fusion occurring within their disk and central regions.

Galaxy - An immense collection of stars orbiting a central core in a whirlpool motion. It fuses hydrogen throughout its disk and disassociates compound atomic nuclei in its core region. Requires the presence of a central black hole to disassociate compound nuclei back into hydrogen in a virtually nonradiative environment.

Gamma luminosity - A weak universal gamma radiation field generated by the annihilation of matter and antimatter and other high-energy events.

Gamma ray - The most energetic type of photon. Generated by the strong acceleration of charged particles or the annihilation of matter and antimatter.

Gammastructure - The largest stable celestial compartmentalization of matter and antimatter. It is the structure necessary to preserve material stability in a universe composed of 50% antimatter. Gammastructure size is currently indeterminant and might not even be finite.

Globular cluster - A spherical, gravitationally bound distribution of as many as a hundred thousand stars, typically found in orbit about a much larger galactic system.

Gly - A billion light years, a distance equal to $9.46(10)^{24}$ m.

Gpc - A gigaparsec, or billion parsecs, a distance equal to $3.084(10)^{25}$ m.

Gravitational force - *Gravitational fields* arise from energy's intrinsic internal hypervolume. *Gravitational phenomena* are caused by the superposition of gravitational fields. This induces an asymmetrical distortion of particles' core boundaries.

Gravitational index, $\kappa_{\parallel}$ - The constant ratio of internal hypervolume to energy content in all physical objects.

Gravitational veneer - The surface of a compact massive object with a gravitational potential near negative unity. Light emitted by this surface experiences an extraordinarily large redshift.

Group - A small collection of gravitationally bound objects containing between 10 and 20 members.

Gyr - A billion years, equal to $3.155(10)^{16}$ s.

h - Planck's Constant. Used to express the constant product of energy and wavelength in a photon, $h = (E\lambda)/c$. It is proportional to and originates from the universe's finite four-dimensional size.

H_0 - Hubble Constant. The exponential time constant for the gravitationally induced decay of electromagnetic energy.

H I - Neutral hydrogen. Detected in deep space by its emission of 21 cm radiation.

H II - Ionized hydrogen.

Hubble constant - The lumetic decay constant of the universe. It is named after Edwin Hubble, who first discovered the proportional loss of energy in ancient photons.

Hypercore - The mathematical depiction of a particle core as a charged hyperspherical hole in space composed of a uniform region of polarvolumetric density.

Hyperdensity - The universe's peak material density, defined by the unit polarvolume distribution and a proton's rest mass, $1.2(10)^{19}$ kg/m^3.

IGM - Intergalactic material. The matter and antimatter sparsely distributed between galaxies in deep space.

Integrated starlight - The universe's averaged optical luminosity, with a spectrum similar to that of an attenuated 10,000 °K blackbody.

Intergalactic - The wide expanses of space between galaxies.

Internal energy - The energy associated with the fourth-dimensional slope of internal deflection.

Interstellar - The space between the stars of a galaxy or cluster.

Intragalactic - The space internal to a galaxy, similar to interstellar space but typically used in reference to the entire interior space.

ISM - Interstellar material. The material sparsely distributed between stars within a galaxy or other large stellar population.

Isoexternal - A three-dimensional surface with uniform fourth-dimensional elevation.

Isointernal - A three-dimensional surface with a uniform magnitude of internal deflection along space.

Isotropic - Having the same characteristics in all directions. The CMB field and the universe's large-scale matter distribution are both isotropic, as viewed from any location within the universe.

J - Abbreviation for the SI unit of energy called the joule.

k - Boltzmann constant. The fundamental relationship, in matter, between kinetic energy and temperature. Typically used to express the energy density of gases at a certain temperature.

KK - Kilokelvin. A unit of temperature equal to one thousand degrees Kelvin.

Kly - A thousand light years, a distance equal to $9.46(10)^{18}$ m.

L - An object's luminosity; the rate that it emits electromagnetic radiation into space.

Lenticular - Galactic morphology combining spiral and elliptical characteristics.

Light year - The distance light travels in one year in free space, equal to $9.46(10)^{15}$ m.

Lumetic decay - The gradual energy loss a photon or neutrino incurs on its journey across vast distances, caused by the gravitationally induced expansion of its wavelength.

Luminosity - The amount of light energy an object radiates per unit time.

Luminous limit - The amount of light energy a luminous object can maintain in space as an equilibrium state between its power output and the rate of gravitationally induced decay in its prior output.

ly - Abbreviation for *light year*.

Main sequence - Stars whose primary energy source is hydrogen fusion.

Matter - A hole in space surrounded by a centralized field of positive or negative fourth-dimensional deflection. Elementary unit charge is quantized by half of the universe's four-dimensional size.

Megaparsec - A unit of distance equal to a million parsecs, or 3.26 million light years.

Metavolume - A magnitude of spatial volume compositionally intermediate between a solitary geometric point and the whole of space.

Milky Way - The spiral galaxy in which our solar system is located, home to approximately 200 billion stars. Our sun is situated in its disk, ~25 Kly from its center.

MK - Megakelvin. A unit of temperature equal to one million degrees Kelvin.

mK - millikelvin. A unit of temperature equal to one thousandth of a degree Kelvin.

Mly - A million light years, a distance equal to $9.46(10)^{21}$ m.

Mpc - A megaparsec, or million parsecs, a distance equal to $3.084(10)^{22}$ m.

Muon - A composite (unstable) elementary particle intermediate in mass between an electron and proton.

Nebula - Celestial collection of gas and dust. *Planetary* nebulae are the remnants of stellar explosions. Others, typically more expansive and diffuse, often contain stellar nurseries where protostars are in the process of condensing and igniting.

Neorealism - The idea that the universe is deterministic and has a single underlying physical structure responsible for all visible phenomena.

Netherspace, δ_3 - The linear resolution and fourth-dimensional thickness of space, $(1^4/\infty^3)$.

Neutrino - A form of neutral radiant energy with an exceptionally small absorption cross-section. Evidence suggests that neutrinos consist of twin photons of virtually equal momentum in a closely bound state. Although contemporary physics postulates the existence of a number of different neutrinos, each with their own antiparticles, this is erroneous. Neutrinos, like photons, are their own antiparticle because they are electrically neutral.

Neutron - The combination of a proton and a heavily compressed (bound) electron in a tightly coupled state.

Neutron star - An ultra high-density supernova remnant, composed of neutronically degenerate matter (material whose orbital electrons have collapsed into their nuclei). Neutron stars are thought to consist of a neutron superfluid core covered by a crust of atomic nuclei of various species, in particular iron.

nm - Nanometer. A billionth of a meter, 10^{-9} m.

Nova - A flaring episode in a star that lasts between a few hours and a few days. The star becomes as much as 100,000 times brighter than its nominal luminosity but returns to its original brightness after several months.

Nuclear condensation - A process resulting in a net increase in nuclear proton count. This includes any type of fusion, from hydrogen to iron or beyond.

Nuclear disassociation - The reduction of compound atomic nuclei to their component protons through the application of kinetic energy and gravitational potential. Galactic cores are the universe's primary site of nuclear disassociation.

Nucleosynthesis - Creation of progressively larger nuclei by repeated combination of protons, neutrons and smaller nuclei. Thought to occur principally in stars off the main sequence, such as red giants and supernovae.

Null Axiom - The premise that existence is a distributed form of nonexistence which sums to nonexistence.

Observable universe - The region of our infinite universe within the range of the most powerful optical and radio telescopes. Its size is limited by the sensitivity of our instruments and the loss of electromagnetic energy due to absorption, scattering, and lumetic decay.

Olbers' Paradox - The erroneous idea that an infinitely large, infinitely old universe would experience an unchecked heating of space due to the light released by stellar fusion.

Omnielement - The material universe's largest, individually varying contiguous component.

Omnipattern - The universal distribution of omnielements.

Oort cloud - The region surrounding our solar system beyond Pluto's orbit. Thought to contain a diffuse amount of material and is probably the source of most comets.

Panspermia - The premise that microbes from space (and/or their by-products) are responsible for the origin of life on Earth and in similar environments throughout the universe.

Parsec - A parallax second. A unit of distance equal to 3.26 light years. The image of an object one parsec from Earth would shift one angular second due to the parallax motion caused by the Earth's yearly orbit around the sun.

Particle - One of the four stable elementary expressions of matter: electrons, positrons, protons, and antiprotons.

Particle core - The central region of an elementary particle. Its size is defined by its rest energy and is devoid of space as well as energy.

pc - Abbreviation for *parsec*, a distance of 3.26 light years.

Photon - The physical representation of unit hypervolume's quantization of momentum. Photon's are neutral and propagate at c.

Planck constant, h - The ratio of a photon's energy and frequency, a direct manifestation of the universe's four-dimensional size.

Planck relation - The ratio of a photon's energy and frequency is constant: $h = E / \nu$.

Planck spectrum - Also called the blackbody spectrum or thermal spectrum. An equilibrium distribution of photon number density as a function of photon energy.

Plasma - A high-energy state of fully ionized matter. Plasma is a mixture of high-speed free electrons and atomic nuclei. This state is by far the most common state for the universe's luminous matter.

Positron - An antimatter electron. It has the same mass as an electron but is positively charged.

Proton - The heaviest stable form of elementary matter. Unless otherwise specified, the term *proton* refers to either protons or antiprotons.

Protostar - An embryonic form of a star, consisting of a rapidly collapsing cloud of dust and gas. This is the stellar state immediately prior to the onset of nuclear fusion.

Pulsar - Pulsating Astronomical Radio source. A star or stellar remnant emitting regularly spaced, short, intense bursts of electromagnetic radiation in the radio wave band.

QSO - Quasi-Stellar Object. QSOs are massive, hot, compact sources of broad spectrum radiation. Evidence suggests that they are naked galactic core material either ejected from an active core or produced by some other stage of galactic evolution. A QSO's remarkably high (typically $|\Phi_g|/mc^2 > \sim 0.4$) surface gravitational potential induces large redshifts in its spectrum, giving the false impression of great astronomical distance.

Quantum - a discrete packet of energy bounded by either unit hypervolume or unit polarvolume, to include photons, neutrinos, and elementary particles.

Quark theory - The illusory concept that protons are composed of more fundamental subunits. It is an ad hoc theoretical response to the variety and complexity of elementary forms of matter.

Quasar - A QSO with pronounced radio wave emission.

Radial velocity - The motion of stars toward a galaxy's center caused by its galactic vortex.

Radiant energy - Any quantized energy that propagates at the speed of light, to include photons and neutrinos.

Red dwarf - A low-mass (typically on the order of a tenth of our sun's mass), low-luminosity star with an extremely long life span.

Redshift - A frequency shift of light toward the low-energy end of the spectrum resulting in a non-scattered energy loss. Although usually cited for visible light it can be found in electromagnetic radiation of any wavelength.

Refraction - The apparent slowing of light as it passes through transparent materials. It is caused by the additional distance light must travel to traverse energy density's extraspatial volume.

Relativistic - A rate of speed in a moving object sufficiently close to c to produce a pronounced increase in mass and decrease in the rate that its internal change passes relative to the rest of the universe. Both effects result because c is the fixed dimensional relationship between distance and time.

SI - International measurement system based on units of meters, kilograms, and seconds.

Space - A three-dimensional distribution of nothingness in the form of geometric points.

Steradian - A unit of angular area. There are 4π steradians in a spherical surface area.

String theory - The entirely fictitious premise that matter and energy are composed of submicroscopic strings that vibrate in seven to eleven (or more) unseen dimensions.

Strong force - The attractive nuclear potential caused by an interaction between the hollow core regions of protons/antiprotons.

Strong veneer binder - The residual nuclear binding energy of the expanded degenerate material of a black hole's gravitational veneer.

Supercluster - An enormous collection of gravitationally bound galaxies, often numbering in the tens of thousands. The spatial distribution of superclusters is the medium- to large-scale topology of the universe, depending on the as yet to be determined size of gammastructures.

Superluminal - A speed faster than light, up to and including infinitely rapid propagation.

Supernova - Explosion of either (a) a massive star that has exhausted its stable fuel or (b) a white dwarf that has accreted too much material from its binary companion (Type Ia). Many supernovae have peak transient luminosities so intense that, under the right conditions, they are visible in distant galaxies.

Surface brightness - Luminosity per angular area of a celestial object.

S-Z effect - Sunyaev-Zel'dovich effect. The slightly reduced CMB radiance associated with rich galactic clusters. It is caused by the galactic power return cycle, where galactic halos absorb enough CMB radiation (through the motion of vast electrical currents) to offset the energy they lose by their fusion luminosity.

Thermal spectrum - Also called the blackbody spectrum or Planck curve. A distribution of photons whose number density and energy are in equilibrium.

Time - A difference of space arising from the intrinsic symmetry of the universe's four-dimensional size. Time is the polar operator responsible for nullifying the net magnitude of space, as is necessary for its contextual existence.

Ultrastasis - The state of universal zero change. Regardless of the appearance of our changing local surroundings, the universe's unbounded material distribution, when viewed in its entirety, is invariant.

Unit hypervolume, $\lozenge_4$ - The universe's finite hypervolume. Infinite three-dimensional space has a finite four-dimensional size. It takes the form of Planck's constant and defines the quantization of photons.

Unit polarvolume, $\lozenge_q$ - Boundary condition for elementary particles' four-dimensional field distribution. It defines unit elementary charge and is equal to half of unit hypervolume.

Veneer - *See* Gravitational veneer.

Veneer capacity - The total flux of material moving in a gravitational veneer.

Veneer flux - The rate material moves out of a black hole's gravitational veneer, governed predominantly by its temperature.

Virial theorem - The total kinetic energy in a gravitationally bound system at equilibrium is equal to half of its time-averaged gravitational potential energy.

W - Watt. The SI unit of power. It is equal to one joule per second.

Weak force - The positive potential required to compress a free electron to nuclear dimensions.

Weak veneer binder - The residual weak binding energy of the bound electrons in the expanded degenerate material of a black hole's gravitational veneer.

White dwarf - The gradually cooling remnant core of a medium mass star ($4M_{sun} - 8M_{sun}$). White dwarfs are produced by supernova events that tear away all of a star's outer layers.

z - The redshift parameter. A z of zero indicates no redshift, a z of 1 represents a doubling of photon wavelength and the corresponding loss of 50% of its energy. $z \equiv (\lambda/\lambda_0) - 1$.

Zero equation - The mathematical expression of the fact that the universe must sum to nothingness in order to be a form of nothingness.

Index

F

G

H

I

L

M

T

U

V